# Progress in the Chemistry of Organic Natural Products

## Volume 130

The volumes of this classic series, now referred to simply as "Zechmeister" after its founder, Laszlo Zechmeister, have appeared under the Springer Imprint ever since the series' inauguration in 1938. It is therefore not really surprising to find out that the list of contributing authors, who were awarded a Nobel Prize, is quite long: Kurt Alder, Derek H.R. Barton, George Wells Beadle, Dorothy Crowfoot-Hodgkin, Otto Diels, Hans von Euler-Chelpin, Paul Karrer, Luis Federico Leloir, Linus Pauling, Vladimir Prelog, with Walter Norman Haworth and Adolf F.J. Butenandt serving as members of the editorial board.

The volumes contain contributions on various topics related to the origin, distribution, chemistry, synthesis, biochemistry, function or use of various classes of naturally-occurring substances ranging from small molecules to biopolymers.

Each contribution is written by a recognized authority in the field and provides a comprehensive and up-to-date review of the topic in question. Addressed to biologists, technologists, and chemists alike, the series can be used by the expert as a source of information and literature citations and by the non-expert as a means of orientation in a rapidly developing discipline.

All contributions are listed in PubMed.

A. Douglas Kinghorn • Heinz Falk
Simon Gibbons • Yoshinori Asakawa
Ji-Kai Liu • Verena M. Dirsch
Editors

# The Diels–Alder Reaction as a Key Step in Natural Product Synthesis

By Heinz Falk

Springer

*Editors*
A. Douglas Kinghorn
College of Pharmacy
The Ohio State University
Columbus, OH, USA

Heinz Falk
Institute of Organic Chemistry
Johannes Kepler University
Linz, Austria

Simon Gibbons
Centre for Natural Products Discovery
School of Pharmacy and Biomolecular Sciences
Liverpool John Moores University
Liverpool, UK

Yoshinori Asakawa
Faculty of Pharmaceutical Sciences
Tokushima Bunri University
Tokushima, Japan

Ji-Kai Liu
School of Pharmaceutical Sciences
South-Central University for Nationalitiy
Wuhan, China

Verena M. Dirsch
Department of Pharmaceutical Sciences
University of Vienna
Vienna, Austria

ISSN 2191-7043 ISSN 2192-4309 (electronic)
Progress in the Chemistry of Organic Natural Products
ISBN 978-3-032-09900-6 ISBN 978-3-032-09901-3 (eBook)
https://doi.org/10.1007/978-3-032-09901-3



This Springer imprint is published by the registered company Springer Nature Switzerland AG
The registered company address is: Gewerbestrasse 11, 6330 Cham, Switzerland

# About This Book

This volume explores the Diels–Alder reaction, a cornerstone of organic chemistry, tracing its evolution from a classical synthetic tool to a versatile platform for constructing complex molecular architectures. It highlights the reaction's historical significance, methodological advancements, and pivotal role in natural product synthesis. From its discovery in the early twentieth century, the Diels–Alder reaction revolutionized organic synthesis by enabling efficient formation of six-membered rings with remarkable stereoselectivity and regioselectivity. Over the decades, innovations such as asymmetric catalysis, organocatalysis, and inverse-electron-demand variants have expanded its scope, allowing precise synthesis of highly functionalized frameworks. The volume delves into biomimetic applications, showcasing how the reaction emulates biosynthetic pathways, bridging chemical synthesis and natural product biogenesis. It also examines the groundbreaking discovery of Diels–Alderases, enzymes that catalyze pericyclic reactions in nature, offering insights into biosynthesis and biocatalysis. Through detailed case studies, the volume illustrates the reaction's strategic importance in tackling synthesis challenges posed by intricate natural products. Looking ahead, the volume discusses emerging applications in sustainable chemistry, photochemical activation, and enzyme engineering, emphasizing the adaptability and enduring relevance of this reaction. By integrating modern computational tools and cross-coupling strategies, the Diels–Alder reaction continues to inspire innovation, solidifying its place as a cornerstone of chemical synthesis and retrosynthesis planning.

# Contents

# The Diels–Alder Reaction as a Key Step in Natural Product Synthesis

**Heinz Falk**

## Contents



H. Falk (✉)
Institute of Organic Chemistry, Linz, Austria
e-mail: heinz.falk@jku.at; heinz@fa.lk



A. D. Kinghorn, H. Falk, S. Gibbons, Y. Asakawa, J.-K. Liu, V. M. Dirsch (eds.),
*The Diels–Alder Reaction as a Key Step in Natural Product Synthesis*, Progress in the Chemistry of Organic Natural Products 130,
https://doi.org/10.1007/978-3-032-09901-3_1

## 1 Introduction

The Diels–Alder cycloaddition certainly is one of the most celebrated reactions in organic synthesis. First described by Otto Paul Hermann Diels and Kurt Alder in 1928 (Fig. 1) [1], this reaction between a diene and an alkene enables the construction of six-membered rings in a single step with remarkable regio- and stereocontrol (Fig. 1). Through the simultaneous formation of two carbon–carbon bonds, the Diels–Alder reaction provides chemists with a reliable strategy to introduce complexity into molecules as needed, particularly in natural product synthesis.

Few other transformations have so profoundly shaped the art of total synthesis. Over the past several decades, the Diels–Alder reaction has enabled many elegant syntheses and cascade sequences, offering numerous and unparalleled solutions to a diverse range of synthesis puzzles provided by Nature. It might be no exaggeration to say that the impact of the Diels–Alder reaction on synthesizing natural products is arguably unmatched by any other reaction in the synthesis repertoire, and in 1950 the

**Fig. 1** The Diels–Alder reaction and the two Nobelists. Photographs, public domain

authors were awarded the Nobel prize in chemistry "for their discovery and development of the diene synthesis".

As early as 1939, Otto Diels summarized the possibilities this extraordinary synthesis, which not only assisted in the generation of organic natural products in these series, but also proved to be a valuable instrument in compound structure elucidation [2]. Thus, it seems timely to provide the reader with a historically oriented "Progress"-contribution, covering the progression of the Diels–Alder reaction from a newly discovered curiosity to a cornerstone of natural product synthesis, since this is a rich narrative. In the early days, the reaction was initially explored on simple model systems, as for example, the original experiments of the authors with cyclopentadiene and quinones or maleic anhydride that demonstrated the facile formation of polycyclic adducts. Visionary chemists soon recognized its potential for constructing complex natural product structures. As early as 1937, Cohen and Warren applied the nascent "diene synthesis" to steroids, using a Diels–Alder reaction to build part of a tetracyclic steroid framework [3]. This pioneering work of forming a steroid C-ring by cycloaddition of 1-vinylnaphthalene and maleic anhydride foreshadowed the utility of this reaction in assembling intricate ring systems. However, it was not until the mid-twentieth century that the Diels–Alder reaction truly came to prominence in natural product chemistry. As discussed subsequently in 1952, R. B. Woodward astonished the world with the first comprehensive uses of the Diels–Alder reaction in natural product synthesis: his formal syntheses of several steroids including cholesterol and cortisone featured a key Diels–Alder cyclization between 1,3-butadiene and an electron-deficient alkene to construct the polycyclic core of the steroids [4]. This achievement was nothing short of revolutionary by constituting a dramatic validation that the Diels–Alder reaction could handle molecules as complex as hormones. Woodward's steroid work, soon followed by his 1956 synthesis of reserpine, which used a Diels–Alder reaction to construct a *cis*-decalin system, heralded a new era: by solving formidable structural puzzles with pericyclic reactions, these syntheses firmly established the Diels–Alder reaction as an indispensable tool in the natural product synthesis toolkit [5, 6].

Following these landmark successes, the subsequent decades saw an explosion in the development and application of the Diels–Alder reaction for natural products synthesis. By the 1960s and 1970s, as targets grew in complexity, chemists refined the scope and sophistication of the Diels–Alder reaction. New methodologies emerged to address challenges of control and versatility. Notably, the synthesis of the prostaglandins $F_{2\alpha}$ and $E_2$ by E. J. Corey in 1969 showcased how a well-orchestrated Diels–Alder reaction could set multiple stereocenters in one step, with, in this case, establishing three contiguous chiral centers in a cyclopentane ring, an achievement that was aided by Lewis-acid activation to improve yield and selectivity [7]. Such advances underscored the ability of this reaction to simplify routes to complex molecular architectures, inspiring chemists to incorporate Diels–Alder steps into an ever-growing list of total syntheses. During this period, theoretical understanding also caught up: the development of the Woodward–Hoffmann rules provided a rigorous explanation for the stereochemical outcomes of pericyclic reactions, including the Diels–Alder reaction, giving chemists confidence in predicting and planning Diels–Alder steps in syntheses [8]. The Nobel Prize in Chemistry in 1981 was awarded to Roald Hoffmann and Kenichi Fukui for this achievement (Woodward had died in 1979 and thus was ineligible for the prize).

The 1970s and 1980s then brought methodological innovations, as exemplified by the introduction of specialized dienes and dienophiles that expanded what structures the reaction could easily build. Danishefsky's diene (1-methoxy-3-trimethylsilyloxy-1,3-butadiene), first reported in 1974, is one such breakthrough reagent [9]. It enabled Diels–Alder reactions to directly yield α,β-unsaturated cyclohexenone adducts after elimination of the 1-methoxy group. Such reagents opened new pathways to construct the six-membered ring motifs common in natural products. Likewise, the scope of the Diels–Alder reaction broadened beyond carbocycles. Thus, the hetero-Diels–Alder variant by using heteroatom-containing π-systems like carbonyls or imines, became a powerful route to heterocycles, widely applied in alkaloid and terpene syntheses. By the 1980s, even cascade sequences involving multiple Diels–Alder steps were being designed. A spectacular example was the approach to polycyclic frameworks by Vollhardt, as applied for the steroid estrone and other polycycles, where sequential Diels–Alder reactions or tandem pericyclic steps built up complex ring systems in just a few operations [10]. These and other "case studies" demonstrated creative strategies that pushed the Diels–Alder reaction into new territory. In many instances, lesser-known contributions proved deeply insightful. For example, Danishefsky's 1979 synthesis of the aromatic precursor prephenate used an intramolecular Diels–Alder to efficiently construct a bicyclic scaffold, illustrating how the reaction could even construct frameworks found in primary metabolism [11]. Dozens of such ingenious applications accumulated in the literature, each extending the boundaries of what the Diels–Alder could accomplish.

Entering the late twentieth and early twenty-first century, the Diels–Alder reaction underwent a renaissance of new techniques and applications. Asymmetric catalysis emerged as a dominant theme. Chemists learned to control not just relative, but absolute stereochemistry in Diels–Alder cycloadditions, thanks to chiral Lewis acids and organocatalysts developed in the 1980s and 1990s. The once "thermal allowed" pericyclic process became amenable to catalysis that could accelerate the reaction under milder conditions and induce high enantioselectivity, crucial for synthesizing natural products in an optically pure form. At the same time, specialized variants of the Diels–Alder blossomed. The inverse electron demand Diels–Alder reaction, essentially flipping the normal electronic pattern, found utility in constructing complex ring systems, especially in cases where electron-rich dienophiles are present. The hetero-Diels–Alder continued to enable formation of oxygen- and nitrogen-containing rings building many bioactive natural products. Other novel variants like the dehydro-Diels–Alder reaction, involving alkynes to form aromatics, the aza-Diels–Alder, forming nitrogen-heterocycles, and even photochemical pericyclic analogues, as, for example, the photoenolization Diels–Alder reaction [12], have been developed. Modern synthesis planning often exploits domino sequences where a Diels–Alder is followed by secondary transformations, or even another cycloaddition, in one pot. These innovations have greatly expanded the scope of the Diels–Alder reaction and kept it at the cutting edge of natural product synthesis. Indeed, even in the last few years, reports abound where Diels–Alder cycloadditions, sometimes in new layouts, continue to deliver efficient routes to complex natural products. The inherent ability of the Diels–Alder reaction to generate complexity with precision remains as invaluable today as ever.

Given this rich background, the present volume will chronicle the historical development of using the Diels–Alder reaction in natural product synthesis, from its inception to its current state. We will revisit the landmark achievements that punctuated this history, from early milestones like the first steroid constructions, through the golden age of pericyclic strategies in the 1960s–1980s, to the advent of asymmetric and biomimetic Diels–Alder tactics in recent decades. However, of course, given the vast amount of material that has accumulated so far, this contribution will avoid being comprehensive and try to describe every one of the Diels–Alder type syntheses, which led to natural products. Understanding this history is more than an academic exercise, it is profoundly relevant to natural product chemists today. The Diels–Alder reaction, now nearly a century old, remains a crucial tool in constructing complex natural product frameworks, and the lessons drawn from its past will guide its future. By examining how each generation of chemists overcame challenges (whether it was controlling stereochemistry, improving yields, or expanding scope) using the Diels–Alder, we gain insight into the general principles of synthesis strategy. This retrospective illuminates how a simple reaction, applied with ingenuity, can unlock solutions to seemingly intractable problems, certainly an inspiring paradigm for modern practitioners. Moreover, appreciating the historical evolution of the Diels–Alder reaction underscores its continuing significance. Even today, new discoveries (e.g., enzymatic Diels–Alderase catalysts in biosynthesis pathways of Nature) hint that we have yet to tap the full potential of this reaction. Of course, a series of reviews are available, which deal with various aspects of the Diels–Alder reaction including natural product synthesis, many of them written to commemorate anniversaries [2, 13–33]. The present contribution is not just a tour through nearly a century of the Diels–Alder reaction and its use as a pivotal step in natural-product synthesis; it is also a journey through the rich diversity and ever-increasing structural complexity of natural products elucidated during these times. In celebrating the legacy and the upcoming centennial of the Diels–Alder reaction, this volume aims to provide both a historically oriented deep understanding of how far chemistry has come and a compelling rationale for why this rich history matters.

## 2 From Discovery to Indispensable Tool (1928–1950s)

In 1928, Otto Diels and Kurt Alder published their investigation of the reaction product between *p*-quinone (**1**) and cyclopentadiene (**2**) [1]. Contrary to prior studies by W. Albrecht [34] and H. Staudinger [35], which proposed the formation of compounds **4** and **5**, or a cyclobutane ring (**6**), using hydrogenation they could show that instead a new six-ring was formed yielding compound **3**.

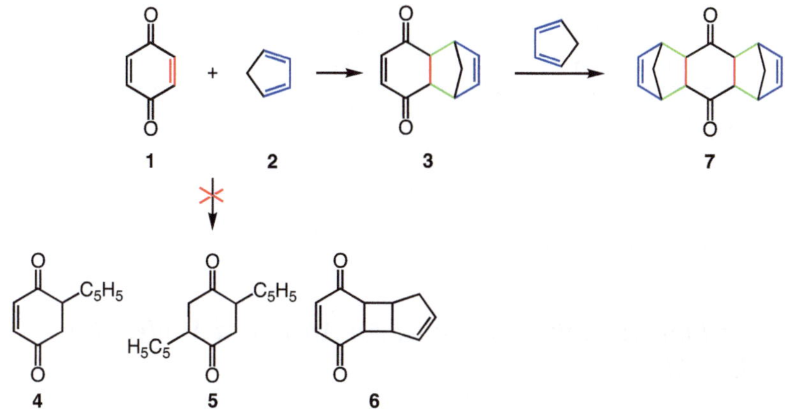


**Fig. 2** The "first" Diels–Alder reaction: *p*-benzoquinone reacts with cyclopentadiene

Formula **5** could be corrected to be the product of a second reaction producing **7** (Fig. 2). Thus, the en-diene cycloaddition, later named the Diels–Alder reaction, was born.

Studying also further educts, like butadiene, naphthoquinone, maleic acid, or acrylic acid, they foresaw even at this early moment: "that this reaction has the possibility to synthesize even complicated natural products, like terpenes, sesquiterpenes, and alkaloids".

In the years following, this reaction was explored varying the dienes (from terpenes to steroids and heterocycles) and dienophiles (olefins, acetylenes, ketenes, quinones) [13], and the stereochemical aspects were investigated mainly by Alder leading to the "*cis* principle" [36].

## *2.1 Model Synthesis of a Steroid Skeleton*

Application of the Diels–Alder reaction to synthesize a natural product was undertaken by A. Cohen and F. L. Warren in 1937 [3]. They demonstrated the first application of this reaction toward a steroid framework (Fig. 3). Thermal cyclization of 1-vinylnaphthalene (**8**), modeling the A/B fused rings of a steroid, with maleic anhydride (**9**) provided a bridged Diels–Alder adduct **10**. Subsequent acid-catalyzed isomerization and ring-forming steps yielded a tetracyclic hydroaromatic steroid nucleus (**11**), proving that the C-ring of a sterol could be assembled via Diels–Alder chemistry (in Fig. 3 and throughout the syntheses a spotlight is used to pronounce the molecular part constructed by the Diels–Alder reaction).

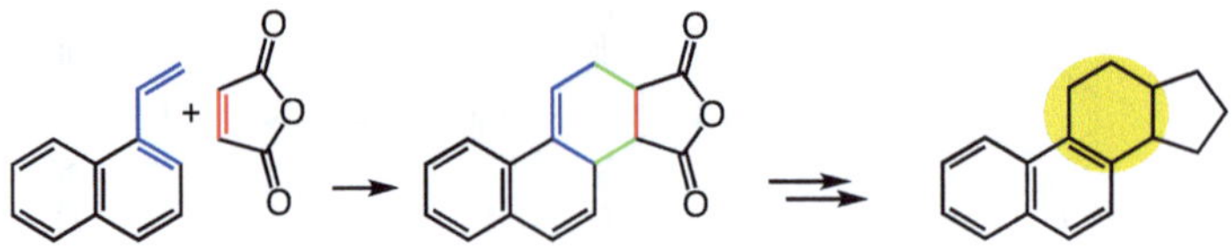


**Fig. 3** Constructing the C/D rings of steroids

## 2.2 Cyclopenta[c]phenanthrene, An Advanced Model of a Steroid

Building on the work of Cohen and Warren described above, Bachmann and Kloetzel thermally reacted (refluxing in xylene for 2 h) naphth-2-ylcyclopentene (**12**) with maleic anhydride (**9**) in a Diels–Alder process to construct a polycyclic system [37] (Fig. 4). This model study of 1938 produced the cyclopenta[*c*]phenanthrene framework (**13**) that essentially constitutes the carbon skeleton of a steroid. Thereby, it validated the Diels–Alder reaction as a tool for assembling complex fused-ring systems relevant to natural terpenoids and steroids.

**Fig. 4** Inserting ring C to yield steroids

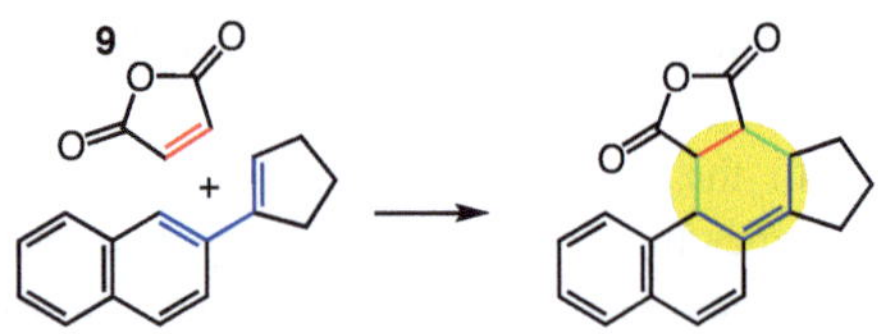


## 2.3 Cholesterol and Cortisone

In 1950, the Nobel Prize in Chemistry was awarded to Diels and Alder, and soon after that in a hallmark paper Woodward et al. described in 1952 the synthesis of cholesterol (**14**), the principal sterol of all higher animals found in all cells of the body tissue, in particular, the brain and spinal cord and in fats and oils, and of cortisone (**15**), a corticosteroid hormone produced by the adrenal glands, which suppresses various elements of the immune system. For this synthesis they used the stereoselective formation of the A/B ring synthon in a Diels–Alder reaction by heating 4-methoxy-2,5-toluquinone (**16**) and butadiene (**17**) to yield the *cis*-decalin **18** thereby producing ring B of the steroid skeleton (Fig. 5) [4]. The latter compound was further elaborated through stereoselective transformations into a fully functionalized steroid skeleton, finally yielding cholesterol (**14**) and cortisone (**15**).

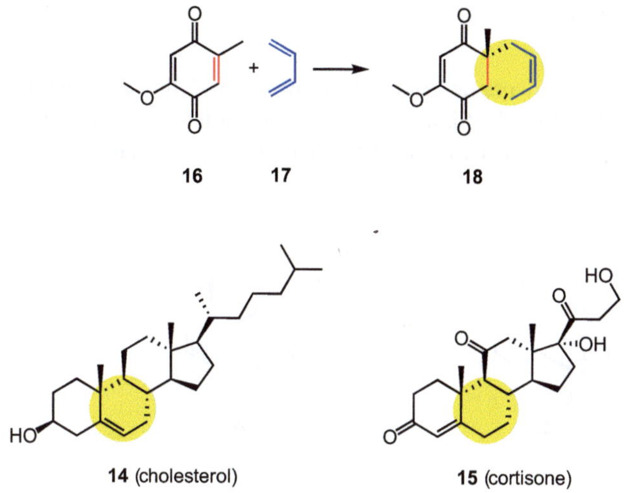


**Fig. 5** The Diels–Alder key step in cholesterol and cortisone synthesis

## 2.4 *Morphine*

Also in 1952, M. Gates and G. Tschudi finally accomplished the first total synthesis of morphine (**22**) [38], the major narcotic analgesic found in the opium poppy, the dark-brown resin produced by drying the latex of *Papaver somniferum* L. (Papaveraceae). They employed one of the earliest Diels–Alder reactions in alkaloid synthesis as described in an earlier paper of 1950 by M. Gates [39]. The key step involved a cycloaddition between the dienophile *o*-naphthoquinone **19** and the conjugated diene **20**, forming a polycyclic adduct **21** that constituted the phenanthrene core of morphine (Fig. 6). This Diels–Alder step rapidly assembled rings A and C of the morphinan scaffold in the correct topology. Subsequent steps introduced the bridge and heterocyclic ring, completing (±)-morphine (**22**). The use of a Diels–Alder reaction to build the tricyclic dihydroanthracene system **21** was thus critical in this milestone synthesis.

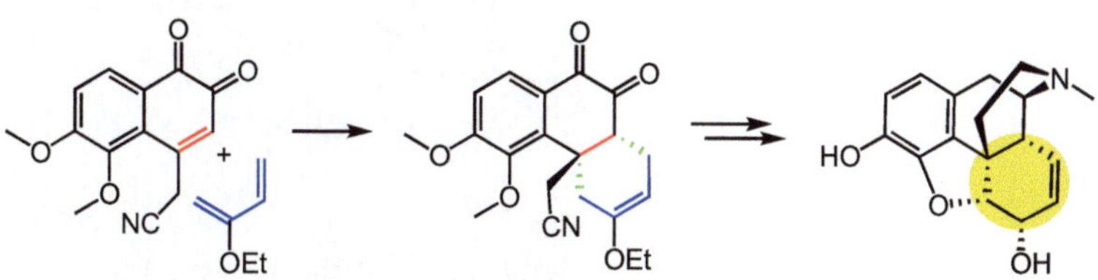


**19** **20** **21** **22** (morphine)

**Fig. 6** The Diels–Alder key step in morphine synthesis

## 2.5 Reserpine

Woodward's steroid work described above was soon followed by his 1956 synthesis of reserpine (**25**) (Fig. 7) [5], an indole alkaloid isolated from the Indian snakeroot *Rauvolfia serpentina* (L.) Benth. [40] (Plate 1), a perennial undershrub from the

1

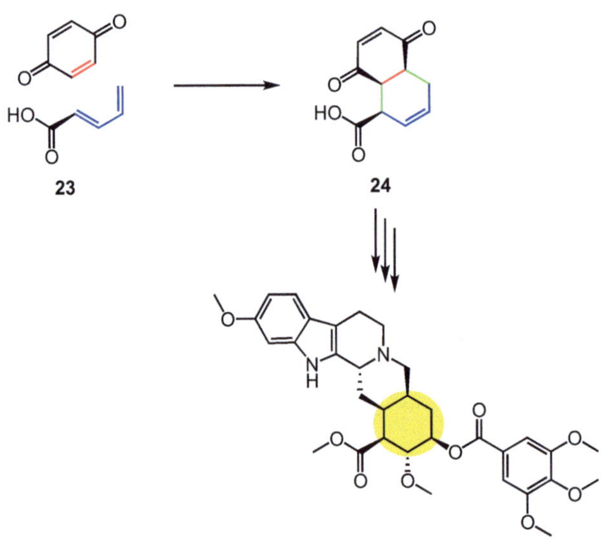

23 24

25 (reserpine)

**Fig. 7** The Diels–Alder key step in reserpine synthesis

**Plate 1** *Rauvolfia serpentina*. Photograph courtesy Vinayaraj, Creative Commons Attribution Share Alike 3.0

milkweed family Apocynaceae, which is used as an antihypertensive and sedative. The ingeniously constructed synthesis route to **25** was based on a Diels–Alder reaction to construct stereoselectively the *cis*-decalin synthon **24** by thermally reacting the dienophile quinone (**1**) and the 1,3-butadiene-derived (*E*)-penta-2,4-dienoic acid (**23**) setting three defined stereocenters (Fig. 7). It thus heralded a new era: by solving formidable structural puzzles with pericyclic reactions, these syntheses firmly established the Diels–Alder reaction as an indispensable tool in the natural product synthesis toolkit.

# 3 Expansion into Complex Natural Products (1960s–1970s)

The 1960s and 1970s marked a transformative era in organic chemistry, characterized by groundbreaking advancements in the synthesis of complex natural products. This section delves into the innovative methodologies and applications of the Diels–Alder reaction, a cornerstone of synthetic organic chemistry, in constructing intricate molecular frameworks. From prostaglandins to sesquiterpenes, alkaloids, and diterpenoids, the section highlights how this reaction enabled the efficient synthesis of biologically significant compounds, many of which play critical roles in medicine, agriculture, and industry. The pioneering efforts of chemists such as E. J. Corey, R. B. Woodward, and others not only expanded the scope of synthesis chemistry but also laid the foundation for future explorations into natural product synthesis. Through detailed discussions of key examples, this section showcases the ingenuity and impact of these synthetic strategies, emphasizing their relevance in addressing challenges in healthcare, fragrances, and agronomy.

## *3.1 Prostaglandins*

Prostaglandins are a group of hormone-like compounds, specifically lipids, produced by the body that play a crucial role in inflammation, pain, and uterine contractions. They are derived biosynthetically from arachidonic acid and are found in nearly every tissue in humans and other animals. Prostaglandins mediate various physiological functions, including regulating blood pressure, blood clotting, immune responses, and induction of labor.

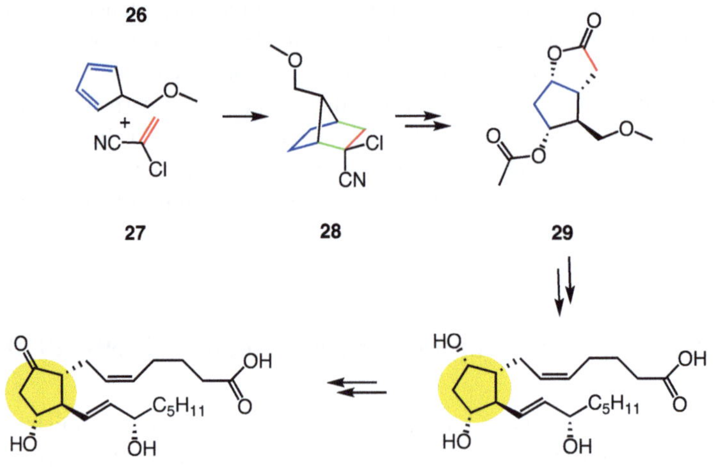


30 (*dl*-prostaglandin $E_2$)

31 (*dl*-prostaglandin $F_{2a}$)

**Fig. 8** The Diels–Alder key step in prostaglandin synthesis

The most notable advancement in this period was certainly the ingenious syntheses carried out by E. J. Corey and coworkers in 1969 and 1970. They prepared prostaglandins $F_{2\alpha}$ (**30**) and $E_2$ (**31**) (Fig. 8).

The synthesis started with the key Diels–Alder reaction between 5-methoxymethyl-1,3-cyclopentadiene (**26**) and 2-chloroacrylonitrile (**27**) to provide the bicyclic synthon **28**. For the first time, such a reaction was catalyzed with cupric fluoroborate, which could be shown to accelerate the addition and allows the reaction to proceed at 0°C in more than 90% yield without appreciable concurrent isomerization of **26** to 1-methoxymethyl-1,3-cyclopentadiene. Compound **28** was then elaborated into the prominent, so-called Corey lactone (**29**), which could then be transformed into the racemic targets, the *dl*-prostaglandins $F_{2\alpha}$ (**30**) and $E_2$ (**31**) [7]. Somewhat later, Woodward and coworkers adapted their approach and succeeded even in the enantioselective total synthesis of naturally occurring stereoisomers [41].

The theoretical insight into the Diels–Alder reaction was strongly furthered by the advent of the Woodward–Hoffmann rules (a ground-state pericyclic change is symmetry-allowed when the total number of $(4q + 2)_s$ and $(4r)_a$ components is odd) published in 1965 [42] and 1969 [8]. According to these rules, it turns out that the Diels–Alder reaction has to be regarded as a synchronous $[_{\pi}4_s + _{\pi}2_s]$-cycloaddition. This indicates that it proceeds via a *suprafacial/suprafacial* interaction of a 4π electron system of the diene component with a 2π electron component of the dienophile. This interaction leads to a transition state without an additional symmetry-imposed energy barrier and permits this reaction to take place easily and is allowed thermally (Fig. 9).

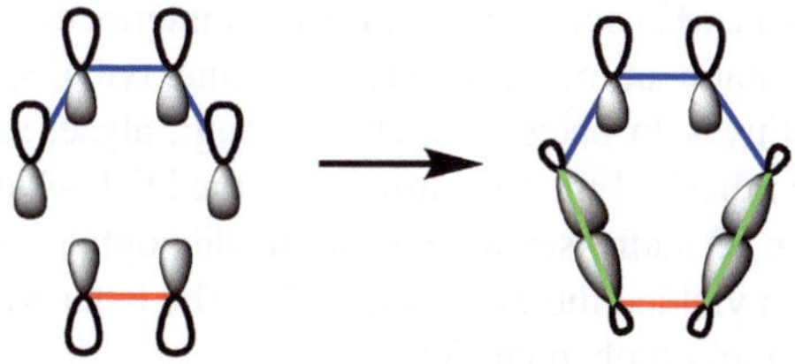

**Fig. 9** The Diels–Alder reaction as a *suprafacial* [4 + 2]-cycloaddition resulting in two σ bonds (green) and one π bond starting from two π bonds of the diene (blue) and one π bond from the dienophile (red)

As mentioned above, the 1981 Nobel Prize in Chemistry was awarded to Roald Hoffmann and Kenichi Fukui "for their theories, developed independently, concerning the course of chemical reactions" two years after Woodward's death.

## 3.2 *Disodium Prephenate*

The 1970s also brought methodological innovations. Thus, for example, the introduction of specialized dienes and dienophiles expanded the range of structures the reaction could easily build. Danishefsky's diene (1-methoxy-3-trimethylsilyloxy-1,3-butadiene (**33**), Fig. 10) first reported in 1974, is one such breakthrough reagent [9]. It enabled Diels–Alder reactions to directly yield α,β-unsaturated cyclohexenone adducts after elimination of the 1-methoxy group. Thus, the synthesis of the aromatic bioprecursor prephenate by Danishefsky in 1979 used an intramolecular Diels–Alder to construct efficiently a bicyclic scaffold, illustrating how the reaction could even construct frameworks found in primary metabolism [11].

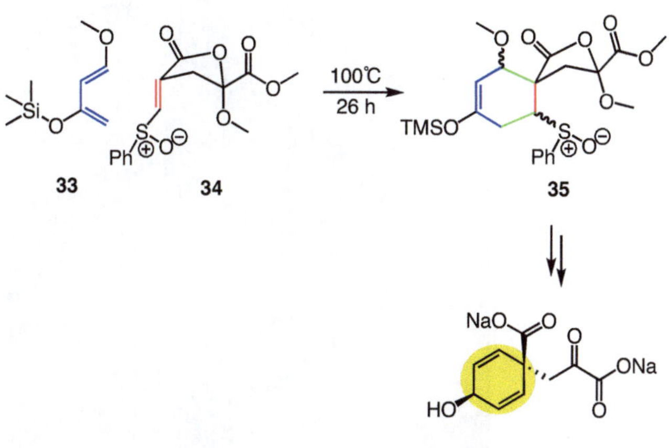


**Fig. 10** The Diels–Alder key step in the synthesis of disodium prephenate (**32**)

Disodium prephenate (**32**) occurs in Nature as an intermediate of the biosynthesis of folates and the aromatic amino acids phenylalanine, tyrosine, and tryptophan via the shikimic acid pathway in bacteria, archaea, fungi, algae, and some protozoans and plants [43]. Its synthesis (Fig. 10) started with the Diels–Alder reaction between the Danishefsky diene **33** addressed above and the dienophile **34** by heating them to 100°C for 26 h, which yielded the cycloadduct **35**. The latter was then elaborated to finally produce disodium prephenate (**32**).

## *3.3 (±)-Cedrol and (±)-Cedrene*

An intramolecular Diels–Alder reaction was used to build the tricyclic cedrenoid sesquiterpenes (±)-cedrol (**36**) and (±)-cedrene (**37**), compounds occurring in the essential oil of cedar trees, *Cedrus libani* A. Rich. in the plant family Pinaceae [44] (Plate 2). They are used in the chemistry of aroma compounds, having antioxidant

**Plate 2** *Cedrus libani* subsp. *atlantica*. Photograph courtesy N. D. Karabelas, Creative Commons Attribution 4.0

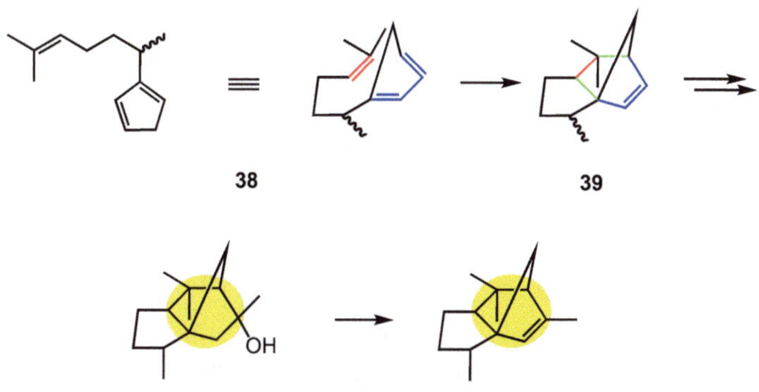


**Fig. 11** The Diels–Alder key step in the synthesis of (±)-cedrol (**36**) and (±)-cedrene (**37**)

effects and also occur globally in traditional medicine and cosmetics. The starting material for the key step, 1-(6-methylhept-5-en-2-yl)cyclopenta-1,3-diene (**38**), was prepared from sodium cyclopentadiene by alkylation with 6-methyl-5-hepten-2-ol tosylate. For the cyclization in an intramolecular Diels–Alder reaction it was heated in hexamethylphosphoramide at 205°C for 7 h to provide the tricyclic olefin **39** in 36% yield. This compound was then transformed using common chemistry to (±)-cedrol (**36**), which upon dehydration yielded (±)-cedrene (**37**) (Fig. 11).

## *3.4 Approaches to Adriamycin: Islandicin and Digitopurpone*

In 1977, a Diels–Alder reaction between a photochemically generated bisketene and an electron-deficient olefin (2-methyl-quinone) (Fig. 12) was shown to produce, although in rather low yields, two adducts (islandicin (**40**) and digitopurpone (**41**)), which could serve as educts for the total synthesis of the antineoplastic agent adriamycin (**42**) [45], also known as doxorubicin, isolated from the bacterium *Streptomyces peucetius* Grein et al., family Streptomycetaceae. Adriamycin (**42**) is a cancer chemotherapeutic medication, which belongs to the anthracycline family and is an antibiotic and topoisomerase inhibitor. Adriamycin (**42**) works by interfering with the growth and reproduction of cancer cells and is used alone or with other drugs to treat many types of cancer, including leukemia, lymphoma, neuroblastoma, sarcoma, Wilms tumor, and certain cancers of the lung, breast, stomach, ovary, thyroid, and bladder.

In the synthesis of **40** and **41**, a rather uncommon diene was thus added to the toolkit of Diels–Alder reactions.

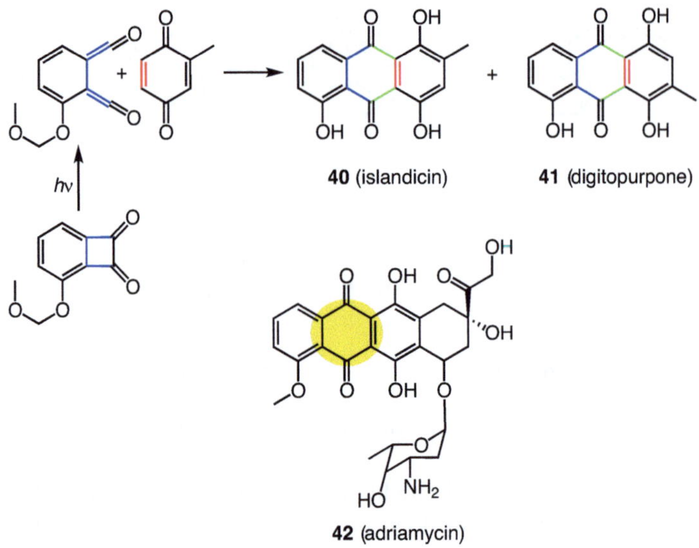


**Fig. 12** The Diels–Alder key step in the synthesis of islandicin (**40**) and digitopurpone (**41**)

## *3.5 Gibberellic Acid*

Gibberellic acid (**43**) (Fig. 13), also called gibberellin $A_3$ or $GA_3$, is a pentacyclic diterpenoid of high stereochemical complexity. It is a hormone found in plants and fungi and promotes the growth and elongation of cells. Thus, it is of agronomic industrial interest in being used with respect to seed germination, stem elongation, flowering, fruit development, and the regulation of gene expression in the cereal

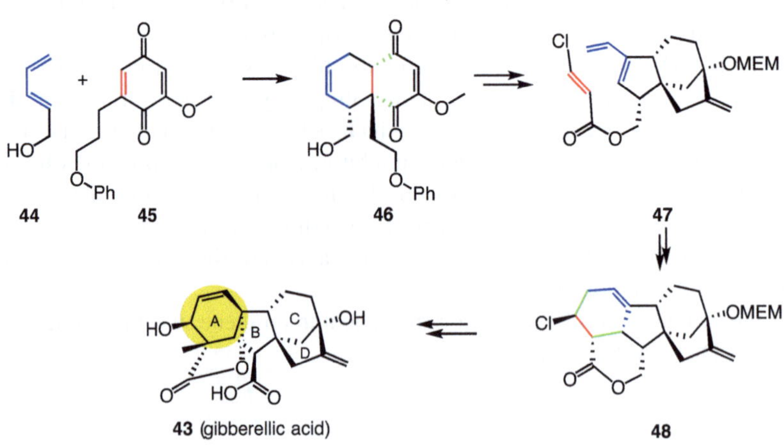


**Fig. 13** Double Diels–Alder reaction to synthesize gibberellic acid (**43**)

aleurone layer. Its biosynthesis originates from geranylgeranyl diphosphate, a common $C_{20}$ precursor for diterpenoids [46].

Again, Corey was finally successful with a total synthesis of gibberellic acid (**43**) after about 150 published papers from 25 laboratories tried in vain to achieve this goal (Fig. 13). In 1978, two consecutive Diels–Alder reactions, one intermolecularly generating the *cis*-fused B/C ring unit, the other intramolecular providing ring A together with the pentacyclic lactone ring, paved the way to construct this complicated skeleton. Thus, in the first step, the diene **44** was added to the quinone **45** providing the *cis*-decalin system **46**, having the proper stereochemistry of the ring B/C fusion [47]. Further elaboration of **46** led to the tricyclic compound **47** [48], which underwent an intramolecular Diels–Alder reaction to yield the pentacyclic precursor **48** displaying a pentacyclic system. The latter compound **48** could then be transformed stereospecifically into the target product gibberellic acid (**43**).

## *3.6 D-Homoestrone, Estrone*

Of course, a special fascination was accorded to the synthesis of steroids, in particular, those of estrone-type compounds. In 1976, Kametani and his group were successful in applying their synthesis principle, which consists of the use of an *o*-quinodimethane prepared in situ from a benzocyclobutane, as the educt for an intramolecular Diels–Alder cyclization. This reaction yielded the B ring of the steroid skeleton, which could then be elaborated along with common methodology to D-homoestrone (**49**) and finally estrone (**50**) [49]. Thus, the starting material **51** (Fig. 14) was easily transformed in boiling *o*-dichlorobenzene to the *O*-methyl-D-homoestrone, via the Diels–Alder system **52**, which after demethylation provided

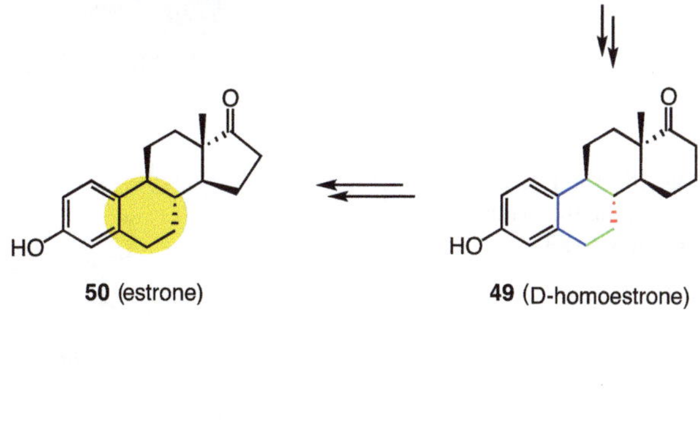


**Fig. 14** The Diels–Alder key step in D-homoestrone (**49**) and estrone (**50**) synthesis

D-homoestrone (**49**). Applying established synthesis methodology, this could then be converted into estrone (**50**).

## *3.7 Khusimone*

The oil of the plant *Vetiveria zizanioides* (*Chrysopogon zizanioides*) L. Nash (Plate 3), from the family Poaceae, native to Africa, China, and India, constitutes a valuable raw material for refined fragrances. The characteristic scent of the essence is in part due to khusimone (**53**), which is a tricyclic *nor*-sesquiterpene ketone with a zizaene skeleton. An intramolecular thermal Diels–Alder reaction of 5-methyl-1-(4-methylcyclohexa-1,3-dien-1-yl)hex-4-en-1-one (**54**) by G. Büchi and his co-workers resulted, after some further elaboration, in khusimone (**53**) (Fig. 15) [50].

**Plate 3** *Vetiveria zizianioides*. Photograph courtesy Ratel, Creative Commons Attribution Share Alike 4.0

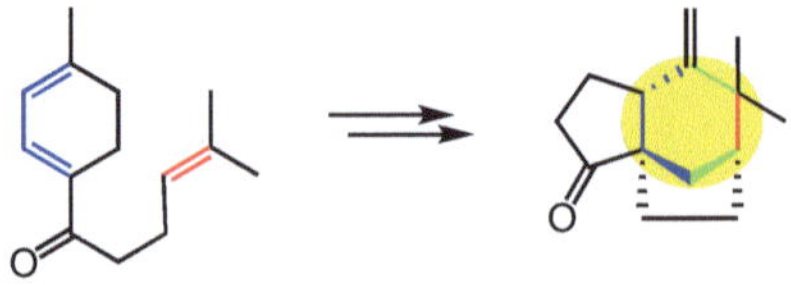

**54**

**53** (khusimone)

**Fig. 15** The Diels–Alder key step in the synthesis of khusimone (**53**)

### *3.8 Selina-3,7(11)-diene*

The oil obtained by steam distillation of hops (*Humulus lupulus* L., family Cannabaceae; Plate 4), which can be added at various points in the beer brewing process to improve aroma yields, contains selina-3,7(11)-diene (**55**), among about 15 sesquiterpene hydrocarbons present [51].

In 1978, an easy entry to this group of eudesmane sesquiterpenoids using an intramolecular Diels–Alder reaction was described by S. R. Wilson and D. T. Mao [52]. Accordingly, the O-trimethylsilyl derivative **56** of the corresponding 1,3,9-decatriene was smoothly cyclized to yield **57**, which could then be elaborated in the usual way into the final product **55** (Fig. 16).

**Plate 4** *Humulus lupulus.* Photograph courtesy H. Zell, Creative Commons Attribution Share Alike 3.0

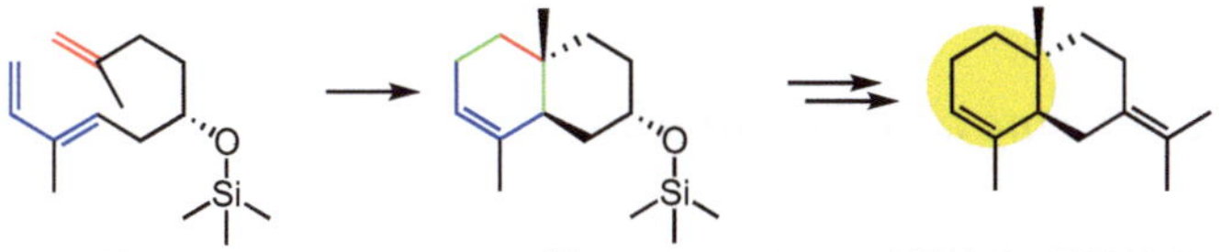


**56** **57** **55** (sela-3,7(11)-diene)

**Fig. 16** The Diels–Alder key step in the synthesis of sela-3,7(11)-diene (**55**)

## 3.9 (±)-Dendrobine

The Chinese herbal drug "Chin-Shi-Hu" (*Dendrobium moniliforme* L., family Orchidaceae; Plate 5), which is used as a tonic and antipyretic, contains as a major component the alkaloid dendrobine (**58**). This compound is structurally and pharmacologically similar to picrotoxinin, which has a role as γ-aminobutyric acid antagonist and serotonergic antagonist [53].

Among several syntheses, an outstanding one of W. R. Roush described in 1978 is based on an intramolecular Diels–Alder reaction [54]: the open-chain triene **59** was cyclized in refluxing toluene to yield after some further elaboration the racemic product, (±)-dendrobine (**58**) (Fig. 17).

**Plate 5** *Dendrobium moniliforme.* Photograph courtesy Naoki Takebayahi, Creative Commons Attribution Share Alike 2.0

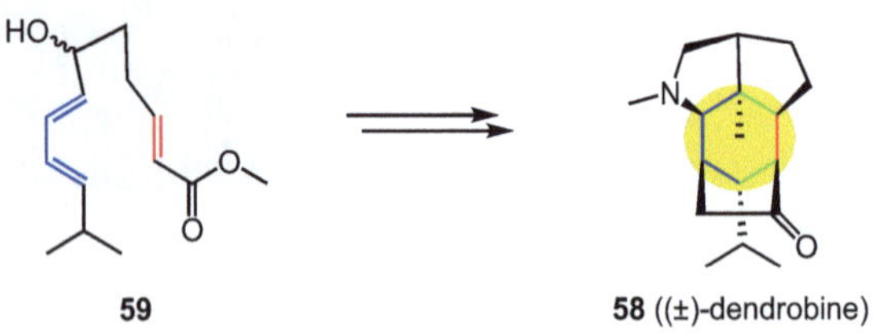


**Fig. 17** The Diels–Alder key step in the synthesis of (±)-dendrobine (**58**)

## 3.10 δ-Coniceine and Tylophorine

A final example of alkaloid syntheses by means of an intramolecular Diels–Alder reaction in the late 1970s, as described in 1979 by S. M. Weinreb and coauthors [55], is the synthesis of the two indolizidine alkaloids δ-coniceine (**60**), found in the poisonous hemlock (*Conium maculatum* L., family Apiaceae; Plate 6) and tylophorine (**61**) found in the Indian Ipecac plant *Tylophora asthmatica* Wight & Arn. (syn. *T. indica* Merr.) of the Asclepiadaceae family (Plate 7).

**Plate 6** *Conium maculatum*. Photograph courtesy W. and W. Follette, Public domain

**Plate 7** *Tylophora indica*. Herbarium JCB, Creative Commons Attribution Share Alike 4.0

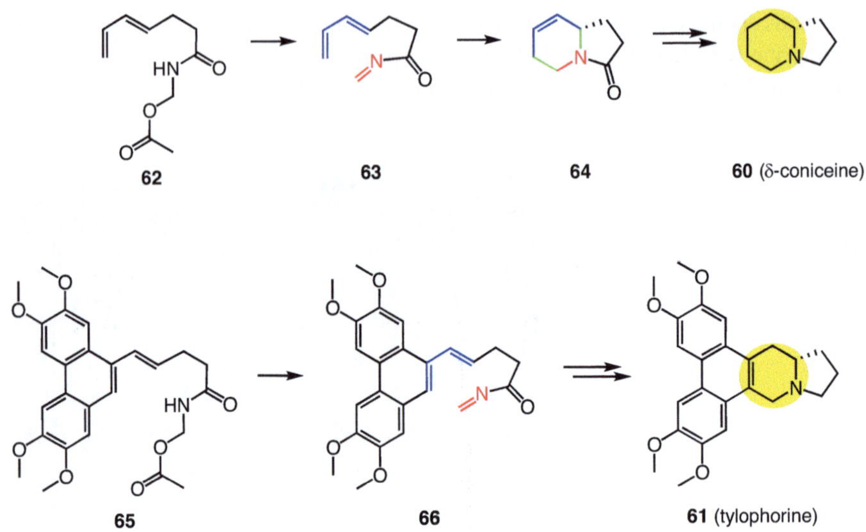


**Fig. 18** The Diels–Alder key step in the synthesis of δ-coniceine (**60**) and tylophorine (**61**)

Conversion of the acetate **62** upon heating a toluene solution by rapidly passing it over a column of glass helices maintained at 370–390°C resulted in the acylimine **63**, which then cyclized in an intramolecular hetero-Diels–Alder reaction into the unsaturated lactam **64**. The latter could then be reduced conventionally to the alkaloid δ-coniceine (**60**) (Fig. 18).

The same principle was applied to the acetate **65**, which upon pyrolysis yielded **66**. Cyclization of **66** and the usual reduction steps then provided tylophorine (**61**) (Fig. 18).

## *3.11 δ-Cadinol*

In 1979, a final example of an intramolecular Diels–Alder reaction was encountered, and was used to construct the skeleton of the sesquiterpene alcohol (±)-δ-cadinol (**67**), formerly called (±)-torreyol [56]. Both enantiomers of **67** are produced by several plants, fungi, or algae, and even act as a pheromone in the butterfly *Plebejus argyrognomon* Bergsträsser (syn. *Lycaeides argyrognomon* Bergsträsser) of the Lycaenideae family (Plate 8) [57].

The intramolecular Diels–Alder reaction started from the trieneol **68**, which cyclized spontaneously to a mixture of two ketones with a *cis/trans* positioned

**Plate 8** *Plebejus argyrognomon* (female). Photograph courtesy Rosenzweig, Creative Commons Attribution Share Alike 3.0

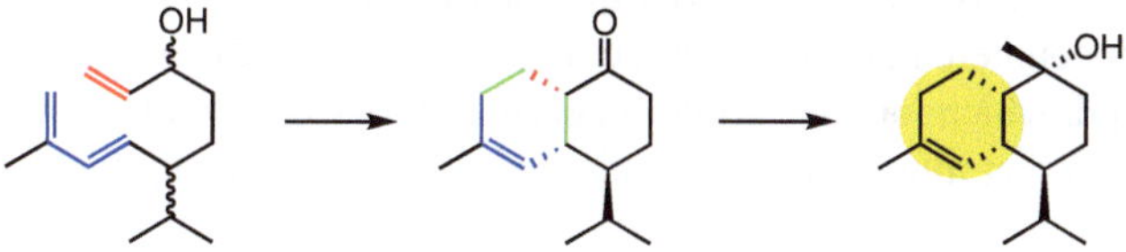


**68** **69** **67** (δ-cadinol)

**Fig. 19** The Diels–Alder key step in the synthesis of δ-cadinol (**67**)

isopropyl group. From the preponderant isomer **69** the desired racemic δ-cadinol (**67**) could then be derived by conventional methylation (Fig. 19).

# 4 Advancement in Strategies (1980s)

This section highlights significant progress in synthesis methodologies during the 1980s, particularly focusing on the application of Diels–Alder reactions in the production of complex organic compounds. This era saw the development of innovative approaches to construct intricate molecular structures, including steroids, alkaloids, terpenoids, and other biologically significant compounds. The section explores various examples where intramolecular and intermolecular Diels–Alder reactions were employed to achieve stereoselective and efficient syntheses, showcasing advancements in reaction conditions, catalysts, and strategies to streamline multi-step processes. These methodologies not only enhanced the synthesis of natural products but also contributed to the development of pharmaceuticals and materials science.

By the 1980s, even cascade sequences involving multiple Diels–Alder steps were being designed. A spectacular example was the approach to polycyclic frameworks by Vollhardt, to for example, the steroid estrone (**50**) and other polycycles, where sequential Diels–Alder reactions or tandem pericyclic steps built up complex ring systems in just a few operations [10]. Let us start with the year 1980, which brought about a memorable total synthesis of *dl*-estrone.

## *4.1 dl-Estrone (Ring B)*

Through the years, estrone (**50**) constituted a challenging synthesis target to measure the utility of novel methodologies on the one hand, and on the other one as a relay point en route to contraceptive drugs [58, 59]. In 1980, R. L. Funk and K. P. C. Vollhardt described an efficient sequence to a precursor suited for the intramolecular Diels–Alder reaction [10] (Fig. 20), which considerably improved on that described in Sect. 3.6: compound **70** was cyclized using cyclopentadienyl-cobalt-dicarbonyl as catalyst to yield the intermediate **71**. In a reaction analogous to the one encountered already with the transformation of **51** to **52**, intermediate **72** was produced, which, upon heating in decane, underwent an intramolecular Diels–Alder reaction that after some elaboration of the product finally yielded *dl*-estrone (**50**).

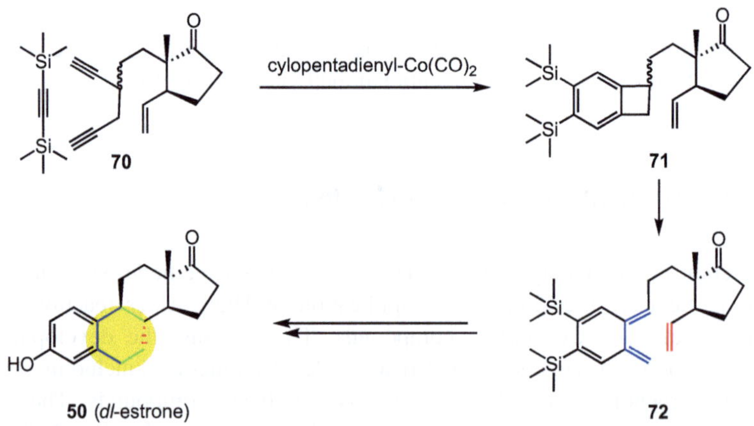


**Fig. 20** The Diels–Alder key step in the synthesis of *dl*-estrone (ring B) (**50**)

## 4.2 Cantharidin

Cantharidin (**73**) is a member of the terpenoid class, which is secreted by many species of blister beetles, as exemplified by the Spanish fly, *Lytta vesicatoria* (family Meloidae) (Plate 9). Besides its historical use as an aphrodisiac, it can be applied in dissolved form as a topical medicine to remove warts and tattoos.

The synthesis of cantharidin (**73**) using an intermolecular Diels–Alder reaction with a heterocycle as the diene and application of high pressure was described by W. G. Dauben and coauthors in 1980 [60]. They used a furan (**74**) as the diene to react with the easily prepared 2,5-dihydrothiophene-3,4-dicarboxylic acid (**75**) in dichloromethane as the solvent and fifteen kilobar pressure at room temperature for 6 h. This reaction resulted in two adduct isomers with **76** as the main component, which could then be easily transformed to the target cantharidin (**73**) by heating with Raney nickel in ethyl acetate (Fig. 21).

**Plate 9** *Lytta vesicatoria.* Photograph courtesy Siga, Creative Commons Attribution Share Alike 3.0

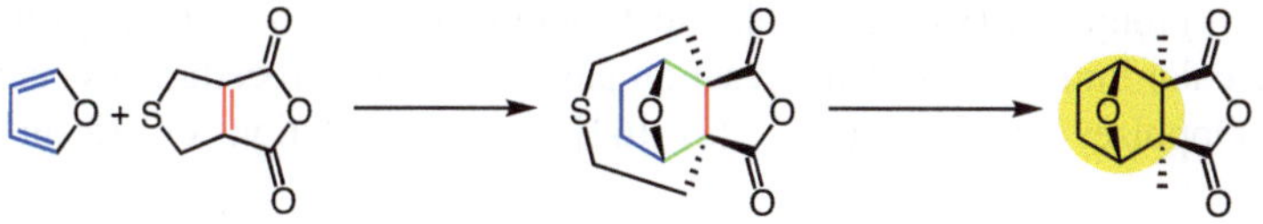


**74** **75** **76** **73** (cantharidin)

**Fig. 21** The Diels–Alder key step in the synthesis of cantharidin (**73**)

## 4.3 (±)-Lysergic Acid

Lysergic acid (**77**) (Fig. 11) is a main constituent of the ergot fungus *Claviceps purpurea* (Fr.) Tul. (family Clavicipitaceae) growing on rye grain (Plate 10). It was identified as the causative agent of the dreaded medieval illness St. Antony's fire or

**Plate 10** *Claviceps purpurea*. Photograph courtesy D. Jacquin, Public domain

Devil's curse, which was characterized by convulsions, muscle spasms, vomiting, hallucinations, and a gangrenous pain where the limbs, fingers, toes, and nose were "eaten up by the holy fire that blackened like charcoal". In modern times, it was recognized as a fundamental unit compound of psychedelic drugs, such as the legendary derivative lysergic acid diethylamide [61].

This ergoline alkaloid lysergic acid (**77**) had already been synthesized by the Woodward group in 1956 using a classical route involving about 15 steps [62]. An intramolecular imino-Diels–Alder reaction paved a shorter procedure, as elaborated by W. Oppolzer in 1981 (Fig. 22) [63]. A 1% solution of **78** in 1,2,4-

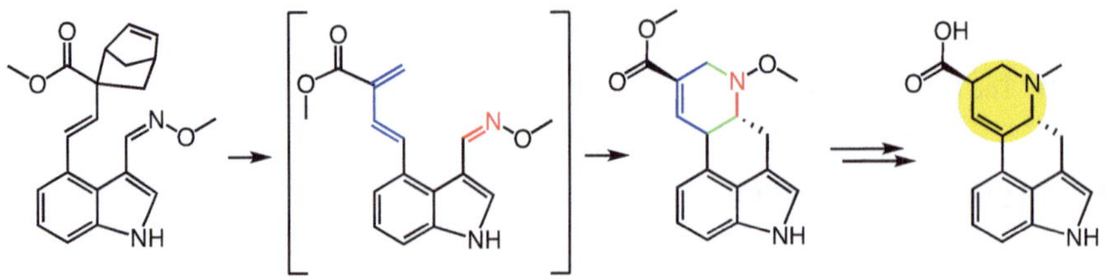


**78** **79** **80** **77** ((±)-lysergic acid)

**Fig. 22** The Diels–Alder key step in the synthesis of (±)-lysergic acid (**77**)

trichlorobenzene was dropped over a time interval of 5 h into this solvent heated to 200°C under argon. In this manner, **78** first underwent a retro Diels–Alder reaction to furnish the desired imino-diene **79**, which then cyclized in an intramolecular imino-Diels–Alder reaction into the 8-ergolene derivative **80**. The latter was then transformed conventionally into the desired product, (±)-lysergic acid (**77**).

## *4.4 11-Ketosteroids*

11-Ketosteroids, like **81**, are typical of cortical hormones occurring in animals and plants, and are thus highly desired synthesis targets. Such a compound can be efficiently synthesized from indanonepropionic acids, like **82**, as shown by the group of G. Stork in 1981 (Fig. 23) [64]. This precursor **82** was prepared in 1982 starting with the intramolecular Diels–Alder reaction between the ketoester **83** and the pyrrolidine enamine of cyclopentanone (**84**) by refluxing in ethanol for 5 min, crossing a type of a transition state like **85**. This furnished the cycloadduct **86**, which could then be transformed into **82** and finally the target **81** using conventional chemistry [65]. In this example, one can clearly see the advancement in thinking about the methodology.

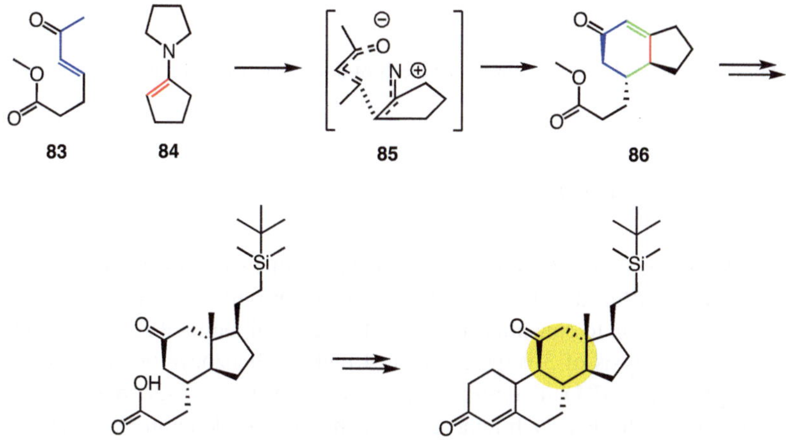


83 84 85 86

82 81 (11-ketosteroid derivative)

**Fig. 23** The Diels–Alder key step in the synthesis of the 11-ketosteroid derivative (**81**)

## 4.5 *dl*-Estrone (Ring C)

The hitherto described syntheses of *dl*-estrone (**50**) using a Diels–Alder synthesis (Sects. 3.6 and 4.1) provided ring B. In 1985, M. E. Jung and K. M. Halweg aimed at the construction of ring C using an intramolecular Diels–Alder reaction [66]: the starting material **87** was heated at 200°C in hexamethylphosphoric amide, which led to the loss of water, and then the resulting triene **88** could cyclize into **89** (Fig. 24). The latter was then easily elaborated to the target compound, *dl*-estrone (**50**).

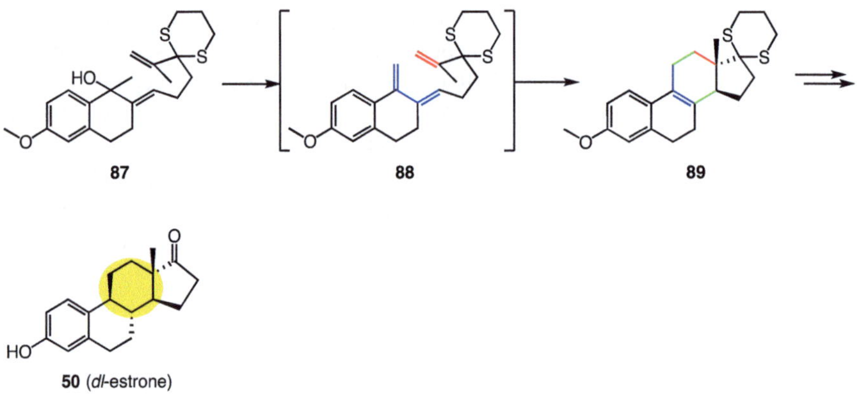


**Fig. 24** The Diels–Alder key step in the synthesis of *dl*-estrone (ring C) (**50**)

## 4.6 $\Delta^9$-Progesterone

In the case of the steroid hormone involved in the menstrual cycle and pregnancy, $\Delta^9$-progesterone (**90**), the steroid skeleton was again built using an intramolecular Diels–Alder reaction to construct ring C [67]. The *o*-quinodimethane (**92**) generated in situ by thermolysis of 3-isopropenyl-5-(4-methoxybenzocyclobutenyl)pentan-2-one-2-ethylene ketal (**91**) underwent an intramolecular Diels–Alder reaction to yield stereoselectively the B/C/D fragment **93** of the steroid skeleton (Fig. 25). Using conventional chemistry, **93** was then elaborated into the target, $\Delta^9$-progesterone (**90**). In addition to their importance as parts of the progesterone family of hormones, compounds like **90** are also of high interest as educts in the synthesis of $C_{11}$-functionalized products.

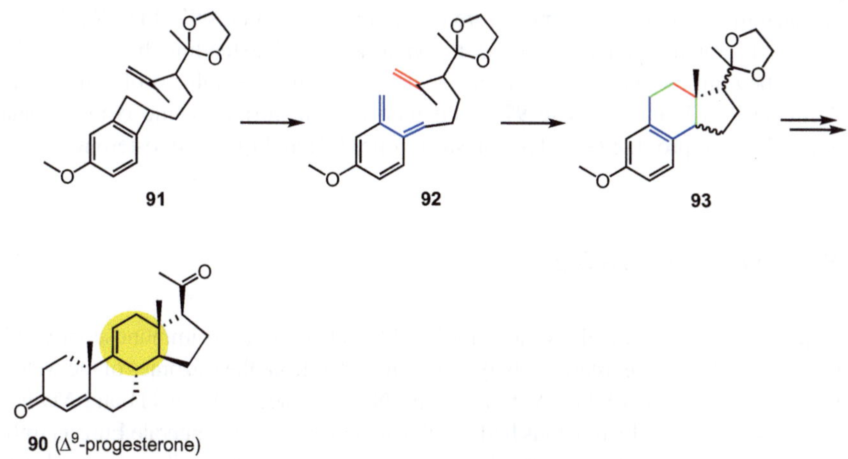


**Fig. 25** The Diels–Alder key step in the synthesis of $\Delta^9$-progesterone (**90**)

## *4.7 (±)-11-Ketotestosterone and (±)-Adrenosterone*

Corticosterones like (±)-11-ketotestosterone (**94**) and (±)-adrenosterone (**95**) (Fig. 26) are challenging targets for total synthesis due to their biological importance in affecting all of the major systems of the body, including the cardiovascular, muscoskeletal, nervous, and immune systems, and playing a critical role in fetal development. A novel strategy to synthesize these compounds involving a D to B/C/D to A/B/C/D skeletal formations was developed by the laboratory of P. J. De Clercq in 1985 [68]. The key step was the high-yield stereoselective intramolecular Diels–Alder reaction of the furan-diene (**96**) by vigorously shaking or with ultrasonic

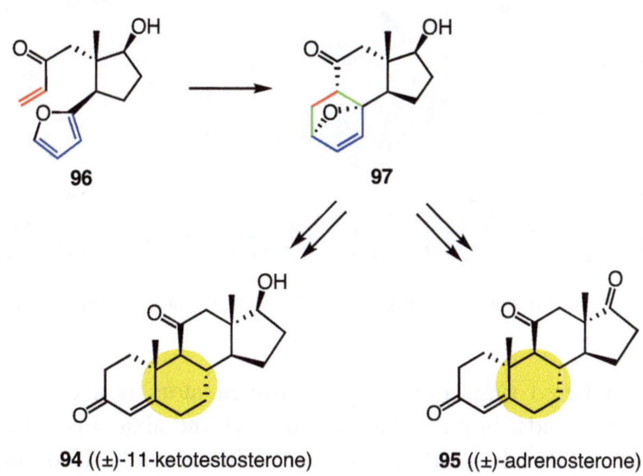


**Fig. 26** The Diels–Alder key step in the synthesis of (±)-11-ketotestosterone (**94**) and (±)-adrenosterone (**95**)

assistance in water at room temperature to yield within minutes **97** (Fig. 26). Here a substituted furan derivative as diene component was used, extending the scope of the reaction beyond the use of furan itself, as described in the synthesis of cantharidin (**73**) in Sect. 4.2. Intermediate **97** could then be transformed using conventional methodology to provide (±)-11-ketotestosterone (**94**) and (±)-adrenosterone.

## *4.8 Deoxynupharidine*

The quinolizidine alkaloid deoxynupharidine (**98**) is known for its immunosuppressive effect and weak anti-metastatic activity. It was reported from the rhizomes of the water-lilies *Nuphar lutea* (L.) Sibth. & Sm., family Nymphaeaceae (Plate 11) and *Nuphar japonicum*, perennial rhizomatous herbs native to subarctic and temperate Eurasia [69], and more recently from the scent glands of the Canadian beaver *Castor canadensis* Kuhl, family Castoridae [70]. Compound **98** is quite distinct from most alkaloids, because its ring system is with respect to its biosynthesis, by being of isoprenoid rather than of amino acid origin. A hydroxy derivative of **98** shows potential by inducing apoptosis in various leukemia, lymphoma, and carcinoma cell lines [71].

**Plate 11** *Nuphar lutea*. Photograph courtesy H. Hillewaert, Creative Commons Attribution Share Alike 3.0

As a six-membered ring containing a nitrogen atom is a common structural feature among alkaloids, in particular of quinolizidine alkaloids, an imine Diels–Alder reaction, as already used within the Weinreb synthesis of the indolizine alkaloid δ-coniceine (**60**) and, as described in Sect. 3.10, can serve as the key reaction. Thus, in the present example, the acylazadiene **100** prepared in situ by the thermal elimination of acetic acid from an hydroxamic acid derivative **99**

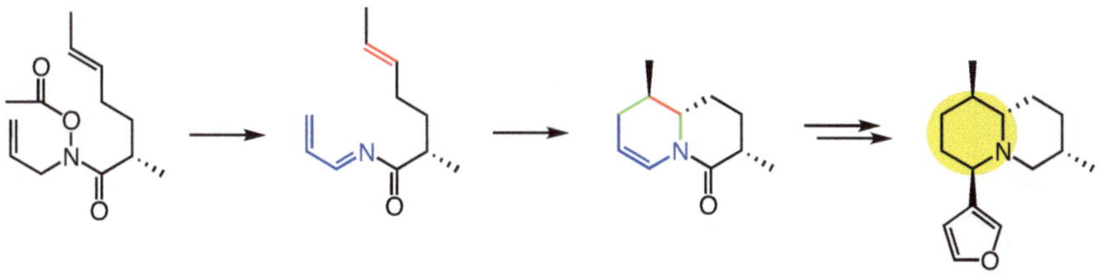

**99** **100** **101** **98** (deoxynupharidine)

**Fig. 27** The Diels–Alder key step in the synthesis of deoxynupharidine (**98**)

underwent the intramolecular aza-Diels–Alder reaction to produce the quinolizidine ring intermediate **101** (Fig. 27) [72]. This intermediate could then be elaborated using conventional chemistry into the target, deoxynupharidine (**98**).

## *4.9 Podophyllotoxin*

Podophyllotoxin (**102**) is a lignan occurring mainly in the roots of *Podophyllum* plants, like the Mayapple *Podophyllum peltatum* L., family Berberidaceae (Plate 12).

**Plate 12** *Podophyllum peltatum*. Photograph courtesy E. Hunt, Creative Commons Attribution Share Alike 4.0

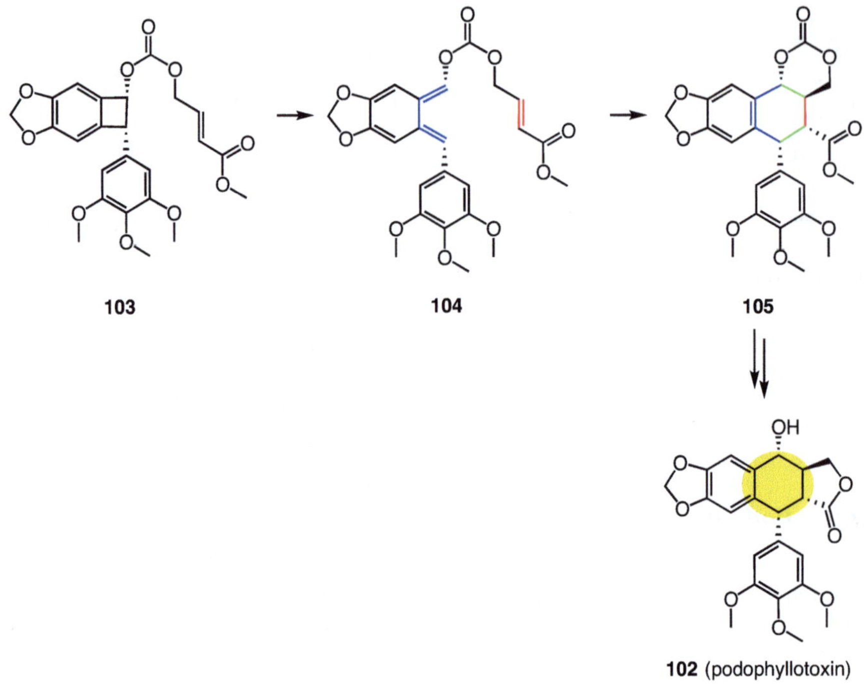

Fig. 28 The Diels–Alder key step in the synthesis of podophyllotoxin (**102**)

It constitutes the antiviral compound contained in medical creams to treat genital warts and the skin infection molliscum contagiosum [73].

The intramolecular Diels–Alder reaction used in situ from the benzocyclobutanol **103** thermally generated *o*-quinodimethane **104**, to yield the tetralin **105**, and provided an efficient approach to the total synthesis of podophyllotoxin (**102**) (Fig. 28) [74]. The tetralin **105** could then be efficiently transformed into the target **102** by common methodology.

## *4.10 Actinobolin*

The broad-spectrum antibiotic actinobolin (**106**), which inhibits protein synthesis by targeting bacterial and eukaryotic ribosomes, was isolated from cultures of the filamentous bacterium *Streptomyces griseoviridis* var. *atrofaciens* [75, 76].

The total synthesis of actinobolin (**106**) was achieved in 1986 starting from the properly substituted diene **107**, which was derived from the chiral pool member L-threonine, and reacting it with methyl propiolate (**108**) in an intermolecular Diels–Alder reaction at six kilobars and room temperature for three days to obtain the

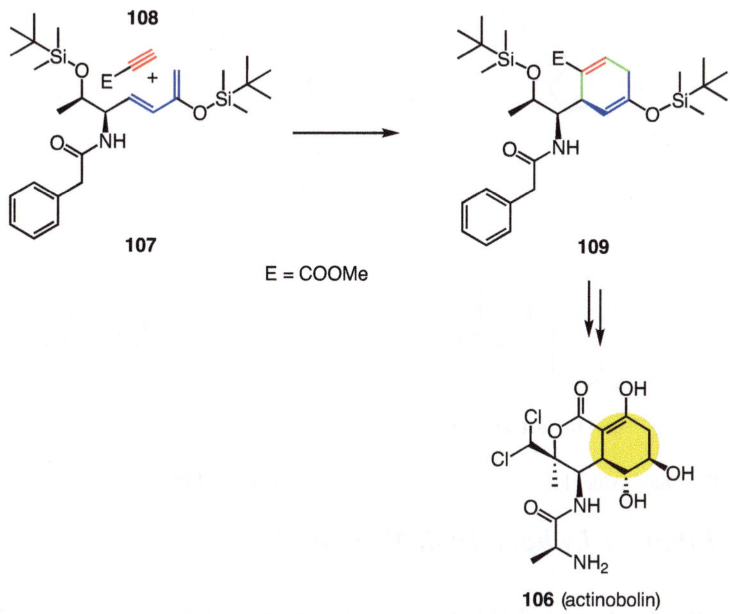


**Fig. 29** The Diels–Alder key step in the synthesis of actinobolin (**106**)

desired stereoisomer of the adduct **109** (Fig. 29) [77]. It is interesting to note that this reaction was very dependent on the reaction conditions. In an experiment running at 220°C, the ratio of undesired to desired diastereomer (**109**) was 1:1.7, whereas in the synthesis described above, which proceeded at room temperature and low pressure, it changed to 1:10! This was rationalized by a transition state favored by the mild conditions for both steric and electronic reasons. Intermediate **109** was then transformed to the target, actinobolin (**106**).

## *4.11 Testosterone and Androsterone (Ring A/B Construction)*

Again, in 1986, steroid synthesis became enhanced by T. Kametani and co-authors, when they achieved an intramolecular Diels–Alder reaction to construct rings A and B in one step [78]. The triene **112** derived from a known optically active indandione could be cyclized by heating a toluene solution in the presence of a catalytic amount of methylene blue at 220°C in a sealed tube for 100 h (Fig. 30). The correct A/B-*trans* isomer **113** dominated the corresponding *cis* isomer by a factor of about four, and their mixture could be separated easily by chromatography to yield pure **113**. By manipulating this latter compound using conventional methodology, the endogenous androgen steroid hormones testosterone (**110**) and androsterone (**111**) were finally obtained.

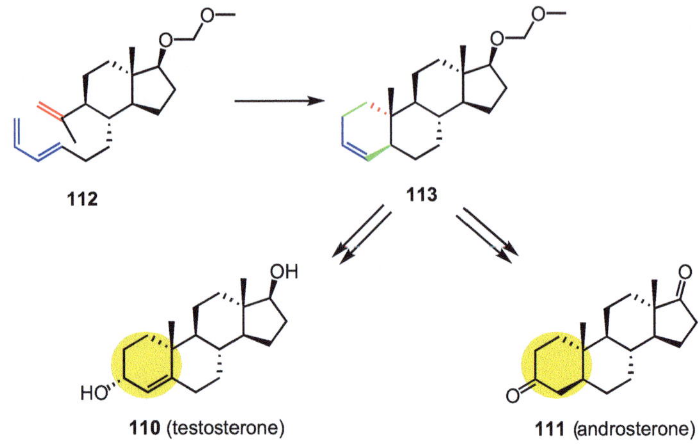


**Fig. 30** The Diels–Alder key step in the synthesis of testosterone (**110**) and androsterone (**111**)

## 4.12 *Inhoffen-Lythgoe Diol: Vitamin $D_3$*

The Inhoffen-Lythgoe diol (**114**) is a synthon commonly used in chiral selective reactions in natural product synthesis [79, 80]. It can also be employed as a catalyst and a ligand, and has a wide range of applications in the preparation of nitrogen- and sulfur-containing natural products, pharmaceuticals, and in material science. The diol plays a very important role in the field of asymmetric synthesis, and, in particular, it was used in the synthesis of vitamin $D_3$ (**115**). The P. A. Grieco group used in 1988 a novel intermolecular Diels-Alder strategy wherein an intact $C_{20}$ stereocenter as part of a diene unit is used to elaborate directly the stereocenters at $C_{13}$ and $C_{17}$ of a latent C/D *trans*-fused hydrindan ring system of **114** (Fig. 31)

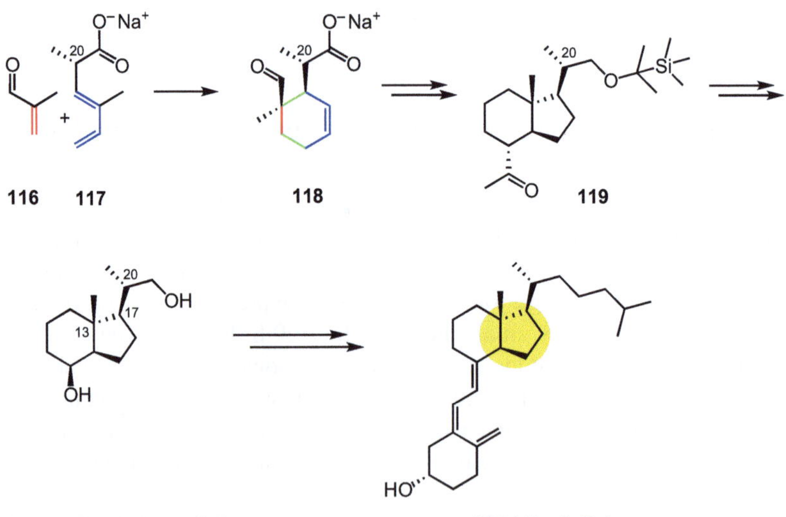


**Fig. 31** The Diels–Alder key step in the synthesis of the Inhoffen-Lythgoe diol (**114**) and vitamin $D_3$ (**115**)

[81]. The starting materials of the intermolecular Diels–Alder reaction, methacrolein (**116**) and the carboxylate **117**, were reacted in aqueous medium to provide the cyclohexene carboxylate **118**. The latter compound was then manipulated by known methodology into the indane derivative **119**, which could then be transformed into the target Inhoffen-Lythgoe diol **114** also using common reaction strategies. The transformation of **114** to vitamin $D_3$ (**115**) was already known through standard methodology.

### *4.13 Androst-6-ene-3,17-dione*

An ingenious route to a steroid skeleton was paved by T. Takahashi by contracting a 14-membered *all-(E)*-configured macrocyclic triene in one intramolecular Diels–Alder step into the A/B/C/D, *cis-anti-trans-anti-trans* steroid skeleton (Fig. 32) [82]. As an example for this methodology, they described the synthesis of androst-6-ene-3,17-dione (**120**), which is a precursor of testosterone (**110**) and estrone (**50**), and is produced in the adrenal gland and the gonads. The cyclopentanone **121** prepared in a conventional way was cyclized with lithium bis(trimethylsilyl)amide in dioxane at 80°C into the macrocycle **122**. The intramolecular Diels–Alder transannular reaction of **122** in xylene at 180°C with a catalytic amount of methylene blue was completed within 1 h to give androst-6-ene-3,17-dione (**120**) in 84% yield. None of its diastereomers could be detected.

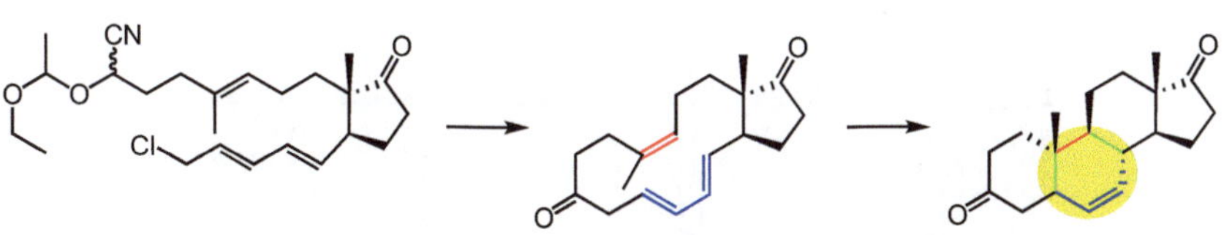


**121** **122** **120** (androst-6-ene-3,17-dione)

**Fig. 32** The Diels–Alder key step in the synthesis of androst-6-ene-3,17-dione (**120**)

## 5 Methodological Advances: The Era of the Intramolecular Diels–Alder Reaction (1990s)

The 1990s witnessed remarkable progress in the field of organic synthesis chemistry, particularly in the application of the Diels–Alder reaction, a powerful tool for constructing complex molecular architectures. This section highlights key methodological advances in the intramolecular Diels–Alder reaction, showcasing its versatility and efficiency in synthesizing biologically significant natural products and challenging molecular targets. From steroids and alkaloids to sesquiterpenes and spiro compounds, the intramolecular Diels–Alder reaction has proven indispensable in achieving stereoselective and regioselective transformations. The examples presented here illustrate the ingenuity of chemists in leveraging this reaction to overcome synthesis challenges, paving the way for advancements in medicinal chemistry, natural product synthesis, and the understanding of biological activity. In this period, advancements in chiral catalysts also enabled enantioselective Diels–Alder reactions [17], facilitating the synthesis of natural products with high enantiomeric purity. Several examples also including the development of tandem reactions [18] should illustrate this point.

## *5.1 (+)-Withanolide E*

Withanolides are a group of at least 300 naturally occurring steroids built on an ergostane skeleton [83]. They occur as secondary metabolites primarily in genera of the Nightshade family. One of the members of this group, (+)-withanolide E (**123**), has an attached δ-lactone and was isolated from the evergreen shrub *Withania somnifera* L. (family Solanaceae) (Plate 13). It displays insect antifeedant, antifungal, and antitumor activity, as already reviewed in this series [83].

**Plate 13** *Withania somnifera*. Photograph courtesy Vinayaraj, Creative Commons Attribution Share Alike 4.0

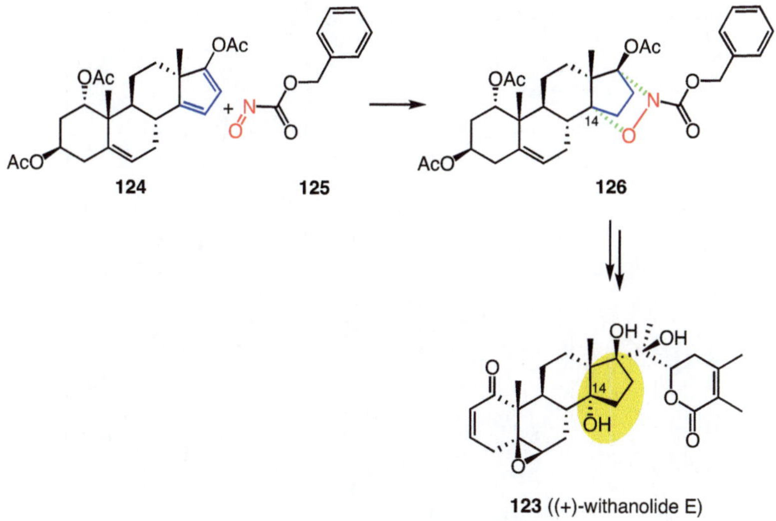


**Fig. 33** The Diels–Alder key step in the synthesis of (+)-withanolide E (**123**)

To synthesize the δ-lactone group pendant on ring D of (+)-withanolide E (**123**) (Fig. 33) A. Perez-Medrano and P. A. Grieco (1991) used a hetero-Diels–Alder reaction between the steroidal dienol acetate **124**, serving as diene, and benzyl nitrosoformate (**125**), as the hetero-dienophile, for the introduction of an α-hydroxy group into the $C_{14}$ position [84]. Treatment of **124** at 0°C for 20 min in dichloromethane solution with benzyl nitrosoformate (**125**), generated in situ from oxidizing benzyl *N*-hydroxycarbamate with tetrabutylammonium periodate, afforded in nearly quantitative yield the adduct **126**. The latter was then elaborated into the target (+)-withanolide E (**123**) using conventional methodology.

## 5.2 *Taxol*

Taxol (**127**), also known as paclitaxel, is a natural compound originally isolated from the bark of the Pacific yew, *Taxus brevifolia* Nutt. from the family Taxaceae (Plate 14), which is a slow-growing conifer native to the Pacific Northwest of the United States. The isolation was first achieved in the 1960s as part of a U. S. National Cancer Institute program to screen plant extracts for their potential antitumor properties. It is a potent anticancer compound notable for its unique mechanism of stabilizing microtubules and preventing cancer cell division [85].

Due to its medicinal potential taxol (**127**) has been an intensively aimed target for synthesis, and a plethora of such investigations have been published [85]. In 1995, an efficient synthesis was reported of the taxane nucleus, in which both the rings A and C were prepared via tandem Diels–Alder cycloadditions from the two simple

**Plate 14** *Taxus brevifolia*. Photograph courtesy W. Siegmund, Creative Commons Attribution Share Alike 3.0

acyclic precursors **128** and **129** (Fig. 34) [86]. The first (intermolecular) Diels–Alder reaction in this tandem reaction proceeded under zinc dichloride catalysis in dichloromethane at 25°C to provide the substituted hexene **130** in about 60% yield. The second (intramolecular) Diels–Alder cycloaddition occurred under boron trifluoride etherate catalysis in toluene to give the tricyclic ketone **131** as a single diastereomer in about 80% yield. Interestingly, neither Lewis acid is capable of catalyzing both Diels–Alder reactions. The diastereomer **132** of **131** could be obtained readily from **131** under reflux in 10% sodium methylate in methanol. With **132**, there was now a useful synthon available to transform it in conventional steps into the desired target, taxol (**127**).

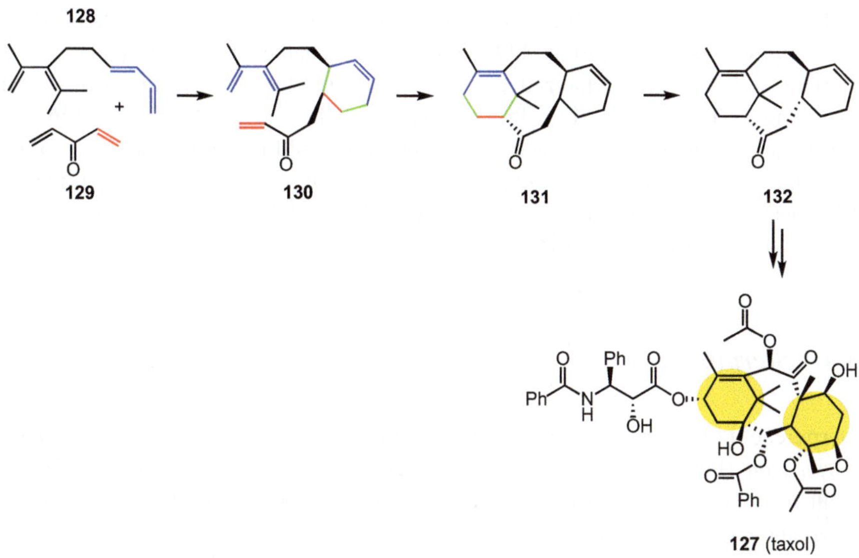


**Fig. 34** The Diels–Alder key step in the synthesis of taxol (**127**)

## *5.3 Cylindrospermopsin*

Cylindrospermopsin (**133**) (Fig. 35), produced by the cyanobacterium *Cylindrospermopsis raciborskii* is recognized globally as one of the most important members of the freshwater algal hepatotoxins [87, 88]. It accumulates in the liver over time, inhibits protein synthesis, disrupts metabolism, and leads to a variety of health issues including chronic liver disease and cancer.

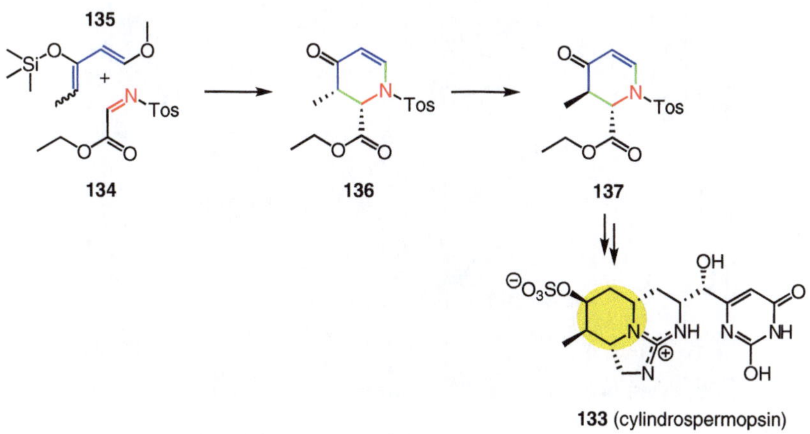


**Fig. 35** The Diels–Alder key step in the synthesis of cylindrospermopsin (**133**)

In 1996, S. M. Weinreb and two co-authors developed an imino Diels-Alder cycloaddition of *N*-tosylimine **134** with the oxygenated diene **135**, under either thermal or Lewis acid-catalyzed conditions, and produced a mixture of *cis*- and *trans*-enones **136** and **137** (Fig. 35) [89]. Although the undesired *cis*-enone **136** was the major product under all reaction conditions, it could be converted easily to the desired *trans*-enone **137** by acid-catalyzed isomerization. This piperidine A-ring unit **137** constituted the desired synthon to proceed along established methodology to the target cylindrospermopsin (**133**).

## 5.4 Sterpurene

Sterpurene (**138**), a sesquiterpene hydrocarbon metabolite of *Chondrostereum purpureum* (Pers.) Pouzar, family Cyphellaceae (Plate 15), constitutes a fungal plant pathogen (silver leaf disease) and potential mycoherbicide [90]. The fungus is a pathogen of mostly deciduous trees.

Sterpurene (**138**) was synthesized using a thermal Diels–Alder cycloaddition of the conjugated ester dienophile **139** to the ketene acetal diene **140** as the key step (Fig. 36) [91]. When **139** and **140** were heated in a sealed tube at 150°C under argon for ten days, the acetal **141** resulted in about 70% yield. The latter intermediate **141** was then transformed into the target compound sterpurene (**138**) using conventional methodology.

**Plate 15** *Chondrostereum purpureum*. Photograph courtesy J. Opioła, Creative Commons Attribution Share Alike 3.0

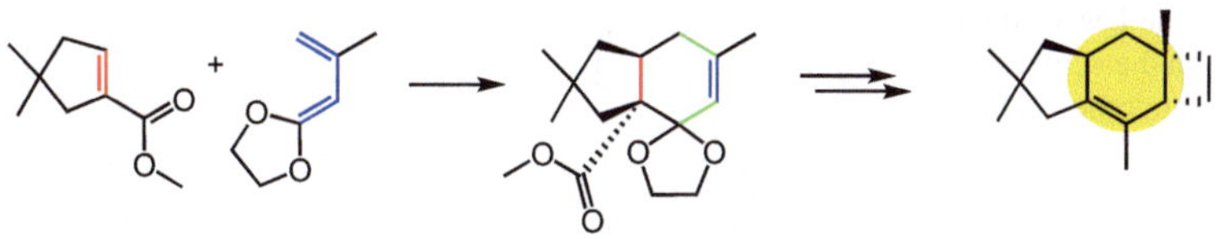


**Fig. 36** The Diels–Alder key step in the synthesis of sterpurene (**138**)

## 5.5 *(−)-Solanapyrone A*

Solanapyrones were isolated as phytotoxic compounds from the phytopathogenic fungus *Alternaria solani* Sorauer, family Pleosporales (Plate 16), which gives rise to the so-called early blight on the stems, leaves, and fruits of potato, tomato, eggplant, and pepper plants of the family Solanaceae, with leaf spots and defoliation symptoms [92].

For (−)-solanapyrone A (**142**), a synthesis was devised that was based on an enzymatic Diels–Alder reaction, resulting in high enantioselectivity and with good exo-selectivity, which are difficult to attain by chemical means (Fig. 37) [93]. To start this synthesis, the decatriene synthon prosolanopyrone (**143**) was prepared using standard chemistry. Execution of a non-enzymatic intramolecular Diels–Alder reaction did not result in satisfactory yields and stereoselectivities even upon variation of reaction conditions (solvent, temperature, and time). However, the use

**Plate 16** *Alternaria solani*. Photograph courtesy Clemson University Slide series, Creative Commons Attribution Share 3.0 United States

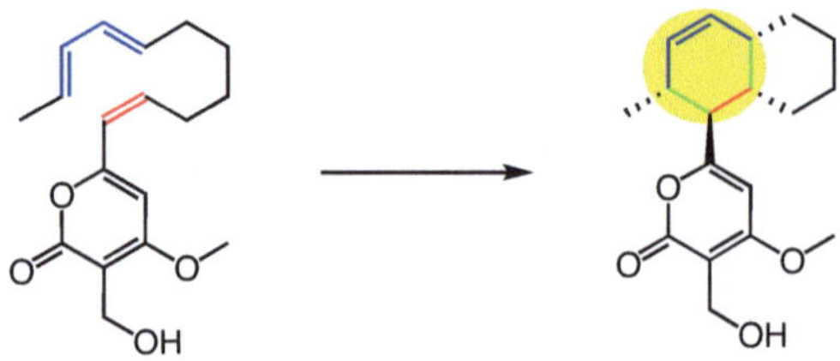


**143** (prosolanopyrone) **142** ((–)-solanopyrone A)

**Fig. 37** The Diels–Alder key step in the synthesis of (−)-solanapyrone A (**142**)

of a crude Diels-Alderase enzyme preparation from *A. solani* resulted in a 50% yield with 98% *ee* and high *exo*-selectivity (6:1) of the desired target, (−)-solanapyrone A (**142**).

## *5.6 (±)-Batrachotoxin A*

Batrachotoxin (**145**), a steroidal alkaloid, is a highly potent neurotoxin found in certain species of poison dart frogs, in particular the golden poison frog *Phyllobates terribilis* (Myers, Daly, and Malkin) from the Dendrobatidae family (Plate 17), as well as in some bird feathers like those of the Pitohui of New Guinea. It is one of the most toxic alkaloids known [94]. The frogs do not produce **145** themselves. Just as in the birds, it is believed that these frogs gain the toxin from batrachotoxin-containing insects that they eat and then secrete it through the skin [95]. Batrachotoxin (**145**)

**Plate 17** *Phyllobates terribilis.* Photograph courtesy W. Berns, Creative Commons Attribution Share 2.0 Germany

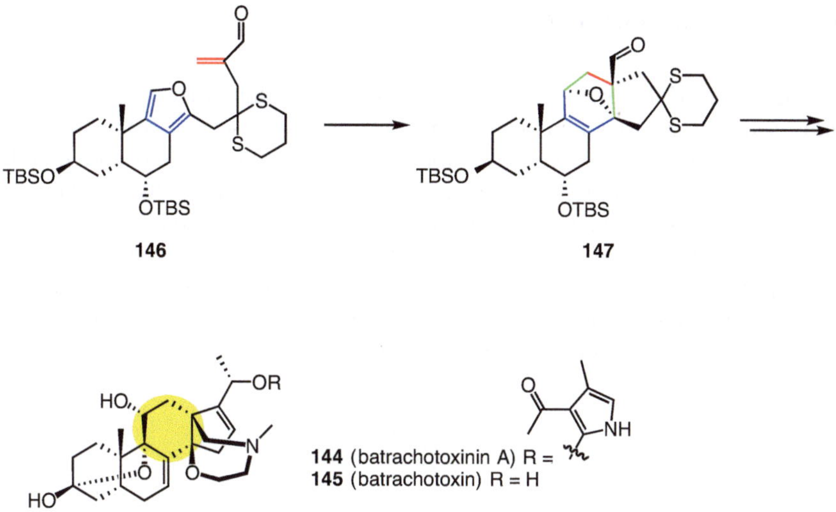


**Fig. 38** The Diels–Alder key step in the synthesis of (±)-batrachotoxinin A (**144**)

targets voltage-gated sodium channels in nerve and muscle cells. It locks them open, causing continuous nerve firing, which can lead to paralysis, cardiac arrest, and death. Extremely small doses can be lethal. It is estimated that as little as 2 micrograms can kill a human. While it has no medical use due to its extreme toxicity, **145** has been very important to better understand ion channel function in neurobiology.

(±)-Batrachotoxinin A (**144**), as a simple derivative of batrachotoxin (**144**) devoid of the pyrrole ring, exhibits a number of unique structural features challenging the synthesis chemist, by containing, for example, a steroid-based pentacyclic core skeleton, an intramolecular 3*β*-hemiketal unit, and a seven-membered oxazapane ring. Its total synthesis included an intramolecular Diels–Alder reaction with a furan ring as the diene partner to construct the C/D rings of the steroid skeleton, as described in 1998 by the group of Y. Kishi (Fig. 38) [96].

The synthon **146** bearing a furan ring as the diene for the intramolecular Diels–Alder reaction was prepared in a conventional way and it underwent smoothly a cycloaddition to yield **147**. The use of a furan ring for this purpose is reminiscent of the synthesis of adrenosterone (**95**) by De Clercq as described in Sect. 4.7. Interestingly enough, the selectivity of this cycloaddition was dramatically influenced by the substituent at $C_6$, with the *O-tert*-butyldimethylsilyl group determined as being optimum. The cycloadduct **147** could then be further elaborated into the target (±)-batrachotoxinin A (**144**), using standard methodology.

## 5.7 (−)-Chlorothricolide

The aglycone of the pyruvate carboxylase inhibitor chlorothricin, which displays some activity against Gram-positive bacteria and was isolated from the Gram-positive bacterium *Streptomyces antibioticus* (Waksman and Woodruff) of the family Streptomycetaceae, is called chlorotricolide (**148**) [97]. Its structure was resolved by the X-ray analysis of its methyl ester [98]. Chlorotricolide (**148**) displays a fascinating structure of a spirotetronic acid involved in a macrocyclic ring system, which has challenged synthesis chemists for many years. The enantioselective total synthesis of **148** was based in an ingenious tandem inter- and intramolecular Diels–Alder reaction described in 1998 (Fig. 39) [99].

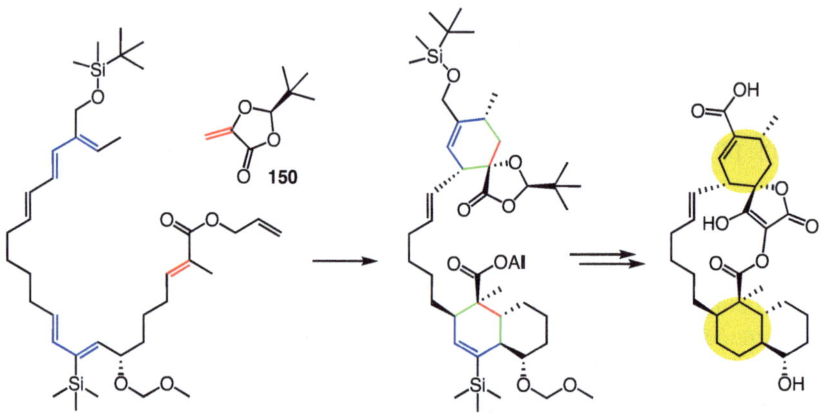


**Fig. 39** The Diels–Alder key step in the synthesis of (−)-chlorothricolide (**148**)

The key tandem inter-intramolecular Diels–Alder reaction was performed by heating a 1 *M* toluene solution of the hexaenoate **149** and two equivalents of the dienophile (*R*)-**150** for 20 h at 120°C with a crystal of 2,3-di-*tert*-butyl-4-methylphenol added as radical inhibitor. This provided stereoselectively the desired adduct **151** in about 40% yield. Conventional synthesis methodology could then be used to transform **151** into the desired target, (−)-chlorothricolide (**148**).

## 5.8 (±)-Bakkenolide A

(±)-Bakkenolide A (**152**) was first isolated from the herbaceous perennial butterbur plant *Petasites japonicus* (Siebold & Zucc.) Maxim. of the family Asteraceae (Plate 18) [100–102]. It is a methylene-substituted sesquiterpene spiro-lactone, which is related structurally to the eremophilanes. (±)-Bakkenolide A (**152**) showed toxicity

**Plate 18** *Petasites japonicus*. Photograph courtesy Joi Ito, Creative Commons Attribution 2.0 Generic

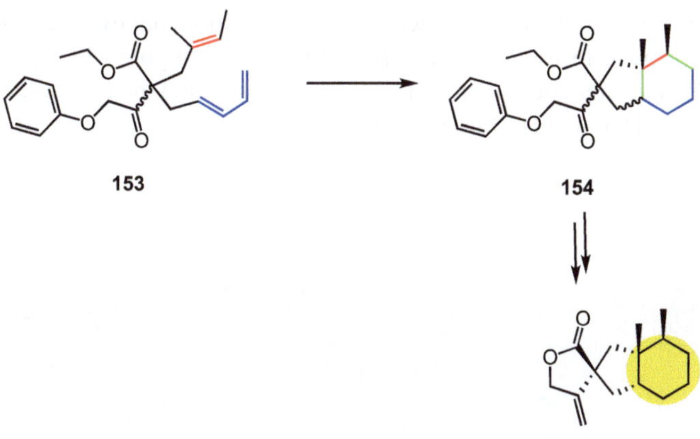


**Fig. 40** The Diels–Alder key step in the synthesis of (±)-bakkenolide A (**152**)

toward several carcinoma cell lines [103] and it also acts as an insect antifeedant [104].

The biological activity and the rather unusual spiro structure of **152** challenged synthesis chemists, and thus, a procedure involving an intramolecular Diels–Alder reaction as the key step was devised (Fig. 40) [105]. The required synthon **153** was prepared by conventional methodology. It underwent the intramolecular Diels–Alder reaction to build the indan fragment intermediate **154** when heated in a sealed tube in toluene at 190°C and in the presence of a catalytic amount of 2,3-di-*tert*-butyl-4-methylphenol as radical inhibitor to suppress polymerization. Again, conventional methodology provided the desired target bakkenolide A (**152**).

## 5.9 (+)-Himbacine and (+)-Himbeline

Himbacine (**155**) and its congener himbeline (**156**) are rather complex piperidine alkaloids, which were isolated from the bark of the northern pigeon berry ash *Galbulimima baccata* F.M. Bailey, of the family Himantandraceae (Plate 19), a rare rainforest species that belongs to the Magnoliales order, which is regarded as a poisonous and hallucinogenic plant [106]. Himbacine (**155**) has been considered as a promising lead in Alzheimer's disease research due to its potent muscarinic receptor antagonist property [107]. The senile dementia associated with Alzheimer's disease is directly correlated with diminished levels of synaptic acetylcholine in the cortical and hippocampal areas of brain, and the current form of therapy addresses this issue by inhibiting cholinesterase, which breaks down acetylcholine. Alternatively, biosynthesis enhancement of synaptic acetylcholine levels could be achieved by selectively inhibiting presynaptic muscarinic receptors ($M_2$), for which its agonist-induced stimulation shuts off acetylcholine release [108].

An early synthesis of (+)-himbacine (**155**) and (+)-himbeline (**156**) in 1999 made use of a diastereoselective intramolecular Diels–Alder reaction (Fig. 41) [109]. The appropriately functionalized tetraene derivative **157** was prepared by conventional means. An intramolecular Diels–Alder reaction was carried out by heating a solution of **157** in toluene at 185°C for 8 h in a sealed tube under argon, which provided the adduct **158**. The latter could then be transformed easily into the two desired targets, (+)-himbacine (**155**) and (+)-himbeline (**156**).

**Plate 19** *Galbulimima baccata*. Photograph courtesy Poyt448 P. Woodard, Creative Commons Attribution Share 3.0

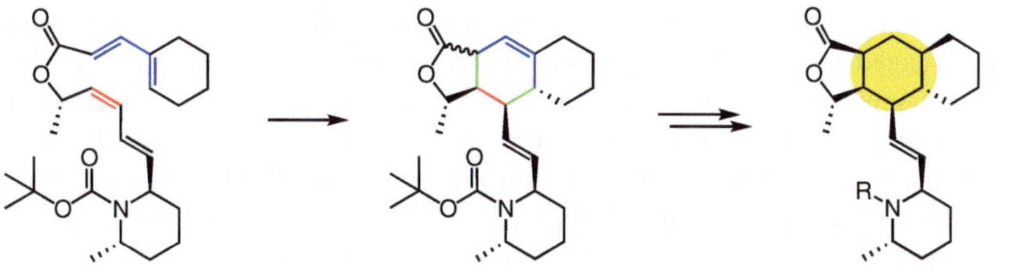

157

158

**155** ((+)-himbacine) R = $CH_3$
**156** ((+)-himbeline) R = H

**Fig. 41** The Diels–Alder key step in the synthesis of (+)-himbacine (**155**) and (+)-himbeline (**156**)

# 6 Complex Architectures and Novel Applications (2000s)

The period between 2000 and 2010 marked a transformative era in the field of organic synthesis chemistry, characterized by the exploration of increasingly complex molecular architectures and the development of innovative methodologies. This section delves into the synthesis of intricate natural products and bioactive compounds, showcasing the pivotal role of Diels–Alder reactions in constructing challenging frameworks with remarkable precision [110, 111]. From marine alkaloids and fungal metabolites to plant-derived terpenoids and bacterial secondary metabolites, the compounds discussed herein represent a diverse array of chemical structures with significant biological and ecological relevance.

Section 6 provides a comprehensive chronicle of modern synthetic applications of the Diels–Alder reaction, emphasizing its centrality in the construction of architecturally complex natural products across diverse biosynthetic classes. Over the past few decades, the Diels–Alder reaction has evolved from a classical tool for building six-membered rings to a multifaceted strategy enabling regio-, stereo-, and enantioselective construction of intricate molecular frameworks, including polycyclic, bridged, spirocyclic, and macrocyclic systems.

What unites the disparate natural products described herein, ranging from alkaloids to elaborate marine-derived terpenoids, is the judicious exploitation of the Diels–Alder reaction to form one or more carbon–carbon bonds in a stereocontrolled manner. These syntheses underscore the power of both intermolecular and intramolecular Diels–Alder reactions, as well as advanced variants such as transannular, inverse electron demand, and hetero-Diels–Alder transformations.

Many cases illustrate biomimetic paradigms, where Diels–Alder reactions emulate proposed biosynthetic steps, effectively bridging chemical synthesis with natural product biogenesis. In others, tandem or cascade strategies integrate the Diels–Alder step with preceding or subsequent transformations, maximizing synthesis economy and complexity generation. Of particular note are examples where the Diels–Alder reaction is embedded within chiral environments, whether via chiral auxiliaries, chiral catalysts, or asymmetric induction from pre-existing stereocenters, to forge contiguous stereocenters with high diastereo- and enantiocontrol.

The structural diversity of the targets spans tricyclic alkaloids, macrocyclic lactones, caged diterpenoids, tropolone-derived systems, and bridged polycycles, many containing multiple contiguous quaternary centers, strained ring junctions, or oxygenated spirocenters. Such architectures pose formidable synthetic challenges, and the success of the Diels–Alder-centered strategies reported herein is a testament to the versatility and adaptability of this reaction. Moreover, several syntheses reveal how subtle variations in tether length, electronic biasing of the diene/dienophile, or the judicious choice of Lewis acid catalysis can dramatically impact reaction course and stereoselectivity.

In compiling these examples, this section not only highlights specific methodological advances but also illustrates a broader conceptual shift: from using the Diels–Alder reaction as a tactical means to assemble six-membered rings to positioning it as a strategic cornerstone for the efficient, selective, and modular assembly of molecular complexity. These case studies exemplify the enduring relevance of the Diels–Alder reaction in total synthesis and the continually expanding frontiers of its application in the construction of functionally rich, stereochemically dense natural products.

## *6.1 (±)-Fasicularin and (±)-Lepadiformine*

From the light-bulb sea squirt *Clavelina lepadiformis* (Müller), family Clavelinidae, which is a shallow-water colonial ascidian (Plate 20) collected in the Mediterranean Sea off Tunisia, the marine alkaloid (±)-lepadiformine (**159**) was isolated, and was

**Plate 20** *Clavelina lepadiformis*. Photograph courtesy R. Strafella, Creative Commons Attribution Share 4.0 International

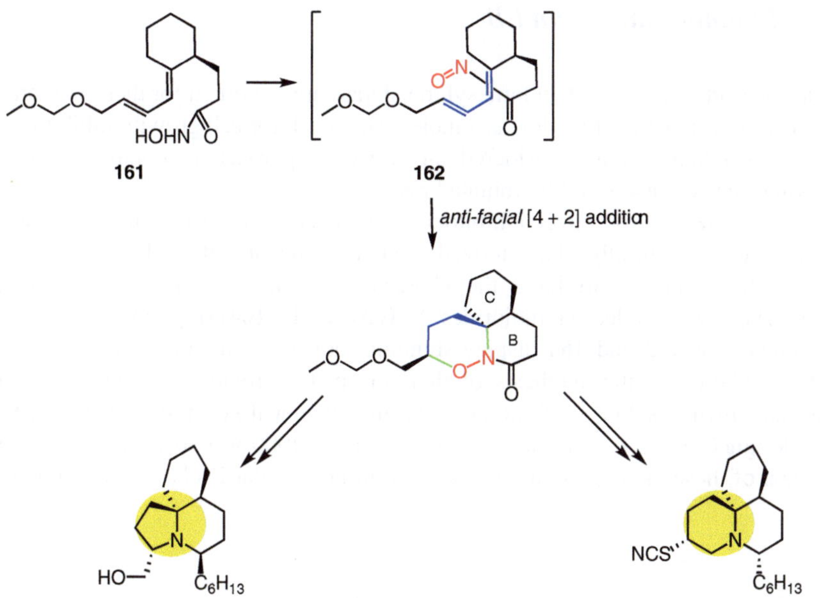


**Fig. 42** The Diels–Alder key step in the synthesis of (±)-lepadiformin (**159**) and (±)-fasicularin (**160**)

found to be moderately cytotoxic towards several tumor cell lines in vitro [112]. In addition to this tricyclic alkaloid, (±)-fasicularin (**160**) of similar structure was recently isolated from the Micronesian ascidian *Nephteis fasicularis*, which displayed selective activity against a DNA repair-deficient organism and was cytotoxic to Vero cells [113].

To construct the rather complex skeleton of **159** and **160**, a stereo-controlled intramolecular hetero-Diels-Alder reaction of an *N*-acylnitroso moiety to an exocyclic diene (as provided by **162**, derived by a periodate oxidation of the conventionally synthesized hydroxamic acid **161**) was applied by C. Kibayashi et al. in 2000 (Fig. 42) [114]. Thereby, mainly the *anti-facial* selectivity leading to the *trans*-B/C-fused decahydroquinoline ring system was achieved by varying the solvent to the optimum water/dimethyl sulfoxide (5/1), involving the simultaneous introduction of the nitrogenated quaternary center in a single step. On further elaboration of the six-membered or five-membered ring A by means of common methodology, the *trans*-fused adduct provided either (±)-fasicularin (**160**) or (±)-lepadiformine (**159**).

## 6.2 Phomoidrides A and B

Squalene synthase and protein farnesyl transferase are important for their membrane localization where they function as a molecular switch for cell growth. Inhibition of ras farnesylation is thus considered an effective approach to control abnormal growth of cells transformed by mutated ras.

Upon screening for novel inhibitors of these enzymes two compounds were isolated and structurally characterized from an unidentified coelomycetous soil fungus that appeared to have the characteristics of a sterile *Phoma* species [115]. These were called phomoidrides A (**163**) and B (**164**) (Fig. 43).

Compounds **163** and **164** display stunning molecular structures, and an asymmetric variation of the synthesis of their racemates provided an opportunity to determine their absolute configurations through chemical synthesis [116]. The key step designed by K. C. Nicolaou and his group in 2000 was to generate the core skeleton of these molecules, and is a type-II intramolecular Diels–Alder reaction.

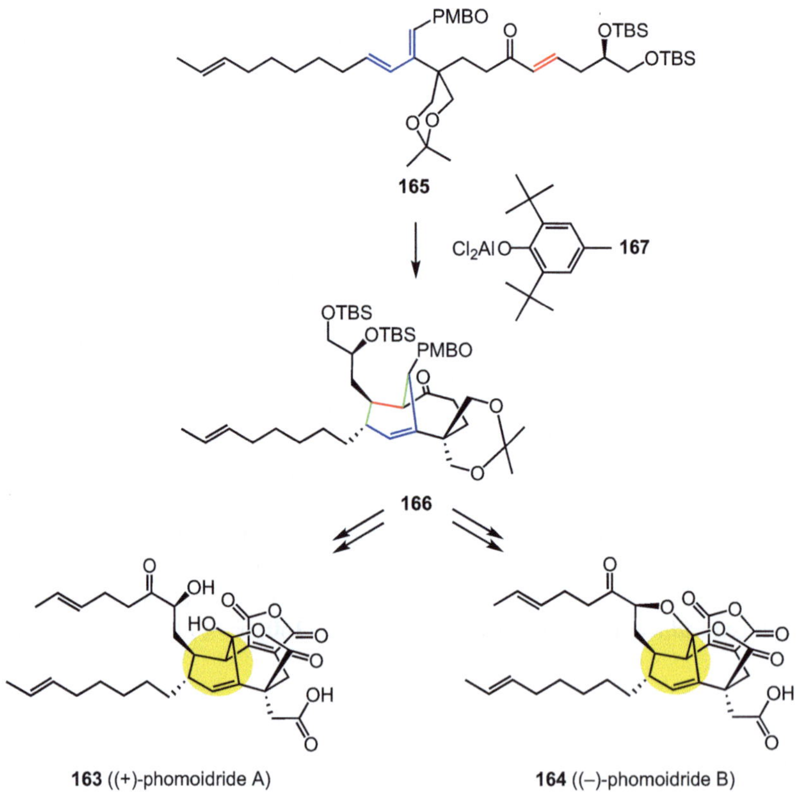


**Fig. 43** The Diels–Alder key step in the synthesis of phomodride A (**163**) and B (**164**)

To comment on the nomenclatural aspects, such a type-II intramolecular Diels-Alder reaction forms a bicyclic ring system with a bridgehead double bond (anti-Bredt alkene). This specific type of reaction involves a diene and dienophile that are connected at the second carbon of the diene, resulting in a bridged bicyclic product. This reaction is valuable for constructing complex molecules, particularly six-membered rings and can also be used to synthesize seven- and eight-membered rings in a single step. In a type I intramolecular Diels-Alder reaction, the diene and dienophile are connected by a tether that is linked to one of the terminal carbons (C-1 or C-4) of the diene. This contrasts with type II reactions where the tether is attached to a non-terminal carbon (C-2 or C-3) of the diene.

The chiral synthon **165** having the necessary diene and ene substructures was elaborated in conventional ways and the reaction was optimized using 20 mol% of the Lewis acid catalyst **167** at −80°C in toluene, which resulted in the intermediate **166**. The latter then could be elaborated into the target phomoidrides A (**163**) and B (**164**) by conventional methodology.

## 6.3 *Rubrolone Aglycone*

Rubrolone (**168**) is a naturally occurring red pigment produced by the soil bacterium *Streptomyces echinoruber* Palleroni et al. of the Streptomycetaceae family, first described in 1978 [117]. This compound belongs to a unique class of glycosylated tropolone alkaloids and has garnered interest for its distinctive chemical structure and potential applications. The rubrolone aglycone (**169**), constituting the primary pigment produced by *S. echinoruber*, features the complex tricyclic aglycone azuleno[2,3-*c*]pyridine-2,5,13-trione composed of a tropolone ring, a cyclopentanone unit, and a pyridine ring. This aglycone (**169**) is glycosylated with the deoxysugar D-fucose, linked via a unique carbon–carbon bond to the tropolone ring. The structure of **168** was elucidated through single-crystal X-ray analysis [117]. The biosynthesis of rubrolone involves a type-II polyketide synthase pathway, which assembles the polyketide backbone. Subsequent oxidative rearrangements led to the formation of the tropolone ring. Notably, the pyridine ring is formed through a non-enzymatic condensation of a 1,5-dione intermediate with ammonia, highlighting a rare example of non-enzymatic pyridine ring formation in natural product biosynthesis [118]. Rubrolone (**168**) exhibits low toxicity, but lacks significant antibiotic activity, making it a candidate for non-pharmaceutical applications. Its vibrant red color and water solubility suggest a potential use as a natural food coloring agent.

For the synthesis of the aglycone **169**, D. L. Boger et al. developed an intramolecular Diels–Alder reaction of an *O*-alkylated α,β-unsaturated oxime carrying an activated alkyne fragment, as encountered in compound **170** [119] (Fig. 44). Optimized conditions for the cyclization were the use of the benzyl residue for the leaving alcohol group, triisopropybenzene as the solvent, a temperature of 185°C, and 36 h reaction time, which yielded about 70% of the desired synthon **171**. The latter was then elaborated along conventional steps into the diene intermediate **172**,

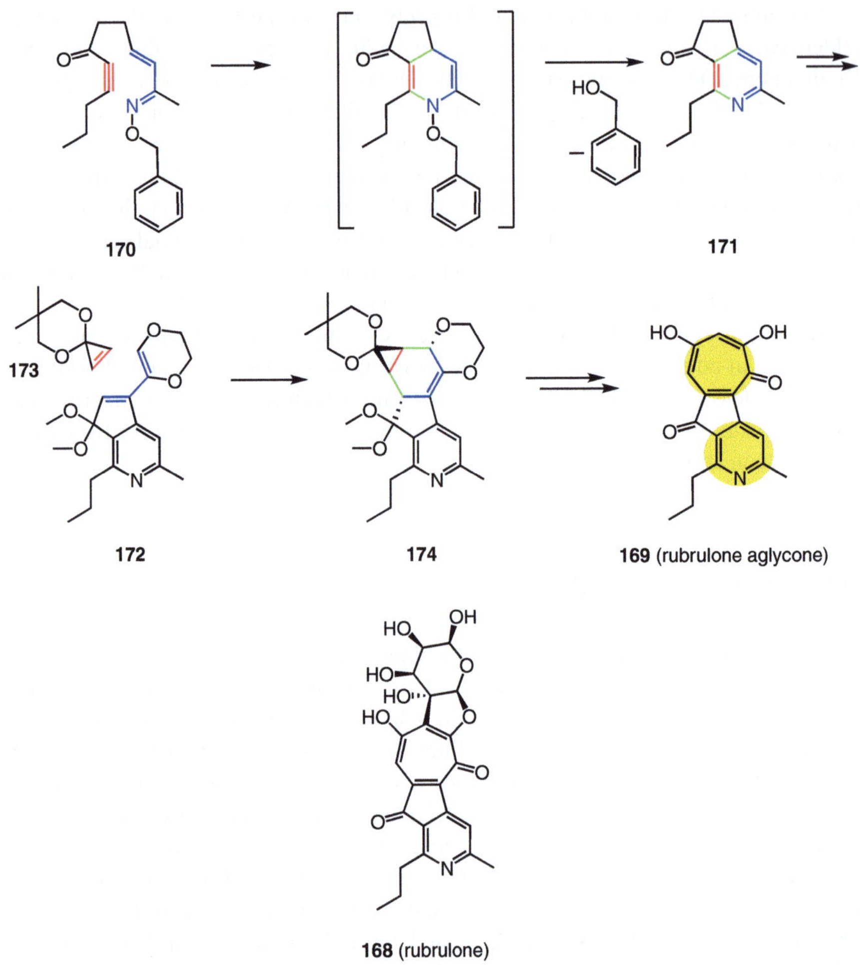


**Fig. 44** The intra- and intermolecular Diels–Alder key steps in the synthesis of the rubrolone aglycone (**169**)

which was subjected to an intermolecular Diels–Alder reaction using the cyclopronenone ketal **173** as the ene partner. This reaction was complete within 45 min at room temperature, providing the product **174** in 97% yield. Synthon **174** was then transformed into the target aglycon **169** along a conventional path.

## 6.4 *Nicandrenones*

The antifeedant and insect (particularly the whitefly) repellent properties of the apple of Peru, or shoo-fly plant, *Nicandra physalodes* (L.) Gaertn. (Plate 21) from the

**Plate 21** *Nicandra physaloides*. Photograph courtesy Flobbadob, Creative Commons Attribution Share 4.0 International

Solanaceae family, are due to the structurally complex, steroid-derived nicandrenones [120]. Interestingly, these compounds are related structurally to the withanolides encountered already in Sect. 5.1.

As targets for a synthesis using an intermolecular Diels–Alder reaction, specifically nicandrenone-1 lactone (**175**), nicandrenone-1 (**176**), and nicandrenone-10 (**177**) were chosen by E. J. Corey and co-workers in 2000 [121]. The key reaction between the cyclohexenone **178** and the diene component **179** turned out to exhibit remarkable *exo*-selectivity for its product **180**, which was unusual and critical for the stereo-controlled generation of all four rings (Fig. 45). Thus, addition of diene **179** (1.05 equiv.) to a mixture of the chiral α,β-enone **178** and methylaluminum dichloride (1.05 equiv.) in dichloromethane at −78°C over 2.5 h, resulted in formation of the *exo* adduct **180** (85%, *exo-endo* selectivity >15:1). This key adduct **180** was then elaborated conventionally into the intermediate **181**, from which also along conventional paths the desired targets nicandrenone-1 lactone (**175**), nicandrenone-1 (**176**), and nicandrenone-10 (**177**) could be obtained, as illustrated in Fig. 45.

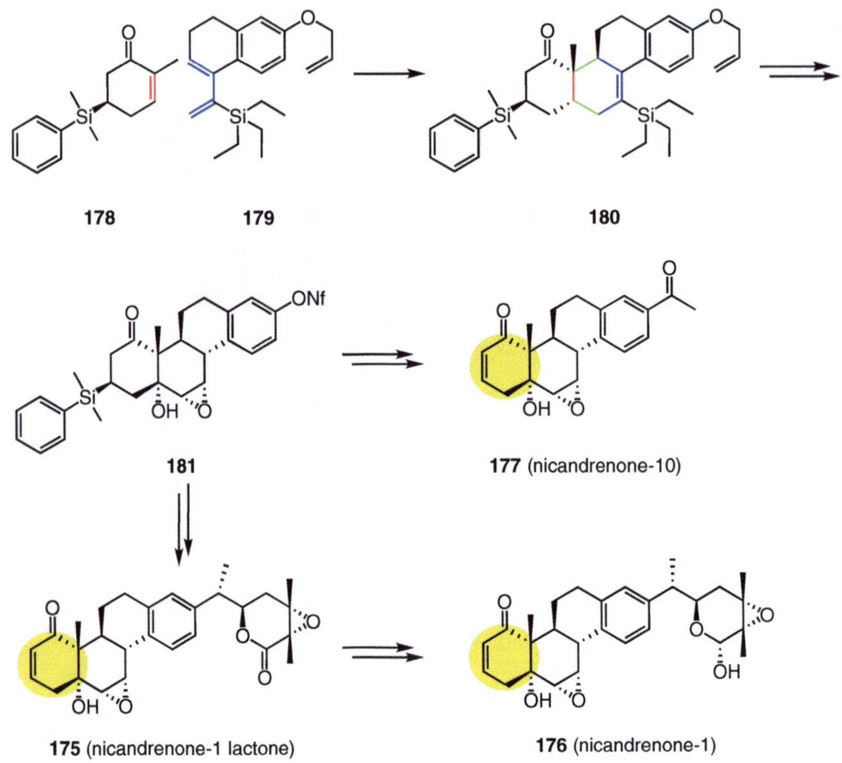


**Fig. 45** The Diels–Alder key step in the synthesis of nicandrenone-1 lactone (**175**), nicandrenone-1 (**176**), and nicandrenone-10 (**177**) (Nf = nonaflate = perfluorobutane-sulfonate)

## *6.5 (−)-Stemoamide*

*Stemona tuberosa* Lour. (Plate 22) is a flowering plant from the family of Stemonaceae, native to China, India, southeast Asia, and Guinea. It is one of the 50 fundamental herbs of traditional Chinese medicine and is used for managing respiratory disorders and as antiparasitic agent. From this plant a number of *Stemona* alkaloids have been isolated and characterized, among them (−)-stemoamide (**182**) [122].

The structurally challenging feature of the *Stemona* alkaloids is their perhydroazaazulene skeleton. For the synthesis of (−)-stemoamide (**182**), P. A. Jacobi devised an intramolecular Diels–Alder reaction, which took advantage of the exceptional reactivity of alkyne oxazoles as present in the educt **183** [123] (Fig. 46). When a solution of **183** in diethylbenzene was refluxed at 182°C, the single cycloadduct **186** was obtained after acid treatment in about 50% yield. Initially, the cyclization of **183** formed a transient intramolecular adduct **184**, which underwent loss of acetonitrile, thus providing the furan derivative **185**. Mild hydrolysis led to **186**, which could be transformed conventionally using a highly

**Plate 22** *Stemona tuberosa*. Photograph courtesy Yercaud-elango, Creative Commons Attribution Share 4.0 International

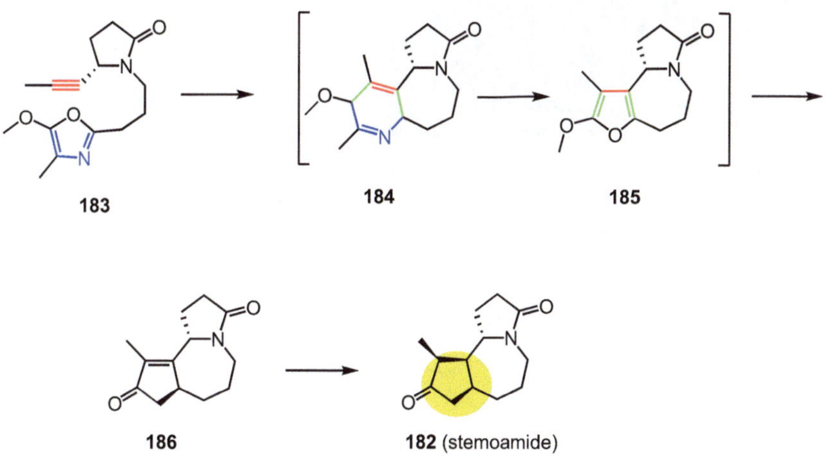


**Fig. 46** The Diels–Alder key step in the synthesis of stemoamide (**182**)

selective reduction with $NaBH_4/NiCl_2$ into the target **182**. The reaction was carried out first to yield the racemate (±)-stemoamide, but could also executed beginning with L-pyroglutamic acid from the chiral pool to yield the natural product (−)-stemoamide.

## 6.6 (±)-2-Isocyanoallopupukeanane

The isocyanopupukeananes constitute a unique class of marine sesquiterpenes, which are further subdivided into three types, according to their tricyclic skeletons. The initial observation that the sea slug *Phyllidia varicosa* of the Phyllidiidae family has acquired defense substances against predatory fish from sponges of the *Ciocalypta* sp., such as *C. penicillus* Bowerbank from the family Halichondriidae (Plate 23) led among other compounds of this type to the subsequent isolation of 9-isocyanoallopupukeanane (**187**) [124].

To synthesize this compound (**187**), T.-L. Ho and coauthors devised a route, with an intramolecular hetero-Diels–Alder reaction as a key step (Fig. 47) [125]. When a toluene solution of the conventionally prepared bicyclo[3.2.1]oct-2-ene **188** was heated in a sealed tube to 225°C for two days, the cycloadduct **189** displaying the

**Plate 23** *Phyllidia varicosa* (left). Photograph courtesy Nhobgood, Creative Commons Attribution-Share Alike 3.0. *Ciocalypta penicillus* (right). Photograph courtesy X. Rufray, Attribution-NonCommercial 4.0 International

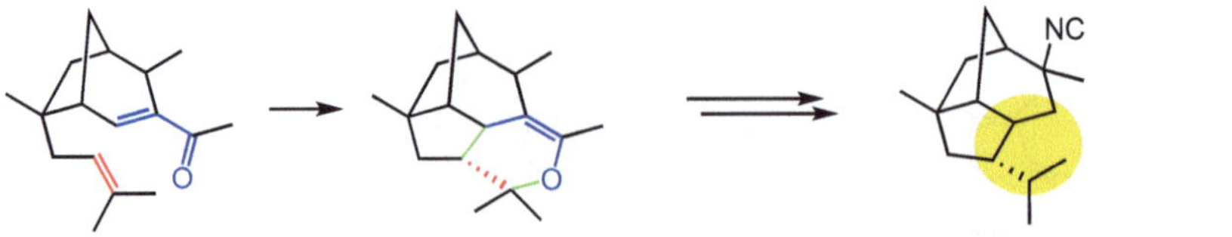


**188** **189** **187** (9-isocyanoallopupukeanane)

**Fig. 47** The Diels–Alder key step in the synthesis of 9-isocyanoallo-pupukeanane (**187**)

tricyclic framework required for the target **187** was formed. 9-Isocyanoallopupukeanane (**187**) could then be synthesized from **189** using conventional methodology.

## 6.7 *Paraherquamide VM55599*

Paraherquamide VM55599 (**190**) is a minor secondary metabolite produced by the fungus *Penicillium* sp. IMI332995 [126]. It belongs to the paraherquamide family of indole secondary metabolites, which are known for their molecular complexity and intriguing biogenesis. The compound VM55599 (**190**) was co-isolated with other paraherquamides, such as paraherquamide A, VM54158 (paraherquamide G), VM54159 (paraherquamide E), and VM55594 (paraherquamide F), as well as several new metabolites of the same type (Fig. 48). Compound **190** shares with these a diketopiperadine-based 2,5-diazabicyclo[2.2.2]octane ring system. Unlike other paraherquamides, **190** has the methyl group in the β-methylproline ring *syn* to the bridging isoprene moiety, making it unique within the family. Compound **190** provides also insights into the biosynthesis pathways of fungal metabolites, particularly the stereochemical outcomes of Diels-Alder cyclizations. It is considered a "dead-end shunt" metabolite, suggesting it may not undergo further biosynthetic transformations like other paraherquamides. While **190** itself does not have any reported biological activity, other paraherquamides are known for their potent antiparasitic properties, making the study of VM55599 relevant for understanding and potentially harnessing the bioactivity of related compounds.

The synthesis of the paraherquamide **190** was achieved by R. M. Williams and coworkers applying an intramolecular Diels–Alder reaction on the azadiene **191**,

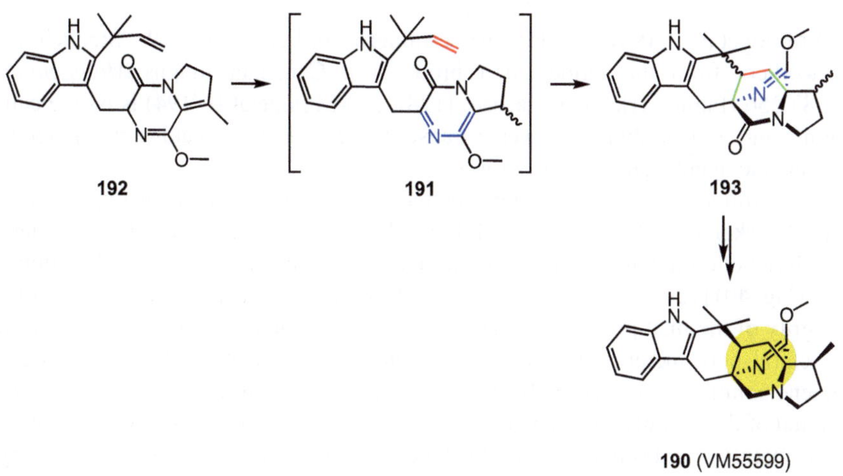


**Fig. 48** The Diels–Alder key step in the synthesis of VM55599 (**190**)

synthesized before by conventional means as the key step (Fig. 48) [127]. Treatment of the conventionally prepared precursor **192** with KOH in aqueous methanol effected tautomerization to the labile incipient azadiene **191**, which underwent spontaneously an intramolecular Diels-Alder cycloaddition at room temperature to give a mixture of all four possible racemic cycloadducts **193** in 78% combined yield in a 3.7/2.6/1.6/1 ratio. The stereoisomer mixture **193** was then transformed to the target **190** by treatment with hydrochloric acid, by boiling in toluene, and reduction with diisobutylaluminum hydride.

The reaction shows modest diastereoselectivity, favoring the *syn*-relative stereochemistry at C-20 with a ratio of 2.4:1. However, the stereochemical disposition of the methyl group in the β-methylproline ring was unexpected, as the major diastereomers displayed the methyl group *syn* to the bridging isoprene moiety, contrary to the anticipated *anti*-disposition. The intrinsic facial bias of the Diels–Alder cycloaddition is modest, with a slight preponderance (1.47:1) of cycloaddition from the same face as the methyl group in the β-methylproline ring. This suggests that the biological system may rely on protein organization of precyclization conformers to achieve higher selectivity. Metabolites possessing the *anti*-relative stereochemistry at C-20 have not been detected in paraherquamide-producing fungi, highlighting a complete facial exclusivity in the biosynthetic system that is not replicated in the laboratory synthesis. As the laboratory cyclization occurs spontaneously at room temperature in water, this indicates that the thermodynamics of the reaction are amenable to cytosolic constraints. This suggests that biosynthetic catalysis may not be required, but protein organization of precyclization conformers could influence the stereochemical outcome.

## 6.8 (−)-Mniopetal E

(−)-Mniopetal E (**194**) is a drimane sesquiterpenoid isolated by W. Steglich and co-workers from the fermentation broth of a Canadian fungus *Mniopetalum* sp. 87,256 (family Tricholomalaceae) [128]. (−)-Mniopetal E (**194**) is biologically significant as it inhibits the reverse transcriptase of HIV-1, making it a potential therapeutic agent against RNA viruses.

An intramolecular Diels–Alder reaction was crucial for constructing the core tricyclic skeleton of (−)-mniopetal E (**194**), which includes a 6–6-5 angularly fused tricyclic framework with multiple stereogenic centers and oxygen functionalities (Fig. 49) [129, 130]. This step ensured the correct stereochemistry and structural integrity of the final product. The cyclization of the conventionally prepared (using the chiral pool member D-mannitol as enantio-source) synthon **195** proceeded in toluene solution in a sealed tube at 180°C for about 7 h in the presence of a trace amount of di-*tert*-butyl-*p*-cresol as antioxidant, and showed high stereoselectivity to form mainly the desired *endo*-cycloadduct **196** with the correct stereochemistry. This was critical for constructing the *trans*-fused octahydronaphthalene core structure (A/B ring) of mniopetal E (**194**). The stereochemical outcome of the reaction is

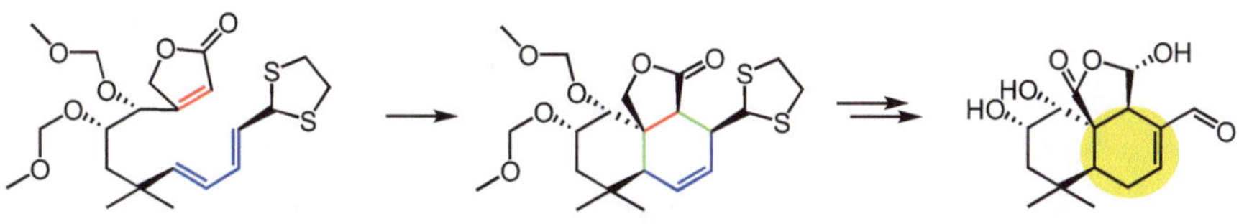

195 196 194 ((−)-mniopetal E)

**Fig. 49** The Diels–Alder key step in the synthesis of (−)-mniopetal E (**194**)

obviously influenced by the transition state. *endo*-Transition states are generally favored due to electronically favorable secondary orbital interactions, while *exo*-transition states are hindered by steric repulsion. Intermediate **196** could be transformed into the targeted (−)-mniopetal E (**194**) by conventional means.

## *6.9 (+)-Maritimol*

(+)-Maritimol (**197**) is a member of the stemodane diterpenoids and was isolated from *Stemodia maritima* L. (Scrophulariaceae) (Plate 24), also known as the Obeah bush or Seaside Twintip, which is a perennial shrub native to the Caribbean region and northeast Brazil [131]. It has been used as a Caribbean folk medicine for the treatment of venereal diseases, and it has represented a long-standing synthesis challenge with respect to its unique tetracyclic stemodane framework and the construction of its seven chiral centers, particularly the two central, adjacent quaternary carbons at positions C-9 and C-10.

The key step for the synthesis of (+)-maritimol (**197**) by P. Deslongchamps and coauthors was a transannular Diels–Alder strategy they and others had followed

**Plate 24** *Stemodia maritima*. Photograph courtesy A. Popovkin, Creative Commons Attribution 2.0 Generic

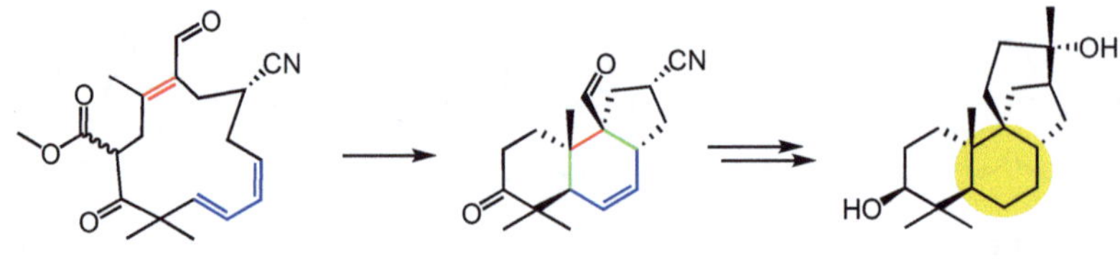


198 199 197 ((+)-maritimol)

**Fig. 50** The Diels–Alder key step in the synthesis of (+)-maritimol (**197**)

previously [132]. Thus, the 13-membered (*E*,*Z*,*Z*)-configured cyclotriene system of **198** (Fig. 50) was reacted either with dimethylaluminum dichloride as Lewis catalyst at room temperature followed by thermal demethoxycarbonylation, or heated in dimethylsulfoxide at 155°C for 4 h directly in a stereoface- and diastereospecific manner to the transannular product **199**, which could then be elaborated conventionally into the targeted (+)-maritimol (**197**).

## 6.10 Colombiasin A

On reaching 2001, we encounter several wonderful syntheses all exploring Diels–Alder reactions as the key steps. Thus, for example, colombiasin A (**200**) (Fig. 51) is a diterpene isolated from the gorgonian octocoral *Pseudopterogorgia elisabethae* Bayer of the Gorgoniidae family (Plate 25) collected in Colombian waters, which was shown to be biologically active against *Mycobacterium tuberculosis* H37Rv [133]. Its structure displays a tetracyclic carbon atom framework of which the periphery is substituted with four methyl groups, two carbonyl functions, two double bonds, and one hydroxy group. In addition, this unprecedented molecular architecture includes six stereogenic centers, two of which are adjacent quaternary carbon atoms, thus, representing a formidable synthesis challenge.

The total synthesis of (±)-colombiasin A (**200**) of the K. C. Nicolaou group started with an intermolecular Diels–Alder reaction of 2-silyloxy-1,3-pentadiene (**201**) and 2-methoxy-3-methyl-1,4-quinone (**202**) in ethanol at 25°C, providing the desired *endo*-cycloadduct **203** as the only product (Fig. 51) [134]. The resulting **203** was elaborated into **204**, a bicyclic intermediate carrying a cyclic sulfone tether. Heating **204** in toluene at 180°C led to a single *endo*-adduct **207** in 89% yield, through the cheletropic extrusion of $SO_2$, which was followed by a [4 + 2] Diels–Alder cycloaddition via **205**, the presumed intermediate possessing an overlapping diene-quinone arrangement, **206**. Removal of the extra hydroxy group of **207**, followed by deprotection of the phenolic methyl ether, provided the target ((±)-colombiasin A (**200**).

To address the asymmetric synthesis of **200**, the Nicolaou group devised an initial Diels-Alder reaction of **201** and **202** in an asymmetric version, using the chiral catalyst (*S*)-BINOL-$TiCl_2$. The adduct **203** was obtained in 94% *ee*. The natural

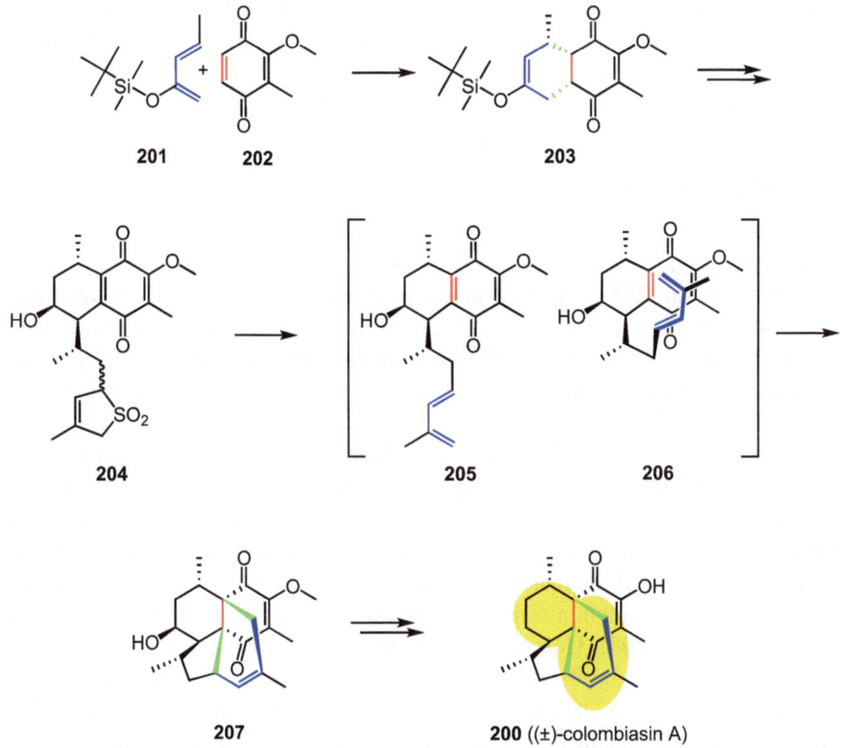


**Fig. 51** The inter- and intramolecular Diels–Alder key steps in the synthesis of (±)-colombiasin (**200**)

**Plate 25** *Pseudopterogorgia elisabethae*. Photograph courtesy Mary Carmen, Reef2Reef

enantiomer, (−)-colombiasin A (**200**), was synthesized from enantiomerically enriched **203**, thereby also establishing the absolute stereochemistry of the natural product.

## 6.11 *Pseudopteroxazole*

As with colombiasin (**200**) discussed in Sect. 6.10, pseudopteroxazole with the suggested structure **208** is another antitubercular marine diterpenoid isolated from the soft coral *Pseudopterogorgia elisabethae* Bayer (syn. *Antillogorgia elisabethae*) of the Gorgoniidae family (Plate 25), provoking a synthesis challenge [135]. Again, an intramolecular Diels–Alder reaction provided the key step in the construction of the skeleton (Fig. 52).

Starting from the conventionally prepared diene amide **209**, treatment with two equivalents of lead tetraacetate in ethyl acetate at room temperature generated the corresponding quinone monoimide **210** that underwent an internal Diels-Alder addition as it was formed, to give the *endo* adduct **211** in nearly 70% yield along with a small amount of the diastereomeric adduct (ratio 8/1). The intermediate **211** could then be elaborated in a series of conventional steps into the target formula of pseudopteroxazole (**208**). Unfortunately, neither this compound, nor its diastereomer, which was also synthesized were identical with the natural product, thus mandating a revision of its structure. This was be achieved much later in 2019 using conventional synthesis methodology, leading to the final structure **212**

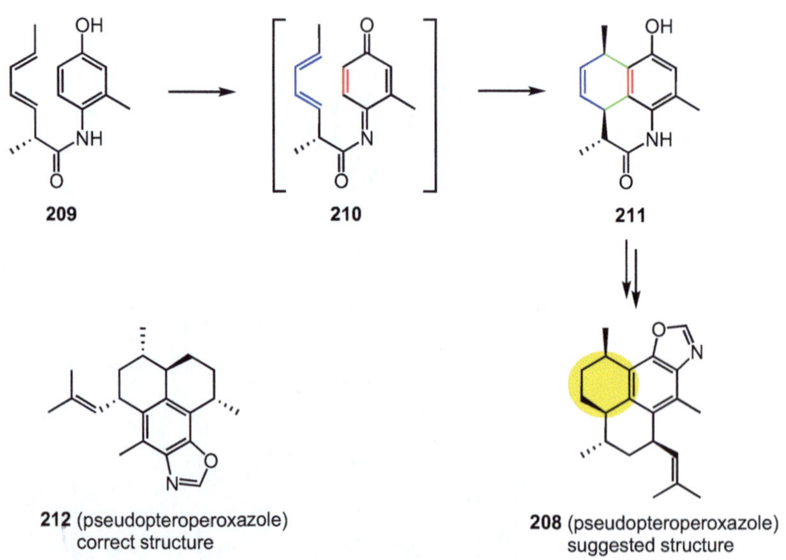


**Fig. 52** The intramolecular Diels–Alder key step in the synthesis of pseudopteroxazole (**208**)

[136]. Nevertheless, the intramolecular Diels–Alder reaction described above helped to pave the road to the correct structure for this natural product.

## *6.12 Eunicenone A*

Eunicenone A (**213**) is an intriguing tetraprenylated 2-cyclohexenone isolated from *Eunicea* species, similar to the soft coral *E. mammosa* Lamouroux, family Plexauridae, illustrated in Plate 26.

For its synthesis (Fig. 53) [137], a catalytic asymmetric Diels–Alder approach was chosen. Thus, on the addition of **214**, synthesized using common methodology, to five equivalents of 2-bromoacrolein (**215**) in the presence of 0.5 equivalents of oxazaborolidine **216** as chiral auxiliary catalyst in dichloromethane at −78°C and standing at this temperature for two days, the reaction proceeded smoothly to form the bromo aldehyde **217** displaying 97% *ee* in 80% yield (*endo*/*exo* ratio > 98/2) (85% yield based on recovered diene component). The latter could be transformed conventionally into the targeted eunicenone A (**213**).

**Plate 26** *Eunicea mammosa*. Photograph courtesy Q. Muñoz, GNU General Public License

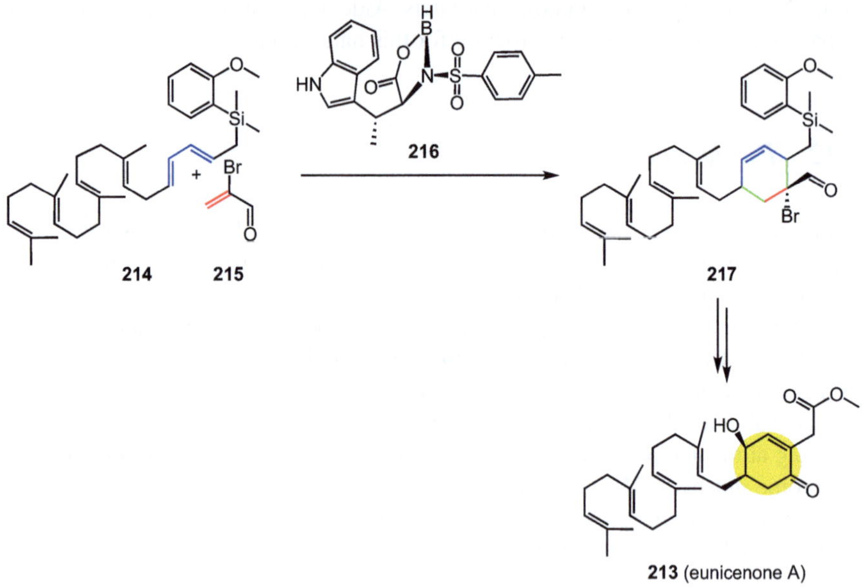


**Fig. 53** The intermolecular Diels–Alder key step in the synthesis of eunicenone A (**213**)

## *6.13 Imerubrine, Isoimerubrine, and Grandirubrine*

The tropoisoquinoline alkaloids imerubrine (**218**), isoimerubrine (**219**), and grandirubrine (**220**) were isolated from *Abuta imene* (Mart.) Eichler (Plate 27), *A. rufescens* Aubl., and *A. grandifolia* (C. Martins) Sandwith, flowering plants from the family Menispermaceae, which are used to prepare arrow poisons by indigenous tribes in South America [138, 139].

These alkaloids pose considerable synthetic challenges, which are in part due to the paucity of general methods for preparing the fused tropolone ring. The key step for their synthesis of these alkaloids (Fig. 54) was the application of an intramolecular Diels–Alder reaction of the acetylene-tethered oxazole **221**, which was prepared following common steps [140]. It was reacted in refluxing dichlorobenzene at 181°C and proceeded via a retro Diels–Alder reaction under the elimination of hydrogen cyanide and simultaneous removal of the *tert*-butoxycarbonyl group to 90% yield of the tetracyclic compound **222**. The latter compound gave the intermediate **223** through a further notable sequence, using an oxallyl [4 + 3] addition with α,α,α′-trichloroacetone. The intermediate **223** could then be elaborated into the three targets **218–220** by applying conventional methodology.

**Plate 27** *Abuta imene*. Photograph courtesy RBG Kews, Creative Commons 3.0

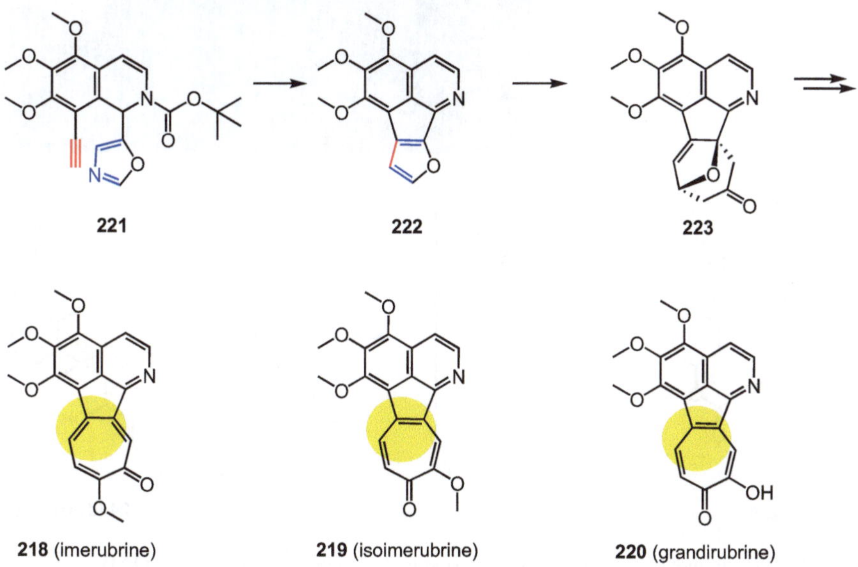


**Fig. 54** The intramolecular Diels–Alder key step in the synthesis of imerubrine (**218**), isoimerubrine (**219**), and grandirubrine (**220**)

## *6.14 Dysidiolide*

Dysidiolide (**224**) is a unique sesterterpenoid that was isolated from the Caribbean sponge *Dysidea etheria* Laubenfels, family Dysideidae (Plate 28). It was

**Plate 28** *Dysidea etheria*. Photograph courtesy G. M. Sandoval, 

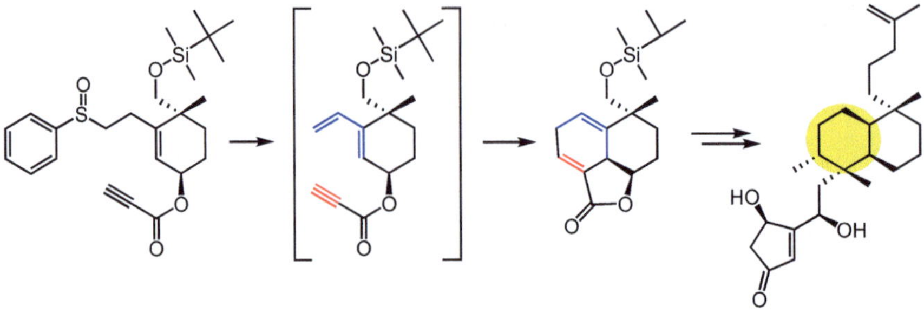

225 226 227 224 (dysidiolide)

**Fig. 55** The intramolecular Diels–Alder key step in the synthesis of dysidiolide (**224**)

documented for the first time to inhibit the action of the protein phosphatase, cdc25A, which is essential for cell proliferation [141].

To synthesize dysidiolide (**224**) (Fig. 55), a solution of ester **225** in toluene was refluxed in the presence of ethyl propiolate and pyridine to generate the diene **226** by elimination of phenyl sulfoxide. This was followed by an intramolecular Diels-Alder reaction to provide the decalin derivative **227** in 89% yield as the sole product

[142]. Interestingly, in the absence of pyridine, decalin **227** was not obtained. The desired target product, dysidiolide (**224**), could then be prepared following standard methodology.

## *6.15 Gummiferolic Acid and Gibberellin $A_{12}$*

Gummiferolic acid (**228**), which was isolated from *Margotia gummifera* (Desf.) Spreng. (Plate 29), a species of flowering plant in the family Apiaceae, native mainly to the western Mediterranean region, is a diterpenoid possessing six contiguous stereogenic centers and a bicyclo[2.2.2]octane ring system as the C/D ring part [143]. In addition, **228** displays plant growth-regulatory activity like or greater than that displayed by the standard compound, gibberellic acid [144]. Gibberellin $A_{12}$ (**229**), is another diterpenoid isolated from the fungal plant pathogen *Gibberella fujikuroi* (Sawada) Wollenw. (Plate 30), which causes the so-called bakanae disease in rice, has a *trans*-hydrindane AB ring system and a spirofused bicyclo[3.2.1]octane moiety that comprises the C and D rings [145].

**Plate 29** *Margotia gummifera* (syn. *Thapsia gummifera*). Photograph courtesy J. Moreira, Creative Commons Attribution Share 4.0 International

**Plate 30** *Gibberella fujikuroi.* Photograph courtesy Clemson University, Creative Commons Attribution 3.0 Unported

An intramolecular Diels–Alder reaction was used to construct ring A of the skeleton of gummiferolic acid (**228**) and gibberellin $A_{12}$ (**229**) (Fig. 56) [146]. The common precursor **230**, synthesized by common methodology, was converted into the two diene–ene synthons **231** and **233**, which were heated in toluene at 200°C in a sealed tube. The intramolecular cyclization products **232** and **234**, obtained in nearly quantitative yields, could then be elaborated following conventional steps into the desired targets, gummiferolic acid (**228**) and gibberellin $A_{12}$ (**229**).

## *6.16 Strychnine*

Strychnine (**235**) is a highly toxic terpene indole alkaloid that occurs in the seeds of *Strychnos nux-vomica* L. (Plate 31), a deciduous tree from the family Loganiaceae, native to southeast Asia. It was first isolated by the French chemists P.-J. Pelletier and J.-B. Caventou in 1818 from the then so-called Saint-Ignatius bean [147].

Owing to its level of structural complexity with respect to its rather small size, strychnine (**235**) has presented a most formidable synthesis challenge, and many

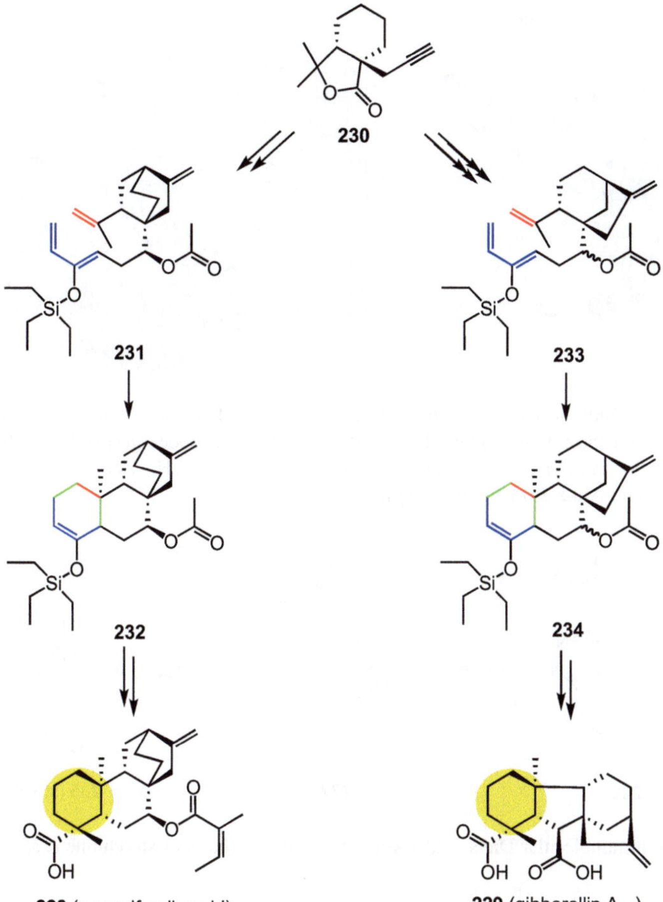


**Fig. 56** The intramolecular Diels–Alder key steps in the synthesis of gummiferolic acid (**228**) and gibberellin $A_{12}$ (**229**)

studies were published since its structure had been derived in detail [148–150]. Among them, a highly ingenious synthesis made use of a transannular inverse electron demanding Diels–Alder reaction (Fig. 57) [151], again expanding the possibilities offered by this reaction. The required cyclophane **236** could be obtained using proper conventional methodology. In heated *N,N*-diethylaniline, it underwent a transannular cycloaddition, which was followed by the expulsion of nitrogen from the initially formed product to give the strychnine synthon **237** in quantitative yield. The latter, already displaying the A/B/C/E/G rings of the desired natural product could be transformed into the racemic target **235** using known techniques.

**Plate 31** *Strychnos nux-vomica* plant, from Köhler's Medizinal-Pflanzen in naturgetreuen Abbildungen mit kurz erläuterndem Texte: Atlas zur Pharmacopoea germanica. Public Domain (left). Seeds of *Strychnos nux-vomica*. Photograph courtesy H. Zell, Creative Commons Attribution-Share Alike 3.0 Unported (right)

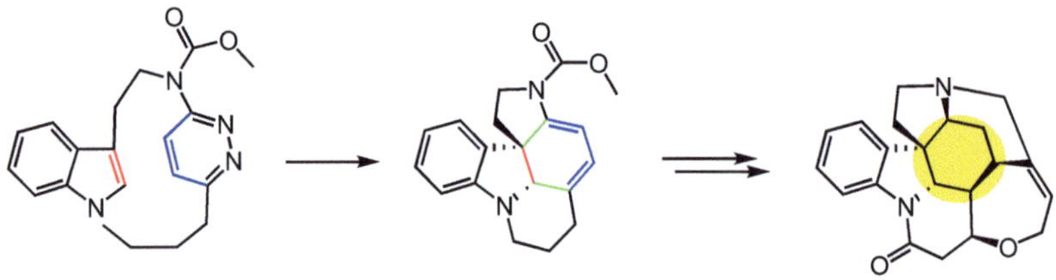

**236** **237** **235** (strychnine)

**Fig. 57** The intramolecular Diels–Alder key step in the synthesis of strychnine (**235**)

This work proved to be a milestone in applying a transannular inverse electron demanding Diels–Alder reaction to natural product synthesis, as the key cyclization proceeded thermally in the absence of added catalysts. It highlighted how enforcing a reactive conformation in a macrocycle can trigger pericyclic ring closures that rapidly generate bridging complexity, representing a formidable shortcut in a historically challenging alkaloid synthesis.

## 6.17 (±)-Stenine

As described in Sect. 6.5, the synthesis of one of the *Stemona* alkaloids, (−)-stemoamide (**182**) was already encountered. in 2000. In 2002, there were two reported processes to synthesize another alkaloid of this type, (±)-stenine (**238**), also occurring in the flowering plant *Stemona tuberosa* Lour. of the Stemonaceae family (Plate 22), which illustrate nicely the extension and variation of the Diels–Alder application to complex molecules. Compound **238** consists of a multifunctionalized perhydroindole skeleton containing seven stereocenters,and thus has challenged synthesis chemists.

An initial approach was applied by J. E. Golden and J. Aubé [152] who first prepared the triene **239** needed for the intramolecular Diels–Alder reaction using standard methodology (Fig. 58). Treatment of **239** with methylaluminum dichloride

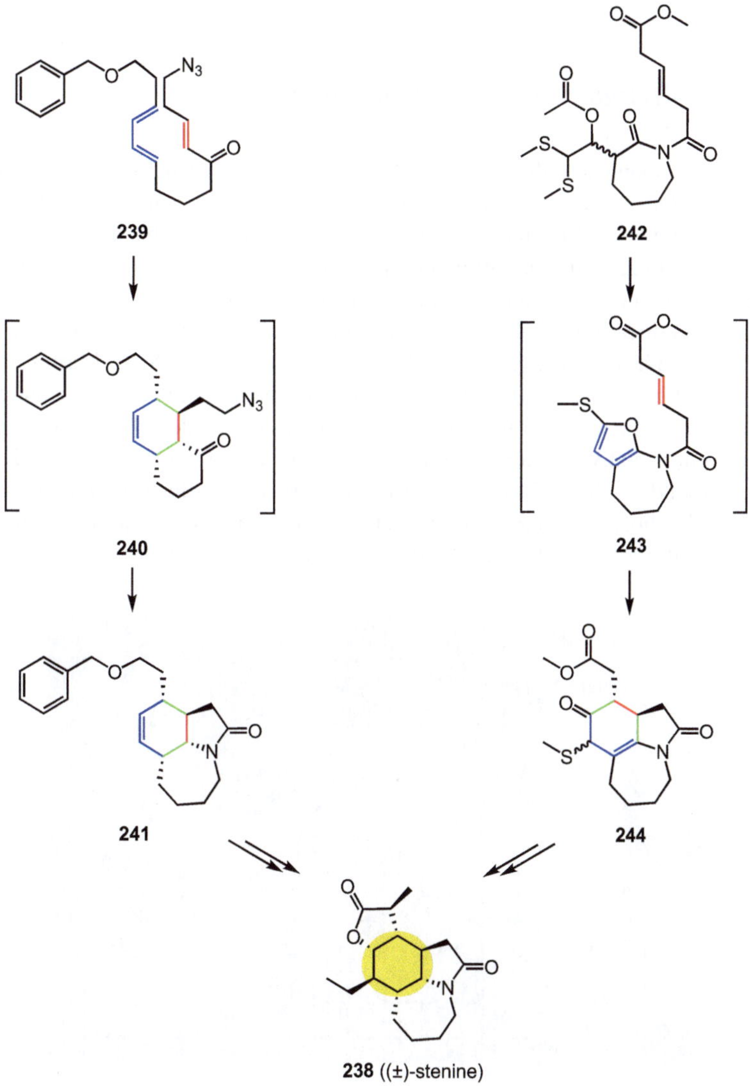


**Fig. 58** The intramolecular Diels–Alder key steps in the synthesis of (±)-stenine (**238**)

in refluxing dichloromethane started the cycloaddition to intermediate **240**, with a subsequent intramolecular Schmidt reaction to afford three stereoisomeric tricyclic lactams, with the desired diastereomer **241** as the major component. The latter was converted to the target (±)-stenine (**238**) in conventional steps.

In the second approach by J. D. Ginn and A. Padwa [153] (Fig. 58) an intermediate non-isolable 2-methylthio-5-amidofuran derivative **243** generated from the lactam **242** was used in an intramolecular Diels–Alder reaction to create the azepindole skeleton-containing synthon **244**, which could then be elaborated in 11 conventional steps into the target (±)-stenine (**238**).

## *6.18 (+)-Ircinal A and (+)-Manzamine A*

From marine sponges of the genera *Haliclona* Grant of the Chalinidae family (Plate 32) and *Pellina*, severally complex indole alkaloids were isolated [154]. Among them manzamine A (**245**) has attracted considerable attention because of its potent antitumor, antimalarial, and antituberculosis activities [155, 156]. Ircinal A (**246**), stemming from the same source, is an intermediate of the biogenesis of **245**.

Again, a critical intramolecular Diels–Alder reaction step opened the way to the synthesis of **246** and finally **245** in the elegant approach of S. F. Martin and his colleagues (Fig. 59) [157]. To attain the educt for the intramolecular cycloaddition, they prepared the correspondingly functionalized chiral dehydropyrrolidine **247** from a commercially available chiral pool compound. Due to the high reactivity of the corresponding diene **248**, this had to be prepared in situ in refluxing toluene, using vinyl tributylstannane in the presence of a palladium(0) catalyst in an intermediary Stille coupling that resulted in the diene intermediate **248**, which then

**Plate 32** An example of encrusting marine sponge of genus *Haliclona*: *Haliclona permollis*. Photograph courtesy jkirkhart35, Creative Commons Attribution 2.0 Generic

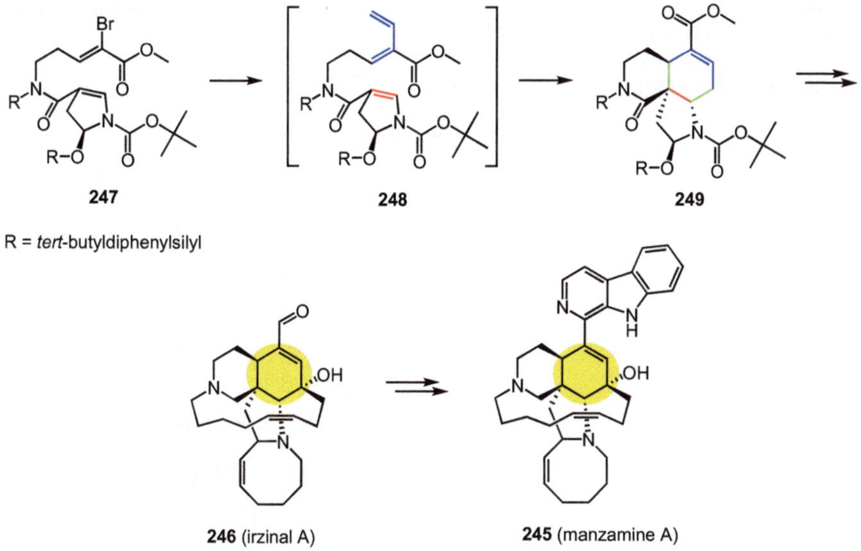


**Fig. 59** The intramolecular Diels–Alder key step in the synthesis of irzinal A (**246**) and manzamine A (**245**)

cyclized spontaneously to **249**, thus building the required A/B/C ring core. Conventional synthesis steps could then be applied to form ircinal A (**246**), and then manzamine A (**245**).

## *6.19 (−)-Longithorone A*

The cyclophane-type quinone (−)-longithorone A (**250**) (Fig. 60) is a cytotoxic marine natural product first isolated in 1994 by F. J. Schmitz and co-workers from the tunicate *Aplidium longithorax* Monniot F. of the Polyclinidae family (Plate 33), which were collected near the island of Palau in the western Pacific Ocean [158]. It is a molecule with an unusual and unique heptacyclic structure that poses quite a challenge for total synthesis.

The synthesis of (−)-longithorone A (**250**), described by D. Shair in 2002, used two Diels–Alder reactions, as derived from its proposed biosynthesis. The first step was intermolecular, whereas the second step was intramolecular (Fig. 60) [159]. Thus, the paracyclophanes **251** and **252**, prepared along conventional routes, were treated with 1.35 equivalents of dimethylaluminum chloride as Lewis acid catalyst in dichloromethane at −20°C, affording the product **253** of the intermolecular Diels–Alder reaction. Removal of the protecting groups with tetra-*n*-butylammonium fluoride and directly oxidizing the intermediate with iodosylbenzene afforded the bisquinone **254**. At room temperature and in the course of 40 h, **254** underwent an intramolecular transannular Diels–Alder reaction to

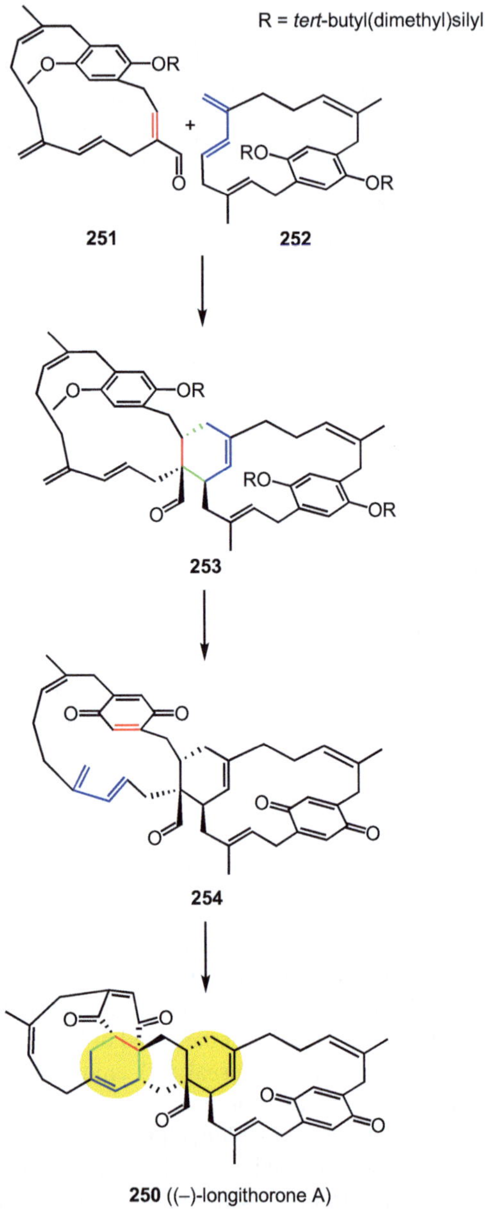


**Fig. 60** The inter- and intramolecular Diels–Alder key steps in the synthesis of (−)-longithorone A (**250**)

provide directly the target, (−)-longithorone A (**250**), in 90% yield. Interestingly, the reaction required Lewis acid catalysis (e.g., dimethylaluminum chloride) to proceed. Other Lewis acids, such as titanium tetrachloride, boron trifluoride etherate, tin tetrachloride, and ytterbium trifluoroacetate either failed to initiate the reaction,

**Plate 33** *Aplidium longithora* (Turtle Cove, Palau). Photograph courtesy R. Spangler, Creative Commons Attribution-NonCommercial 4.0 International

reduced selectivity, or caused decomposition. This suggests that the reaction conditions are highly sensitive and specific.

## 6.20 *trans*-Dihydroconfertifolin

*trans*-Dihydroconfertifolin (**255**) (Fig. 61) is a terpenoid lactone isolated from the New Zealand marine sponge *Dysidea* sp. [160] — these sponges were encountered already in Sect. 6.14 on dysidolide (**224**) in Plate 28. Compound **255** displayed antimicrobial activity.The 1,2-lactone **255** is a simple representative of a common structural theme among many polycyclic terpenoids, the *trans*-fused angularly substituted 7,8- dialkyldecalin. The polycyclic terpenoids that fall in this class can be subdivided further based on the stereochemical relationships at C-7 and C-8, and natural products in each of the four diastereomeric series are known. It was found that an inactivated intramolecular Diels-Alder reaction allows easy access to two of the four diastereomeric series, thus allowing for the synthesis of **255** (Fig. 61) [161].

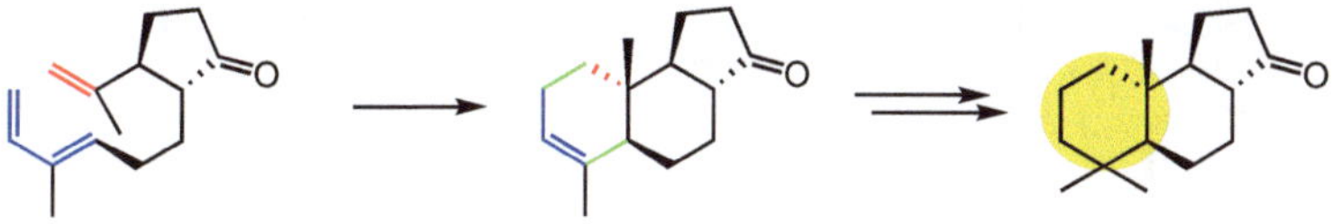

**256** **257** **255** (*trans*-dihydroconfertifolin)

**Fig. 61** The intramolecular Diels–Alder key step in the synthesis of *trans*-dihydroconfertifolin (**255**)

The inactivated intramolecular Diels-Alder reaction was executed by heating a mixture of **256** (prepared earlier by using common methodology) and methylene blue in benzonitrile under nitrogen and reflux for 16 h, which resulted in the predominating diastereomer **257**. This was then elaborated into the target *trans*-dihydroconfertifolin (**255**) via a Simmons–Smith cyclopropanation.

## *6.21 (+)-FR182877*

The bacterial-derived natural product, (+)-FR182877 (**258**) (Fig. 62) was described by scientists of the Fujisawa company in 2000 [162], and it challenged synthesis chemists due to its highly carbon- and oxygen-functionalized hexacyclic core, comprising twelve stereogenic centers and a strained double bond in a bridgehead position. In addition, **258** displayed a remarkable microtubule stabilization propensity like paclitaxel and showed potent cytotoxicity towards several tumor cell lines [163].

Tandem-transannular Diels–Alder reactions inspired by its postulated biosynthesis were thought to provide key steps to synthesize (+)-FR182877 (**258**), and indeed, when the group of E. J. Sorensen oxidized the 19-membered macrocycle **259** by means of phenyl-selenyl bromide and *m*-chloroperbenzoic acid the intermediate **260** formed, which, upon warming up the reaction mixture in chloroform to 40°C, resulted in compound **261** (Fig. 62) [164]. The 19-membered macrocycle of **259** thus provided the complex pentacycle of **261**, with seven new contiguous

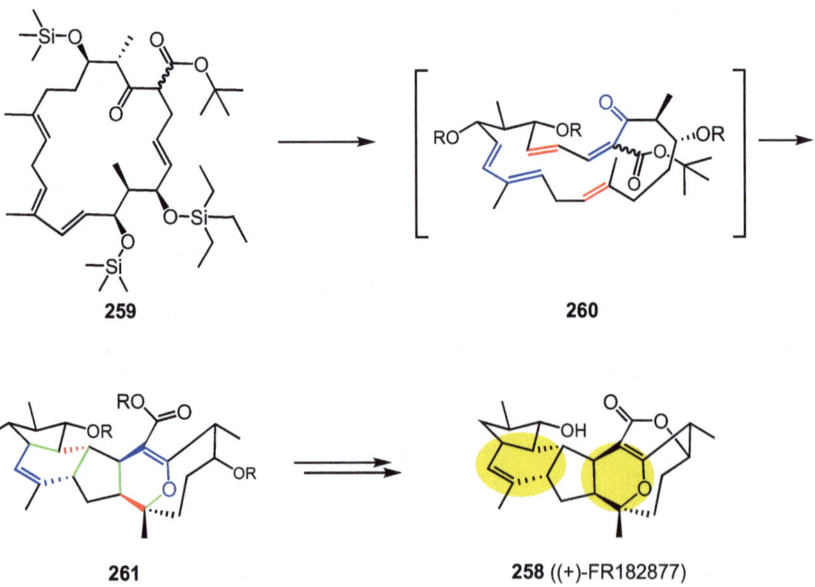


**259** **260** **261** **258** ((+)-FR182877)

**Fig. 62** The double-transannular intramolecular Diels–Alder key steps in the synthesis of (+)-FR182877 (**258**)

stereocenters! Compound **261** was then elaborated following standard methodology to give the target compound (+)-FR182877 (**258**). It should be mentioned that this synthesis was followed up in 2003 with a multigram version, with the possibility to access also its enantiomer (−)-FR182877 [165].

## *6.22 (−)-Elisapterosin B and (−)-Colombiasin A*

In 2003, the interesting and challenging structures of the compounds (−)-elisapterosin B (**262**) and (−)-colombiasin A (**200**), isolated from the West Indian sea whip *Pseudopterogorgia elisabethae* Bayer (Plate 25), stimulated considerably the interest in their enantioselective preparation, after the Nicolaou group had been successful with the synthesis of the colombiasin A racemate (**200**) (and in its enriched enantiomer) along a Diels–Alder route (see Sect. 6.10, Fig. 51). Thus, an ingenious route was devised involving Diels-Alder steps and [5 + 2] cyclization as key features.

A. I. Kim and S. D. Rychnovsky followed steps involving the structure of the serrulatane diterpenoid **263** (Fig. 63), also produced by the same organism [166]. They synthesized the synthon **264**, which easily (25 equivalents boron trifluoride etherate in dichloromethane at −78°C) underwent a [5 + 2] cyclization to yield (−)-elisapterosin B (**262**) (Fig. 63). However, when **264** was heated at 180°C in toluene and treated subsequently with *N*,*N*-dimethylaniline in

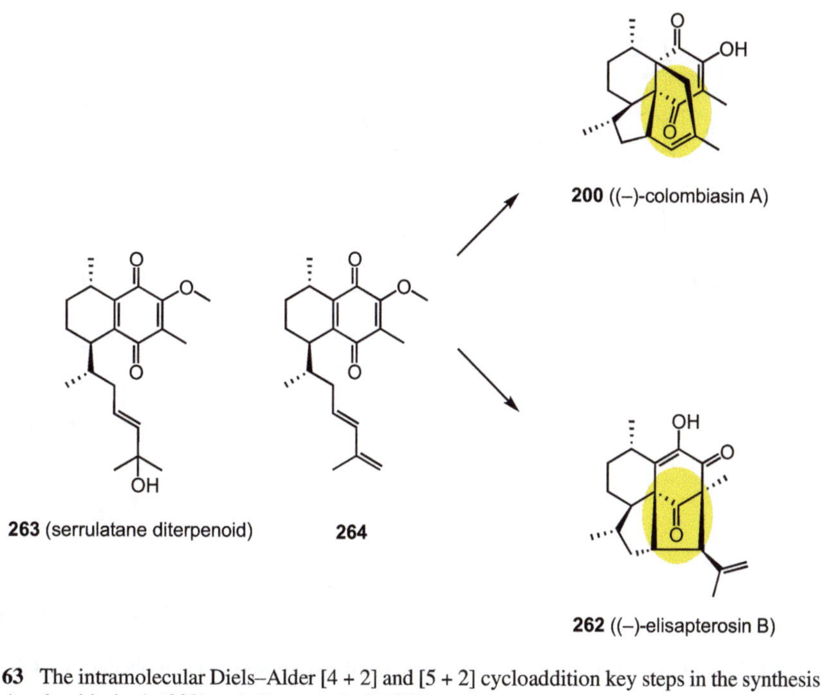


**Fig. 63** The intramolecular Diels–Alder [4 + 2] and [5 + 2] cycloaddition key steps in the synthesis of (−)-colombiasin A (**200**) and elisapterosin B (**262**)

dichloromethane at a temperature range of 0–23°C, (−)-colombiasin A (**200**) was produced by an intramolecular Diels–Alder [4 + 2] cycloaddition, in about 70% yield.

## 6.23 Elisabethin A

In addition to the fascinating compounds (−)-elisapterosin B (**262**) and (−)-colombiasin A (**200**), isolated from the octocoral *Pseudopterogorgia elisabethae* (Plate 25) and discussed in the preceding Sect. 6.22, the organism contains the structurally related elisabethin A (**265**), which also displayed significant activity against *Mycobacterium tuberculosis* [167].

The challenge to synthesize elisabethin A (**265**) was taken up by T. J. Heckrodt and J. Mulzer, and they conceived an intramolecular Diels–Alder cyclization following biomimetic considerations as the key step (Fig. 64) [168]. The chiral *tert*-butyl(dimethyl)silyl-protected precursor **266**, synthesized by conventional means, was easily deprotected with tetra-*n*-butylammonium fluoride and oxidized with aqueous iron trichloride to form the fleeting quinone intermediate, which in situ underwent an intramolecular Diels–Alder reaction to the intermediate quinone **267**. The latter was elaborated conventionally to the enantiopure target (+)-elisabethin A (**265**), thereby also giving proof to the absolute configuration of the natural product.

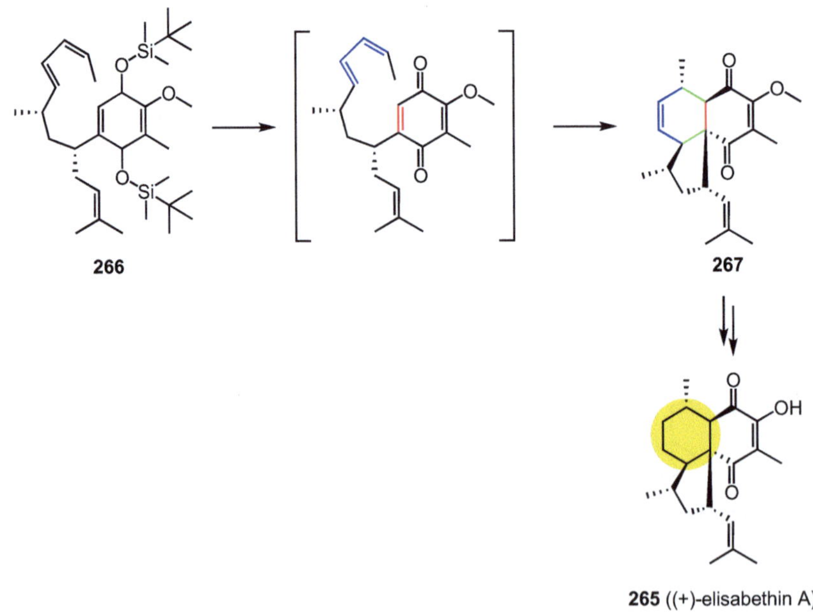


**Fig. 64** The intramolecular Diels–Alder key step in the synthesis of (+)-elisabethin A (**265**)

## *6.24 (+)-Torreyanic Acid*

The dimeric quinone epoxide (+)-torreyanic acid (**268**) (Fig. 65) was isolated among other compounds, like taxol, from the endophytic fungus *Pestalotiopsis microspora* (Speg.) G.C. Zhao & N. Li of the Sporacadaceae family (Plate 34 left), which is associated with the Florida torreya tree or stinking cedar, *Torreya taxifolia* Arn. of the Taxaceae family (Plate 34, right), one of the most endangered conifers in North America [169, 170]. However, the fungus does not seem to be pathogenic to the tree, which is infested by other fungi, mainly *Fusarium torreyae* causing the canker disease, which seems to be the primary culprit behind the population reductions. Interestingly, it turned out that *Pestalotiopsis microspora* is a species capable of breaking down and digesting polyurethane [171]. (+)-Torreyanic acid (**268**) is known to trigger apoptosis and to affect cancer cells through oxidative stress and mitochondrial damage. Its quinonoid moieties are thought to contribute to redox cycling and reactive oxygen species. This natural product **268** was also found to be more potent in cell lines that are sensitive to the protein kinase C agonist, 12-*O*-tetradecanoylphorbol-13-acetate, and showed G1 arrest of G0 synchronized cells [172].

The synthesis of (+)-torreyanic acid (**268**) centered on the biosynthesis-inspired key step of an *endo* Diels–Alder reaction between two 2*H*-pyran monomers with diene *facial* selectivity and an *anti* arrangement of the pentyl side chains, experimenting first with racemic torreyanic acid [173], and then applying the results to the (+)-enantiomer **268**. The biomimetic synthesis involved the rapid conversion of the protected quinone epoxide **269** to the *syn*- and *anti*-pyrans **270** and **271** via an oxaelectrocyclization, using Dess-Martin periodinane oxidation (Fig. 65). Next, a spontaneous intermolecular Diels–Alder reaction between **270** and **271** proceeded to

**Plate 34** Spores of *Pestalotiopsis microspora* (left). Photograph courtesy MSchink, Creative Commons Attribution-Share Alike 3.0 Unported. *Torreya taxifolia* with seeds (right). Photograph courtesy Cbarlow, Creative Commons Attribution-Share Alike 4.0 International

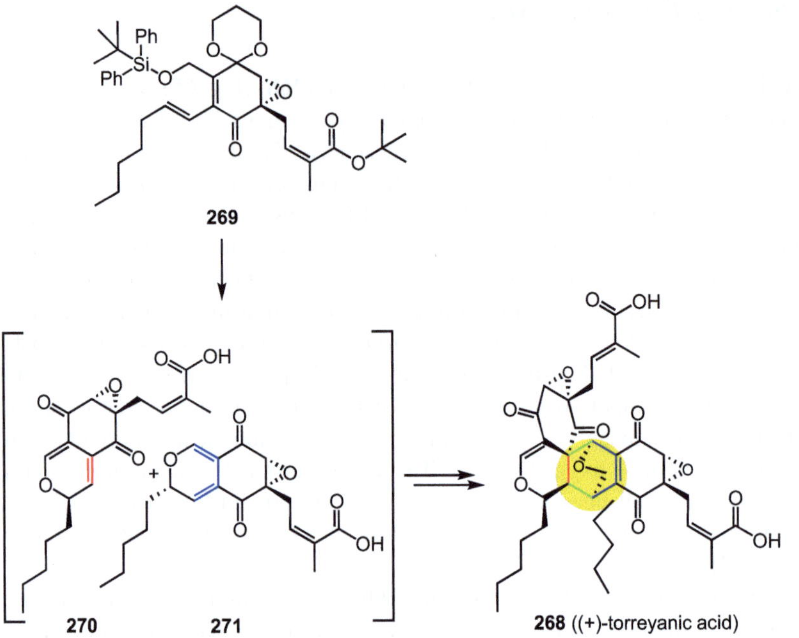


**Fig. 65** The intermolecular Diels–Alder key step in the synthesis of (+)-torreyanic acid (**268**)

yield the diastereomers from which, after appropriate deprotection, the target (+)-torreyanic acid (**268**) could be obtained.

## *6.25 (±)-Pinnatal and (±)-Sterekunthal A*

From the plant family Bignoniaceae a range of biologically active natural products for which their biosynthesis might involve pericyclic steps were isolated. Among these, pinnatal (**272**) (Fig. 66) was characterized from the sausage tree *Kigelia pinnata* (Jacq.) DC. (= *K. africana* (Lam.) Benth.) (Plate 35), a tree occurring throughout tropical Africa and producing poisonous fruits [174]. The related naphthoquinone, sterekunthal A (**273**), has been found in *Stereospermum kunthianum* Cham., also from the family Bignoniaceae (Plate 36), a tree occurring also in central Africa [175]. Decoctions of the fruits, bark, and roots are used in traditional medicine for infections, venereal diseases, diarrhea, and dysentery. Both compounds show antiplasmodial activity, and, accordingly, represent interesting leads for drugs against malaria.

The syntheses of the quinones (±)-pinnatal (**272**) and (±)-sterekunthal A (**273**) were enhanced by the then recent disclosure of the first crystal structure of a Diels–Alderase [176]. The authors presented a 1.70 Å resolution structure of the natural

**Plate 35** *Kigelia africana*. Upper left, tree, upper right, fruits, bottom left, flower, bottom right fruit cut open with seeds. Photographs courtesy E. Balocchi, B. C. Tørrissen, M. Schmidt, Genet; montage by RoRo, Creative Commons Attribution-Share Alike 3.0 Unported

**Plate 36** *Stereospermum kunthianum*. Photograph courtesy G. White, CC BY-NC 4.0

fungal macrophomate synthase complexed with pyruvate, displaying a large hydrophobic active site that can hydrogen-bond to the substrate 2-pyrone moiety, thereby suggesting that the reaction proceeds by a large-scale structural reorganization of the product. J. P. Malerich and D. Trauner conceived such a biomimetic synthesis involving an intramolecular Diels–Alder reaction [177]. Thus, the educts were synthesized by a Knoevenagel condensation of 2,7-dihydroxy-naphthoquinone (**274**) with the geraniol-derived pyranyl-protected aldehyde **275**, resulting in the intermediary naphthoquinone **276**, which underwent spontaneously a 6-π-electrocyclization producing the pyran ring in **277** (Fig. 66). Functional group transformations then led to the biomimetic precursor **278**, which, upon standing at room temperature led to an intramolecular Diels–Alder reaction into the target molecule, (±)-pinnatal (**272**).

Heating (±)-pinnatal (**272**) in benzene at 160°C in a sealed vessel resulted in a retro hetero-Diels–Alder reaction to provide the other target compound, (±)-sterekunthal A (**273**), in 92% yield.

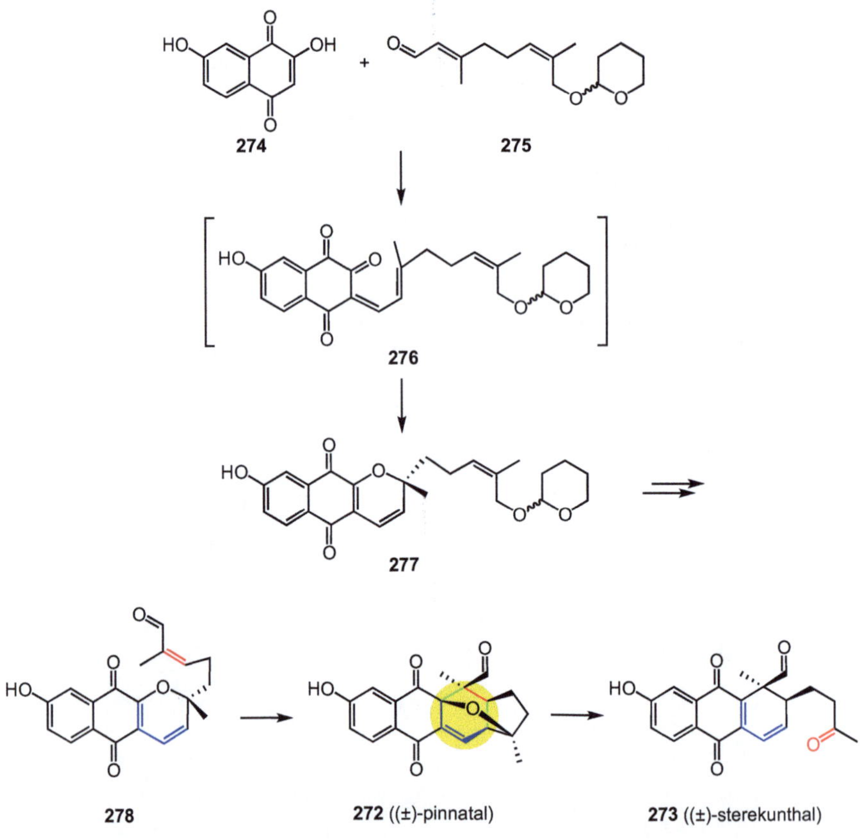


**Fig. 66** The intramolecular Diels–Alder key step in the synthesis of (±)-pinnatal (**272**) and the retro hetero Diels–Alder reaction to give (±)-sterekunthal (**273**)

It is interesting to note that 2003 seems to be the year when it became apparent that the repertoire of biosynthesis transformations in Nature obviously also included Diels–Alder cycloadditions [178].

## *6.26 (+)-Macquarimicins A, B, and C*

(+)-Macquarimicin A (**279**) in 2003, was isolated from cultures of the bacterium *Micromonospora chalcea* Foulerton from the Micromonosporaceae family (Plate 37) by Abbott Laboratories researchers in 1995 [179, 180]. Later, it was found that **279** is a selective inhibitor of membrane-bound neutral sphingomyelinase and exhibits anti-inflammatory activity in vivo [181]. The structure of **279** is characterized by a unique tetracyclic framework, which comprises a *cis*-configured tetrahydroindanone ring, a β-keto-δ-lactone ring, and a ten-membered macrocyclic ring [180]. Among the isolates the microtubule-stabilizing FR182877 (**258**) encountered in Sect. 6.21 above, and the antitumor antibiotic cochleamycin [182] (to be discussed in the next Section) displaying similar structural aspects, were obtained [183].

A biosynthesis-inspired intramolecular transannular Diels–Alder synthesis was envisaged for the key step in ring construction of (+)-macquarimicin A (**279**) by K.-I. Takao and coworkers [184] (Fig. 67). Thus, the authors synthesized a diastereomeric mixture of **280** by conventional methods, and when it was heated

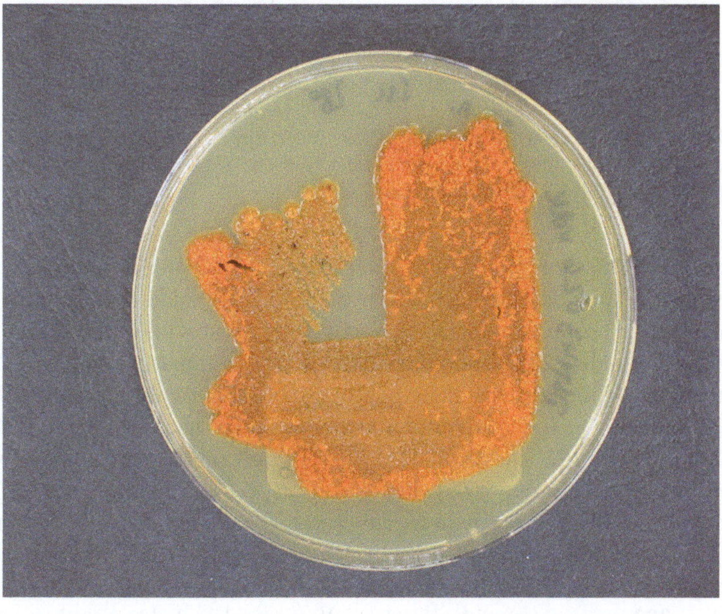

**Plate 37** *Micromonospora chalcea*. Copyright Leibniz-Institut (DSMZ 43026)

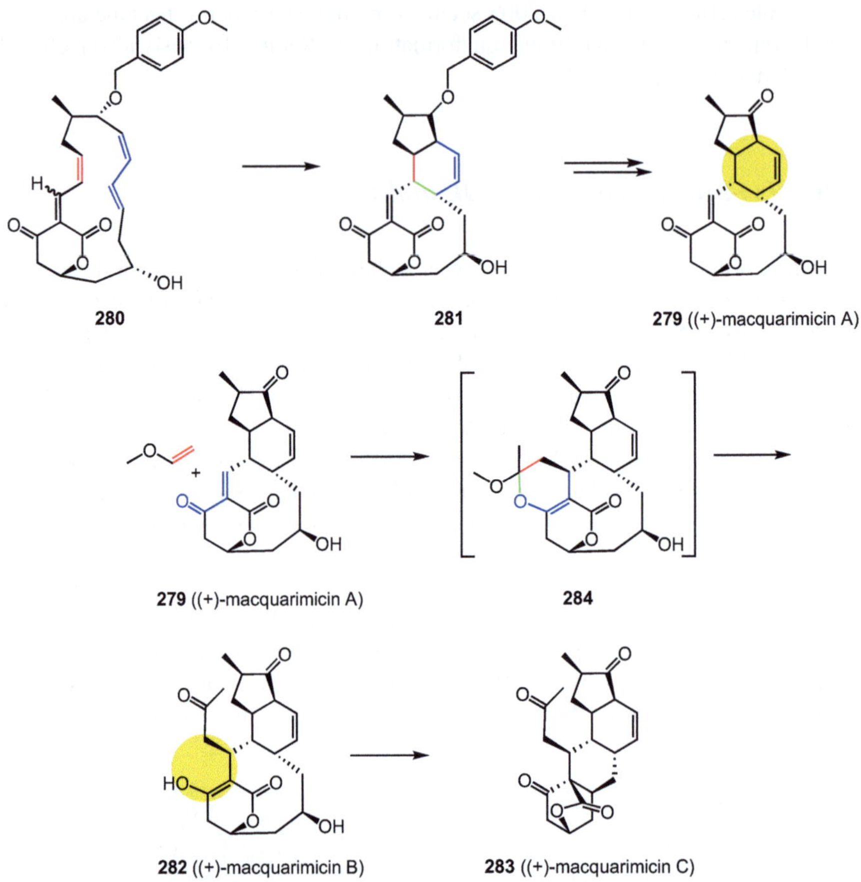


**Fig. 67** The intramolecular transannular Diels–Alder key step in the synthesis of (+)-macquarimicin (**279**), the intermolecular step in the synthesis of (+)-macquarimicin B (**282**), and its further reaction product (+)-macquarimicin C (**283**)

in toluene at 130°C in a sealed tube in the presence of 2,6-di-*tert*-butyl-*p*-cresol it furnished the desired diastereomer **281** as the sole Diels–Alder cycloadduct. In this reaction, the (*Z,E*) configuration of the reacting diene seems to be the origin of *endo*-selectivity. The adduct **281** was converted to the target (+)-macquarimicin A (**279**) in a few more conventional steps, thereby also establishing its absolute configuration.

In the following year, the Abbott Laboratories group succeeded in preparing (+)-macquarimicins B (**282**) and C (**283**) [179, 180], this time using an intermolecular hetero Diels–Alder key step [185] (Fig. 67). Thus, (+)-macquarimicin A (**279**) was reacted with 2-methoxypropene in tetrahydrofuran/water to give smoothly the transient cycloadduct **284**, which underwent spontaneous hydrolysis to afford the target (+)-macquarimicin B (**282**) in 83% yield. To achieve a dehydrative transannular alkylation, **282** was treated with camphorsulfonic acid in dichloromethane to yield quantitatively the desired (+)-macquarimicin C (**283**).

## 6.27 (+)-Cochleamycin A

(+)-Cochleamycin A (**285**) was isolated from a cultured broth of *Streptomyces* sp. and it showed cytotoxicity against P388 leukemia cells and other cell lines, and also exhibited antimicrobial activities [186]. It is analogous structurally to the macquarimicins discussed in the preceding Section, and again, the intramolecular Diels–Alder reaction provided the key to its synthesis (Fig. 68).

There were two main syntheses for the target **285**. For the first synthesis by K. Tatsuta and co-workers [187], the carbinol-protected open-chain triene **286** was produced by classical steps, and when this compound was heated in xylene at 140°C in the presence of ytterbium heptafluorodimethyl-octanedionate and traces of 2,6-di-*tert*-butyl-*p*-cresol for 4 h, the indene derivative **287** was obtained. This could then be transformed by standard methodology into the target molecule, (+)-cochleamycin A (**285**). Since the configurations of the chiral centers of **286** stemmed from the chiral pool, the absolute configuration of the (+)-enantiomer of cochleamycin (**285**) could be confirmed.

In the second synthesis, T. A. Dineen and W. R. Roush [188] followed mainly the macquarimicin strategy (see Sect. 6.26), by first synthesizing the ansa-pyrone **288** by conventional means (Fig. 68). On heating **288** with traces of 2,6-di-*tert*-butyl-*p*-cresol as antioxidant in toluene at 125°C for 21 h, the intramolecular Diels–Alder reaction provided **289** in nearly 70% yield, which contained already the target skeleton. Conventional steps then led to (+)-cochleamycin A (**285**).

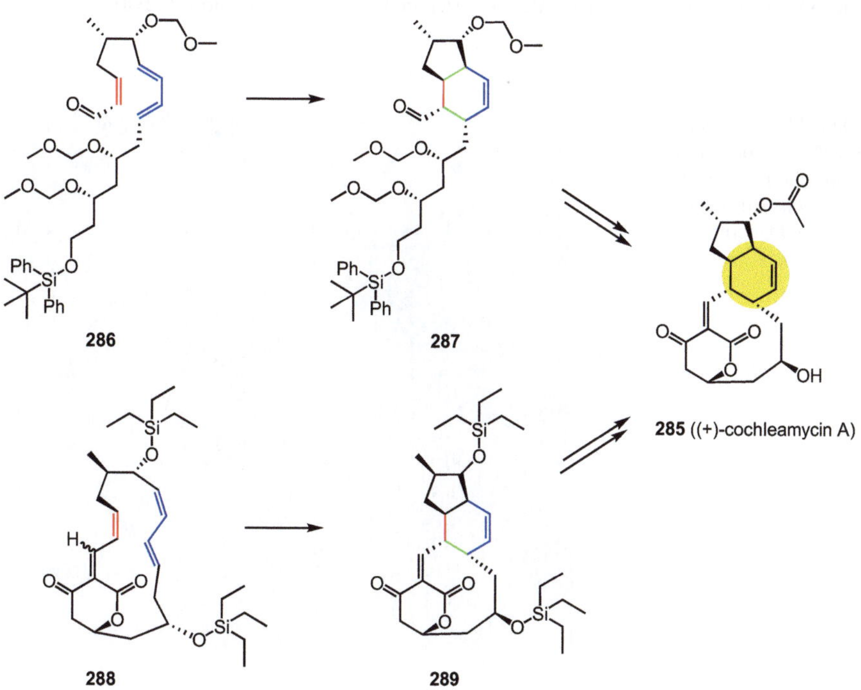


**Fig. 68** The two independent intramolecular transannular Diels–Alder key steps in the synthesis of (+)-cochleamycin A (**285**)

## 6.28 Ophirin B

The eunicellin ophirin B (**290**) (Fig. 69) is a member of the C-2,C-11-cyclized cembranoid diterpenes, which is produced among others as a secondary metabolite of the gorgonian octocoral *Briareum* Blainville sp. from the Briareidae family (Plate 38), found in Caribbean waters and the West Pacific Ocean, and it displays diverse biological activities [189].

The synthesis of ophirin B (**290**) by M. T. Crimmins and B. H. Brown followed an intramolecular Diels–Alder strategy [190] (Fig. 69). Preparation of the Diels–Alder

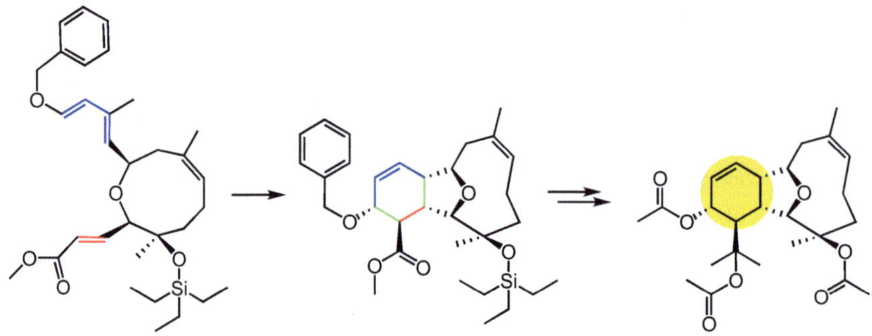

**291** **292** **290** (ophirin B)

**Fig. 69** The intramolecular Diels–Alder key step in the synthesis of ophirin B (**290**)

**Plate 38** *Briareum* sp. Photograph courtesy Haplochromis, Creative Commons Attribution-Share Alike 3.0 Unported

substrate, the oxonon-diene **291**, followed conventional steps. Upon standing at 25°C, **291** was cyclized quantitatively into the oxatricyclic intermediate **292**, which could then be elaborated into the target compound ophirin B (**290**) along standard steps.

## *6.29 (±)-Alantrypinone*

The polycyclic alkaloid alantrypinone (**293**) was isolated from a *Penicillium* species [191] and is characterized by a spiro-bridged pyrazinoquinazolindione unit. An efficient total synthesis of **293** by an intermolecular hetero Diels-Alder reaction of the pyrazine diene **294** with 3-methyleneoxindole (**295**) was described in 2003 (Fig. 70) [192]. The Diels-Alder reaction provided both the desired regiochemistry and *exo* selectivity, and yielded the adduct **296** upon leaving a chloroform solution of the two educts at room temperature for 24 h. The intermediate **296** was then hydrolyzed readily with 1 *N* hydrochloric acid in ethyl acetate for 5 h at room temperature to give the target compound (±)-alantrypinone (**293**) in 85% yield.

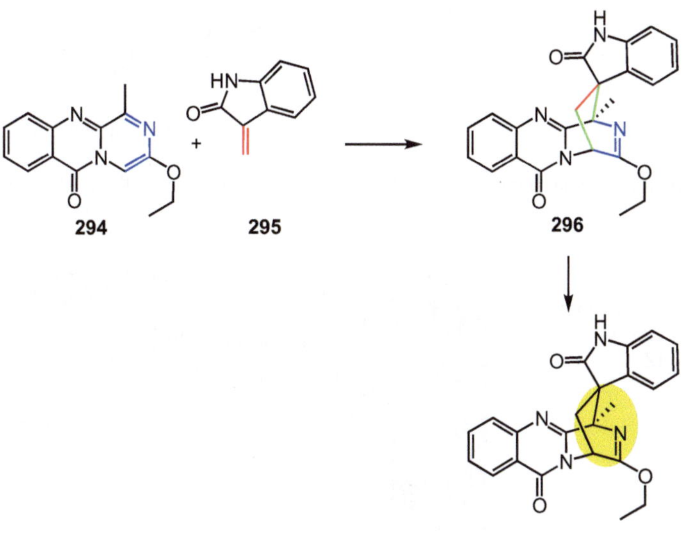


**Fig. 70** The intermolecular hetero Diels–Alder key step in the synthesis of (±)-alantrypinone (**293**)

## *6.30 (−)-Spinosyn A*

The tetracyclic macrolide (−)-spinosyn A, also called A83543A (**297**) (Fig. 71), is an extraordinarily efficient insecticide isolated from the soil actinobacterium *Saccharopolyspora spinosa* Mertz and Yao from the Pseudonocardiaceae family

(Plate 39) [193]. Its structure is comprised of a twelve-membered lactone fused to a 5,6,5-*cis*,*anti*,*trans* carbocyclic ring system.

A biosynthesis-inspired, convergent, highly stereoselective synthesis of **297** was published in 2004 by W. R. Roush and his coworkers, rooted in an intramolecular

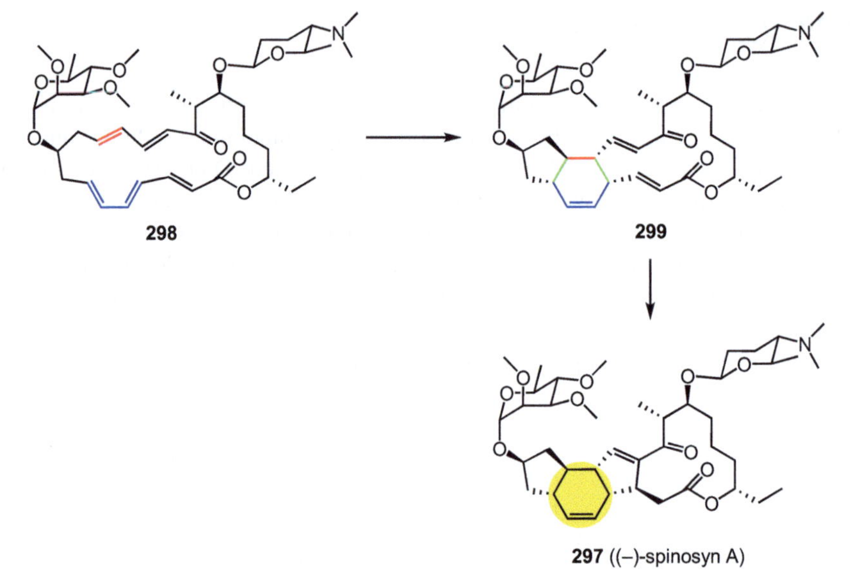


**Fig. 71** The transannular intramolecular Diels–Alder key step in the synthesis of (−)-spinosyn A (**297**)

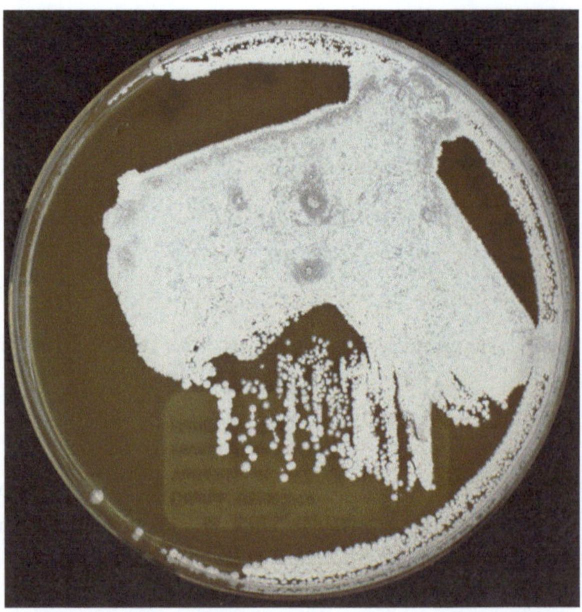

**Plate 39**
*Saccharopolyspora spinosa.*
Copyright Leibniz-Institut DSMZ

transannular Diels–Alder reaction as the key step (Fig. 71) [194]. Thus, the L-rhamnopyranosyl-substituted macrolactone **298** was synthesized along common steps and it underwent the transannular Diels-Alder reaction immediately, under the conditions of its lactone cyclization step at room temperature, yielding the tricyclic system **299**. The final cyclization step consisted of a vinylogous Morita–Baylis–Hillman reaction to produce the targeted compound, (−)-spinosyn A (**297**).

## *6.31 (−)-Norzoanthamine*

From colonies of the marine cnidaria *Zoanthus* aff. *pulchellus* Duchassaing & Michelotti of the Zoanthidae family (Plate 40), (−)-norzonthamine (**300**), a heptacyclic aminal-containing alkaloid, could be isolated [195]. It displays an interesting antiosteoporotic activity and inhibits the growth of leukemia cell lines [196].

For the synthesis of (−)-norzonthamine (**300**), an intramolecular Diels–Alder reaction was the key step to construct the *trans-anti-trans*-fused phenanthrenoid-type part of the target (Fig. 72) [197]. The synthon **301** underwent an intramolecular

**Plate 40** *Zoanthus* aff. *Pulchellus*. Photograph courtesy J. D. Reimer, M. Hirose, P. Wirtz, Creative Commons Attribution-Share Alike 3.0 Unported

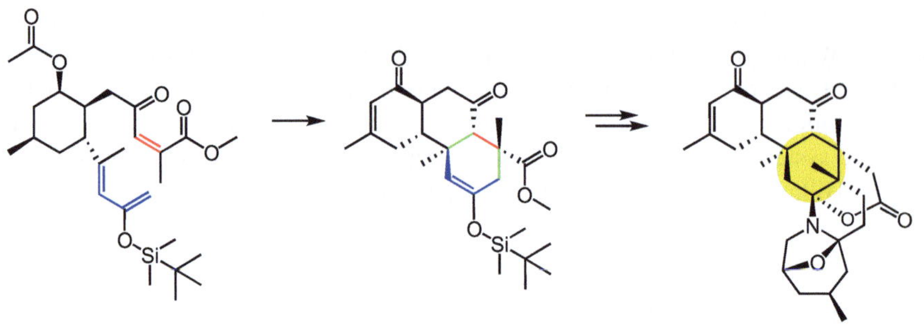


301 302 300 ((−)-norzoanthamine)

**Fig. 72** The intramolecular Diels–Alder reaction key step in the synthesis of (−)-norzoanthamine (**300**)

Diels–Alder reaction in trichlorobenzene at 240°C for 1.5 h to yield the *exo*-configured tricyclic system **302**, with the *endo*-configured adduct as a minor component. The latter was then elaborated through conventional steps into the target, (−)-norzoanthamine (**300**). This synthesis also verified the absolute configuration of the natural product. It is interesting to add that the 41 steps in the synthesis resulted in an overall yield of 3.3% of **300**. This synthesis clearly ranks as one of the landmarks of its type from the turn of the present century.

## *6.32 Solanapyrone D*

As described in Sect. 5.5, solanapyrones were isolated in the 1980s as phytotoxic compounds from the phytopathogenic fungus *Alternaria solani* (Plate 16) [92]. Solanapyrone D (**303**) isolated in the late 1990s from the same source [198] was found to be a valuable target to develop and test a new dimension of Diels–Alder reactions, the enantioselective organocatalytic intramolecular Diels–Alder reaction, as was demonstrated by D. W. C. MacMillan and his group in 2005 (Fig. 73) [199].

The educt for this most elegant Diels–Alder step was the trienal **304**, which was reacted in the presence of 20% of the trifluoroacetate of the second-generation imidazolidinone catalyst **305** in acetonitrile at 5°C to afford the aldehyde **306** in about 70% yield, and with *dr* > 20/1 and 90% *ee*. Importantly, all four stereocenters of the octahydronaphthalene part of solanapyrone D (**303**) were installed efficiently in this single catalytic operation! Conventional steps were then used to elaborate the target, solanapyrone D (**303**), in six steps from the trienal **304**.

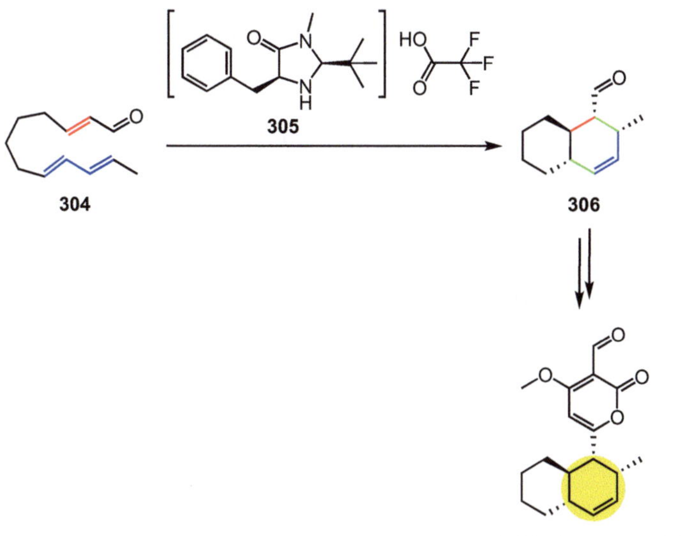


**Fig. 73** The enantioselective organocatalytic intramolecular Diels–Alder reaction key step in the synthesis of solanapyrone D (**303**)

## *6.33 (+)-Tubelactomicin A*

The macrolide antibiotic (+)-tubelactomicin A (**307**) was isolated from the culture broth of the actinomycete strain MK703-102F1, a member of the genus *Nocardia*. It showed potent antimicrobial activity against acid-fast bacteria, including drug-resistant strains [200]. The structure of **307** was elucidated by extensive NMR analysis and confirmed by the single-crystal X-ray analysis of the carboxamide derivative with L-phenylalanine methylester [201]. Accordingly, its absolute stereochemistry also was established.

The structure of **307** is characterized by a *trans*-fused octahydronaphthalene moiety possessing six contiguous stereogenic centers and a 16-membered macrolactone, incorporating an (*E*,*E*)-conjugated diene and an α,β-disubstituted (Z)-acrylic acid moiety.

Starting from the chiral pool member diethyl (*R*)-malate, synthesis of the lower-half segment of **307**, the octalin **308** was the first target (Fig. 74) [202]. Its synthesis involved a highly *endo*- and π-facial selective intramolecular Diels-Alder reaction achieved by using the trisubstituted methacrolein derivative **309** tethering a 10-carbon dienyne unit at the β-carbon, which in turn was prepared from a known allylated malic acid derivative. The reaction proceeded smoothly by heating **309** in toluene at 80°C for 24 h (with some 2,6-di-*tert*-butyl-4-methyl-phenol added to avoid oxidation) in 93% yield, producing mostly the desired *trans*-fused cycloadduct **308** with an *endo* configuration. This product could then be elaborated using conventional methodology into the target compound, (+)-tubelactomicin A (**307**) [203].

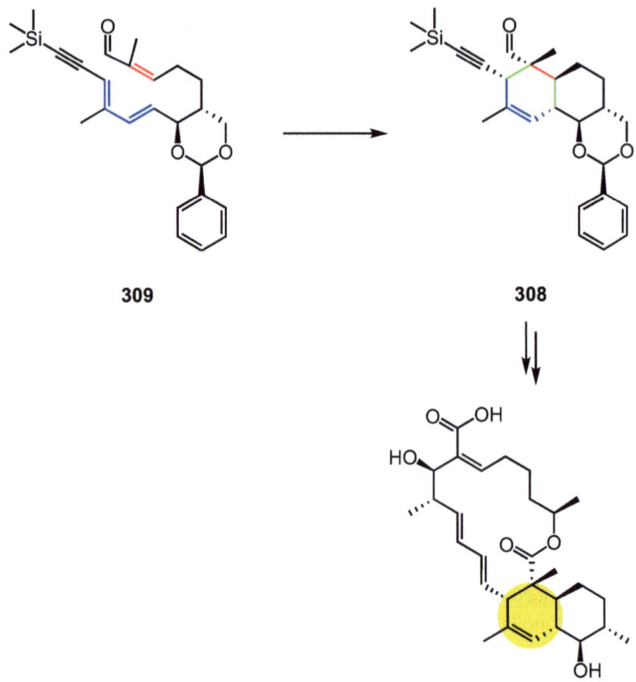


**Fig. 74** The intramolecular Diels-Alder reaction key step in the synthesis of tubelactomicin A (**307**)

## *6.34 Palominol, Dolabellatrienone, β-Araneosene, and Isoedunol*

The highly bioactive dolabellane-type marine diterpenes palominol (**310**), dolabellatrienone (**311**), β-araneosene (**312**), and isoedunol (**313**), isolated from soft corals or brown algae, were targeted by S. A. Snyder and E. J. Corey in 2006, by envisioning an enantioselective Type I intramolecular Diels–Alder macrobicyclization as the key step (Fig. 75) [204].

The precursor **314** to be used in the enantioselective Diels–Alder macrobicyclization was prepared in ten steps from (*E*,*E*)-farnesol. As the absolute stereochemistry-inducing catalyst, the authors constructed the (*S*)-oxazaborolidine **315**, which not only provided the absolute stereochemistry, but also set the relative *trans*-stereochemistry and constructed the challenging eleven-membered carbocycle of the bicyclo[9.4.0]pentadecatrienal **316**. In the proposed transition state complex **315*** (also supported by means of density functional theory at the B3LYP/6-31G(d) level [205]), the catalyst binds to the substrate via a two-point binding motif in which the carbonyl oxygen of the substrate complexes with boron, and a B–O···H–C bond is formed.

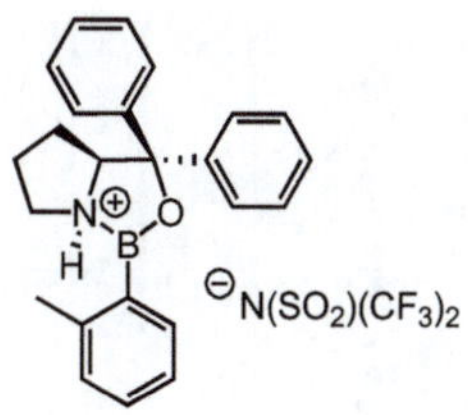


315

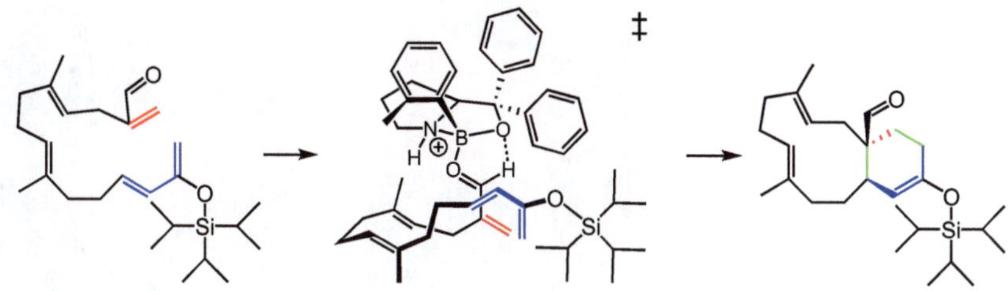


314 315* 316

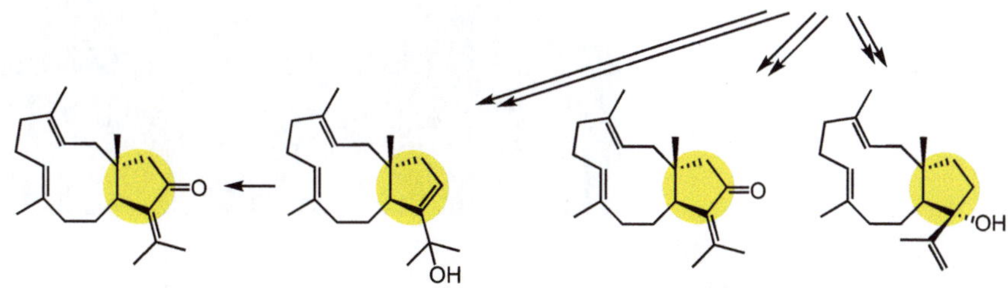


311 (dolabellatrienone) 310 (palominol) 312 (β-araneosene) 313 (isoedunol)

**Fig. 75** The enantioselective Type I intramolecular Diels–Alder macrobicyclization key step in the synthesis of palominol (**310**), dolabellatrienone (**311**), β-araneosene (**312**), and isoedunol (**313**)

Slow addition of a solution of 20 mol% of (*S*)-oxazaborolidine **315** in toluene to a stirred solution of the highly acid-sensitive precursor **314** in toluene at −93°C, followed by 3 h at −93°C and then 10 h at −78°C, resulted in the formation of macrocycle **316** in more than 70% yield and 90% enantiomeric excess. The structure and absolute configuration of **316** were verified by X-ray crystallographic analysis of the tosylhydrazone derivative. Compound **316** was then transformed by the application of conventional steps into the natural compounds palominol (**310**), dolabellatrienone (**311**), β-araneosene (**312**), and isoedunol (**313**).

## *6.35 Kwanzquinone C*

The anthraquinone glucoside kwanzoquinone C (**317**), was isolated in 2002 by M. G. Nair and co-workers [206] from an extract of Kwanzo roots, i.e., the roots

**Plate 41** *Hemerocallis fulva*. Photograph H. F. (author)

of the orange daylily, *Hemerocallis fulva* L., of the Asphodelaceae family (Plate 41), together with a series of other hydroxylated anthraquinones. Plants from the genus *Hemerocallis* are known to be applied in traditional medicine systtems in eastern Asia, for example, as a remedy for schistosomiasis [207]. Kwanzoquinone C (**317**) was demonstrated as an anthraquinone glycoside that displays the growth inhibition of different human tumor cells [208].

The synthesis of **317** proved to be a good example of the progress being made in delicately balanced intermolecular Diels–Alder reactions. L. F. Tietze and his coworkers succeeded in 2006 in preparing the necessary tetrasubstituted 1,3-butadiene **318** (Fig. 76) [209]. This was added dropwise to a solution of the naphthoquinone **319** in dichloromethane at −10°C over 10 min and then stirred at this temperature for 20 h. After solvent evaporation, the residue was treated for 2 min with 1% trifluoroacetic acid, to achieve aromatization of the adduct **320** and quenched with potassium carbonate at 20°C, resulting in the desired anthraquinone **321** after workup. Conventional glycosidation using 2,3,4,6-tetra-*O*-acetyl-*α*-D-glucopyranosyl trichloroacetimidate then afforded the target compound, kwanzoquinone C (**317**).

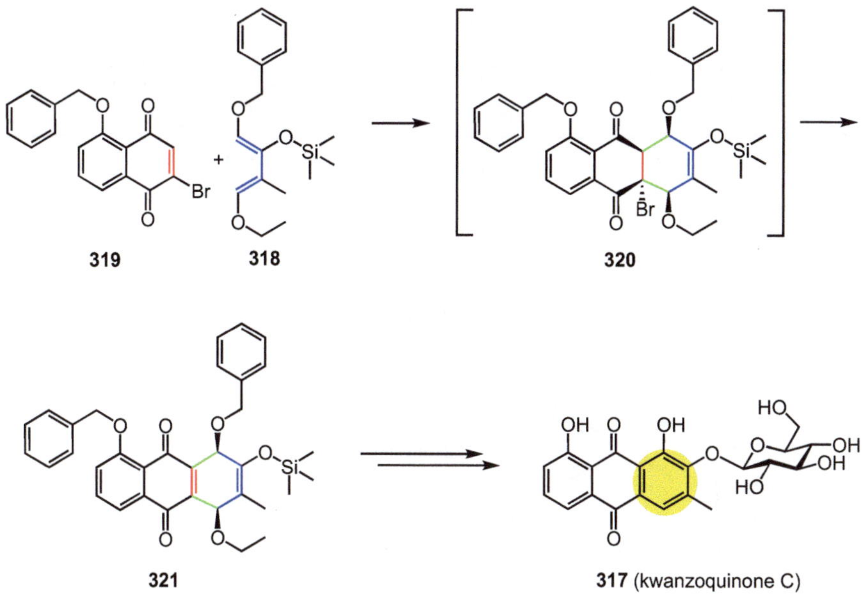


**Fig. 76** The intermolecular Diels–Alder key step in the synthesis of kwanzoquinone C (**317**)

## *6.36 Abyssomicin C and atrop-Abyssomicin C*

Abyssomicin C (**322**) is an antibiotic with an unusual molecular architecture and unique mechanism of action [210]. Isolated from the actinomycete *Verrucosispora* strain AB 18–032, cultivated from a sediment sample collected from the depths of the Sea of Japan, this aptly named natural product inhibits the biosynthesis of *p*-aminobenzoic acid in bacteria, a pathway that is absent in human physiology. The intriguing structural topology of **322**, coupled with its reported activity against both methicillin-resistant *Staphylococcus aureus* and vancomycin-resistant *S. aureus* [211], have made it a fascinating target for synthesis. Again, the key intermolecular Diels–Alder reaction had to be refined considerably to produce the desired molecule, as achieved by K. C. Nicolaou and S. T. Harrison in 2006 (Fig. 77) [212].

After considerable experimentation, it was discovered that mixing one equivalent of the diene **323** with three equivalents of the catalytic ligand 2-(pyrrolidin-1-yl) phenol (**324**) and four equivalents of methylmagnesium bromide in toluene at 0°C, followed by the addition of ten equivalents of the ene component methyl acrylate (**325**) and warming up to 55°C for 24 h, this furnished the desired cycloadduct **327** as a single diastereomer and in 80% yield. The reaction involving the postulated transition state **326** represents a remarkable Lewis acid-templated intermolecular Diels–Alder reaction. The cycloadduct **327** was then elaborated conventionally into the target molecule abyssomicin C (**322**) on the one side, and into *atrop*-abyssomicin C (**327**), which is biologically inactive, on the other side. K. C. Nicolaou and S. T. Harrison further elaborated and improved the various synthesis steps in 2007, and in

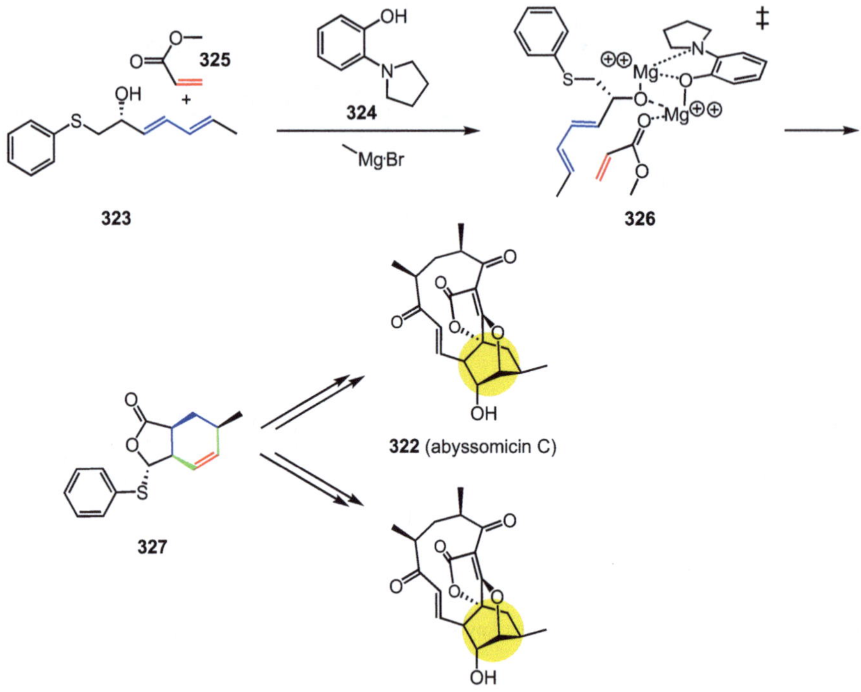


**Fig. 77** The Lewis acid templated intermolecular Diels–Alder key step in the synthesis of abyssomicin C (**322**) and *atrop*-abyssomicin C (**327**)

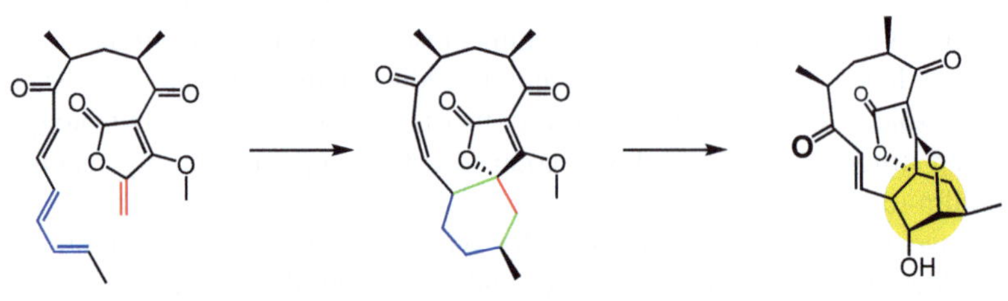


**Fig. 78** The intramolecular Diels-Alder reaction key step in the synthesis of abyssomicin C (**322**)

particular, determined the configuration of the atropisomer **327** using X-ray crystallographic analysis [213].

Due to the extensive interest in this compound, several groups have invested in synthesis efforts for abyssomicin C (**322**). For these, an intramolecular Diels–Alder key step was crucial for the skeleton construction. Thus, in 2005 the group of E. J. Sorensen synthesized conventionally the lactone precursor **328**, which underwent the intramolecular step catalyzed by lanthanum(III) triflate to yield diastereoselectively the tricycle **329**, and was then transformed into the target abyssomicin (**322**) (Fig. 78) [214]. The same principal approach of an intramolecular

Diels–Alder reaction was also followed by B. B. Snyder and Y. Zou in 2005 [215], and the group of E. A. Couloadouros in 2006 [216].

## *6.37 11,12-Diacetoxydrimane*

The catalytic asymmetric transannular Diels-Alder reaction that affords polycyclic products in high enantiomeric excess was the focus of an extended study by E. P. Balskus and E. N. Jacobsen in 2007 [217]. This catalytic system can also alter the inherent diastereoselectivity of cyclizations with substrates containing chiral centers. As an illustrative example, the catalytic enantioselective transannnular Diels-Alder reaction was used as the key step in a total synthesis of 11,12-diacetoxydrimane (**330**).

The drimane sesquiterpene 11,12-diacetoxydrimane (**330**) is characterized by a bicyclic carbon skeleton derived biosynthetically from farnesyl pyrophosphate on the mevalonate pathway. It is found particularly in wood-decaying fungi, like *Polyporus ciliatus* Fries of the Polyporaceae family (Plate 42).

The asymmetric transannular cyclization started with the macrocycle **331** prepared in a conventional way (Fig. 79). It underwent an asymmetric transannular Diels-Alder reaction catalyzed by the specifically tailored oxazaborolidine catalyst **332** (like that discussed in Sect. 6.34) in toluene solution for 20 h at room

**Plate 42** *Polyporus ciliatus*. Photograph courtesy A. Kunze, Creative Commons Attribution-Share Alike 3.0 Unported

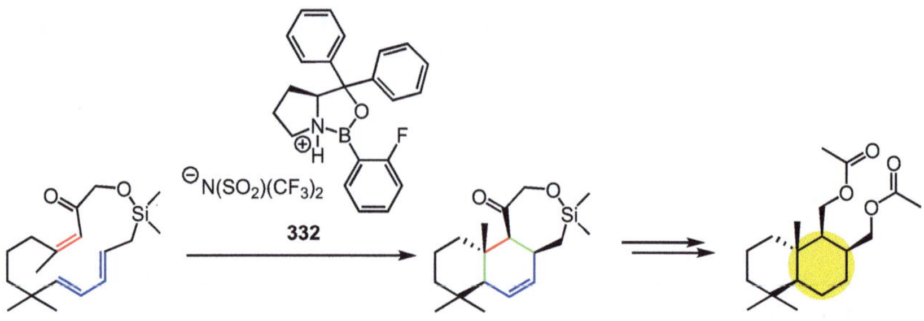


**Fig. 79** The catalytic enantioselective transannular Diels-Alder reaction key step in the synthesis of 11,12-diacetoxydrimane (**330**)

temperature, to yield enantioselectively the tricyclic compound **333**, illustrating wonderfully the high potential of this approach. The latter was then transformed in two further conventional steps into the natural product 11,12-diacetoxydrimane (**330**).

## *6.38 Vinigrol*

The unusual diterpene vinigrol (**334**) from the fungal strain *Virgaria nigra* F-5408 (Plate 43) was isolated by M. Hashimoto and coworkers in 1987 [218] The promising biological properties of **334**, including antihypertensive and platelet aggregation-inhibitory effects [219], combined with its unique terpene framework, have attracted significant attention from the chemical synthesis community.

Vinigrol (**334**) in having eight contiguous stereocenters provides a particularly difficult challenge, as it is the only known natural product to contain a decahydro-1,5-butanonaphthalene carbon skeleton with a strained bicyclo[5.3.1]undecane ring, reminescent of taxol (**127**) (Sect. 5.2). In 2009, P. S. Baran and his group succeeded with a synthesis of the core structure involving two distinct Diels–Alder steps as the key reactions (Fig. 80) [220]. The final complete synthesis of **334** was then achieved in the following year by extensive modifications of the substrates and conventional steps [221].

The first step was a clear-cut and smoothly proceeding *endo*-selective intermolecular Diels–Alder reaction between (*E*)-methyl 4-methyl-2-pentenoate (**335**) and the diene **336**, to yield the bicyclic ketone **337** in the presence of aluminum trichloride. This ketone was then modified into intermediate **338**, which in situ reacted in an intramolecular Diels–Alder reaction at room temperature into the synthon **339** for final manipulation into the target compound vinigrol (**334**). Interestingly, this is the only example of a completely electron-neutral diene and simple olefin taking part in a non-catalyzed Diels–Alder reaction at ambient temperature. Obviously, a strong proximity effect bringing the ene and diene partial structures into an optimal region occurs in this step.

**Plate 43** *Virgaria nigra* on rotten wood. Photograph courtesy E. Rubio, Creative Commons

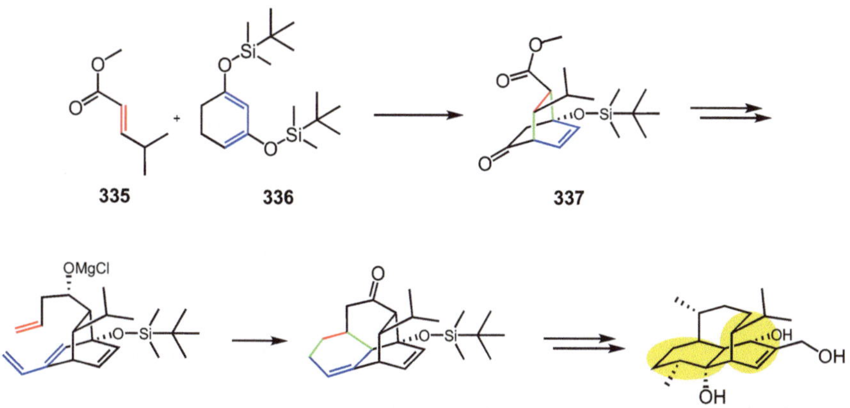


**Fig. 80** Inter- and intramolecular Diels–Alder key steps in the synthesis of vinigrol (**334**)

## 6.39 (+)-Cassaine

A nonsteroidal inhibitor of $Na^+$,$K^+$-ATPase, (+)-cassaine (**440**) (Fig. 81), is known to possess a pharmacological action like that of the digitalis glycosides, such as digitoxin, even though their structures are quite different [222]. In 1935, G. Dalma isolated (+)-cassaine from the bark of the tree *Erythrophleum guineense* G. Don (syn. *E. africanum* Harms), of the Fabaceae family (Plate 44) [223]. The structure elucidation was achieved by R. B. Turner and his group in 1959 [224]. Structurally, **440** features a *trans-anti-trans* tricyclic system and possesses six stereocenters and an *exo*-α,β-unsaturated ester containing a dimethylamino group.

With respect to the synthesis of (+)-cassaine (**440**), a transannular intramolecular Diels–Alder reaction was chosen by P. Deslongchamps and his group in 2008 as the pivotal step (Fig. 81) [225]. The synthon **441** for this reaction was obtained via couplings and macrocylization steps. When dissolved in toluene, **441** was heated in a sealed tube to 123°C and gave the tricycle **442**. Then, the latter could be derivatized in a conventional manner to form the target molecule, (+)-cassaine (**440**).

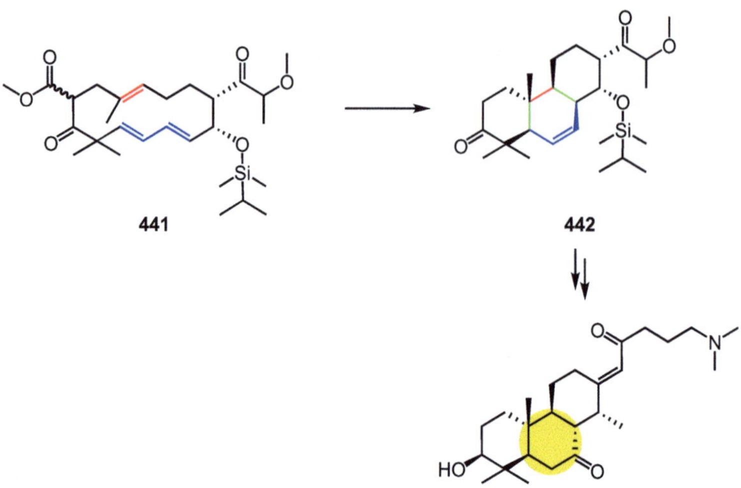


**440** ((+)-cassaine)

**Fig. 81** The intramolecular transannular Diels-Alder reaction key step in the synthesis of (+)-cassaine (**440**)

**Plate 44** *Erythrophleum guinneense*. Photograph courtesy SAplants, Creative Commons Attribution-Share Alike 4.0 International

## 6.40 Triptolide

The isolation of triptolide (**443**) (Fig. 82) from the Asian vine *Tripterygium wilfordii* Hook.f. (Celastraceae) (Plate 45), known as thunder duke vine, was first reported by S. M. Kupchan and his group in 1972 [226]. Plants in this genus have a long history of use in traditional Chinese medicine to treat rheumatoid arthritis and psoriasis. Extracts of *Tripterygium*, along with **443**, its main bioactive constituent, display potent antileukemic and antitumor properties, in addition to anti-inflammatory and immunosuppressive activities, and show potential as a male contraceptive [227].

The tetracyclic system of the oxasteroid triptolide (**443**) was assembled by means of a diene-transmissive double Diels–Alder sequence of the [3]dendralene **444**, with the first step proceeding in toluene with methyl acrylate (**445**) to provide lactone **446**, and the second step with **446** and the quinone **447** in dichloromethane at 19 kbar pressure to yield the synthon **448** (Fig. 82) [228]. The latter was elaborated by known methodology into the target compound triptolide (**443**).

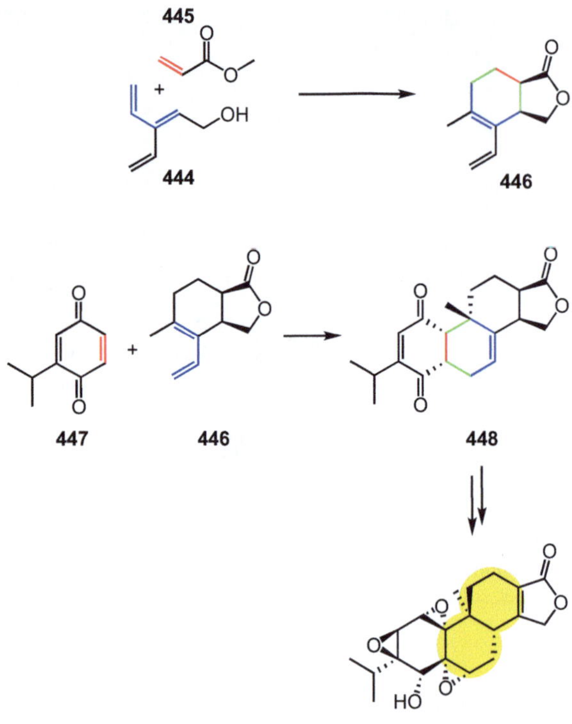


**Fig. 82** The diene-transmissive double Diels–Alder sequence in the synthesis of triptolide (**443**)

**Plate 45** *Tripterygium wilfordii*. Photograph courtesy BotBln, Creative Commons Attribution-Share Alike 3.0 Unported

## *6.41 (+)- and (−)-Versicolamide B*

(+)-Versicolamide B (**449**) and (−)-versicolamide B (**450**) (Fig. 83) are structurally complex spiro-cyclized alkaloids and are versions of prenylated cyclo(L-Trp-L-Pro) derivatives, which possess a unique spiro-fusion to a pyrrolidine ring at the 3-position of the oxindole core together with the bicyclo[2.2.2]diazaoctane ring system. Interestingly, this is one of the rare cases in which both enantiomers are produced by different organisms. Thus, the (+)-enantiomer **449** was isolated from the filamentous fungus *Aspergillus versicolor* NRRL 35600, whereas the (−)-enantiomer **450** was obtained from a marine *Aspergillus* species [229–231].

R. M. Williams and co-workers accomplished the enantioselective total syntheses of both enantiomers **449** and **450** of versicolamide B using an intramolecular Diels–Alder reaction as the linchpin step (Fig. 83) [231]. The carefully crafted enamide precursors underwent the reaction to form the core bicyclo[2.2.2]diazaoctane system with high stereocontrol. Thus, treatment of the two diastereoisomers of the enamide **451** resulted in tautomerization and provided the azadiene **453**, which underwent spontaneously an intramolecular Diels–Alder reaction to afford a mixture of (+)-versicolamide B (**449**) and its diastereomer in a 1.4/1 mixture, which could be separated easily and purified. The enamide diastereomers **452** under identical conditions provided via the azadiene **453a** (−)-versicolamide B (**450**) and its diastereomer in a 1.4/1 mixture. As predicted by ab initio calculations, only the *anti* cycloadducts were obtained. Laboratory validation of the proposed biosynthetic Diels–Alder construction, coupled with the secondary metabolite profile of the producing fungi, reveals that each *Aspergillus* species obviously has evolved enantiomerically distinct indole oxidases, as well as enantiomerically distinct Diels–Alderases.

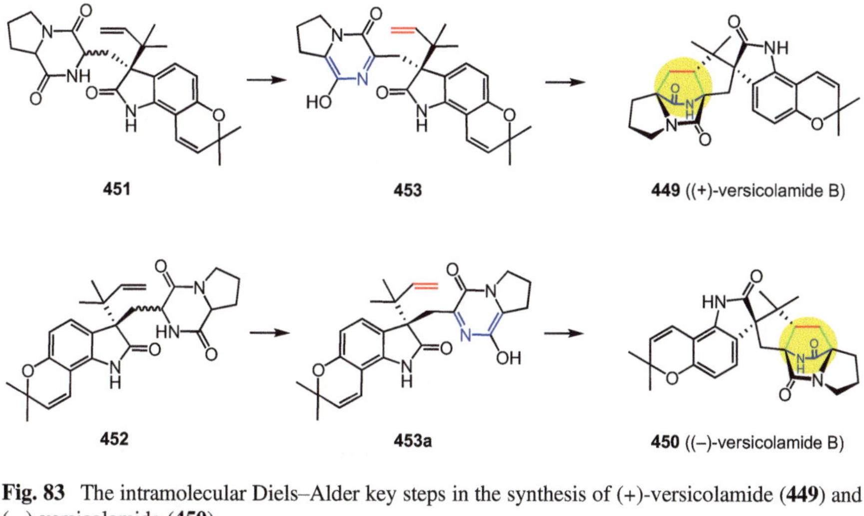


**Fig. 83** The intramolecular Diels–Alder key steps in the synthesis of (+)-versicolamide (**449**) and (−)-versicolamide (**450**)

This work provided an experimental validation of a biosynthesis Diels–Alder hypothesis: it suggests that separate *Aspergillus* species produce versicolamide B of opposite configuration via enantiomerically distinct "Diels–Alderase" enzymes. Methodologically, the paper showcases asymmetric intramolecular Diels–Alder reactions in complex settings and a biomimetic approach to polycyclic construction. It underscores how strategic Diels–Alder cyclizations can be used to probe biosynthesis pathways, while also advancing synthesis efficiency. The Diels–Alder reaction, particularly in the context of the synthesis of (+)- and (−)-versicolamide B (**449** and **450**), presents a notable issue that underscores the challenges in understanding and replicating the stereochemical and mechanistic intricacies of Diels–Alder reactions in natural product synthesis.

## 6.42 (−)-FR182877

To finalize this decade, it is necessary to examine earlier times, which brought inter alia the case of (+)-FR182877 (**258**), so elegantly solved by the Sorensen group (Sect. 6.21). Compound (−)-FR182877 (**454**) (Fig. 84) was isolated from *Streptomyces*

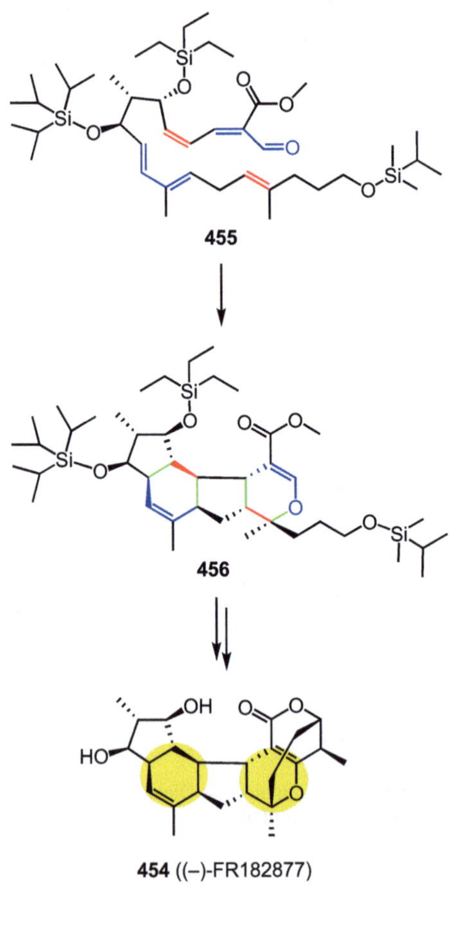


**Fig. 84** The one-pot stereoselective tandem intramolecular–intramolecular hetero-Diels–Alder reaction in the synthesis of FR182877 (**454**)

sp. 9885 [162] and was targeted by the M. Nakada group using an approach from the acyclic synthon **455** prepared by conventional means in situ (Fig. 84) [232]. A one-pot stereoselective tandem intramolecular–intramolecular hetero-Diels–Alder cyclization of **455** in toluene at 80°C led to the tetracyclic compound **456**, with seven newly generated, correct stereogenic centers. This latter compound then could be elaborated into natural product (−)-FR182877 (**454**) by conventional methodology.

## 7 Expanding Frontiers of Diels–Alder Chemistry: Innovative Catalytic and Biomimetic Applications in Natural-Product Synthesis (2010s)

The decade beginning in 2010 witnessed a remarkable renaissance of the Diels–Alder reaction as a strategic workhorse for complex-molecule construction. Building on the asymmetric and tandem paradigms introduced in the 1990s and early 2000s, researchers over these 10 years pushed the [4 + 2] cycloaddition far beyond its classical boundaries. Organocatalysts, chiral Brønsted acids, and metal-free hydrogen-bond networks delivered high levels of enantio-control in transformations once dominated by Lewis-acid complexes, while ruthenium, gold, and even silver-nanoparticle catalysts unlocked difficult intramolecular cyclizations and chemoselective intermolecular unions. At the same time, inverse electron demand, dehydro-, vinyl-quinone, and strain-promoted variants dramatically widened the electronic and topological space accessible through Diels–Alder chemistry.

This methodological breadth translated directly into synthesis power. From the concise organocatalytic access to (+)-estrone and the biomimetic assembly of schefflone, to tandem Diels–Alder cascades that forged the congested frameworks of homodimericin A and (−)-daphenylline, the reaction proved adept at delivering target compounds from small heteroaromatics to highly oxygenated polycyclic scaffolds bearing double-digit numbers of stereocenters. Biomimetic approaches were particularly prominent: chalcomoracin, sorocenol B, paeoveitol, and the brosimones illustrate how carefully designed Diels–Alder unions can recapitulate—or even improve upon—Nature's own cycloaddition logic. Equally striking was the synergy between Diels–Alder reactions and modern cross-couplings, radical cascades, or computational design, enabling multicomponent, one-pot sequences that collapse synthetic routes to a handful of steps.

Beyond their sheer structural complexity, many of the targets covered in this volume exhibit potent biological activity, as exemplified by antimalarial artemisinin analogues, the anti-Alzheimer's drug galanthamine, the neurotrophic scabronine A, and numerous antibacterial or anticancer lead compounds. Their syntheses underscore how advances in Diels–Alder chemistry directly support medicinal discovery by guaranteeing scalable, enantioselective access to scarce natural products and their analogues. Finally, the period saw parallel progress in theory: frontier-orbital, hard and soft acids and bases interactions, and modern density functional theory analyses to now rationalize and predict stereochemical outcomes, guiding catalyst development and substrate design with unprecedented precision.

Collectively, the 44 case studies that follow chart the evolution of the Diels–Alder reaction from a venerable text-book transformation into a versatile, programmable engine for contemporary synthesis. They highlight not only the continued relevance of this reaction, but its expanding role at the interface of organocatalysis, transition-metal chemistry, and biomimetic strategy, an expansion likely to accelerate in the decades ahead.

## *7.1 (+)-Estrone*

In Sects. 4.1 and 4.5, the synthesis of *dl*-estrone (**50**) was encountered where constructing ring B or C by means of a Diels–Alder reaction was the key step. With advancing methodology, one can now discuss the synthesis of (+)-estrone (**457**) (Fig. 85), aiming at the enantioselective construction of ring D using a chiral catalyst in the hydrogen bond-promoted Diels–Alder reaction, as conducted by the group of M. W. Göbel in 2010 [233]. Contrary to prior approaches that used liganded titanium as catalyst [234], the catalyst employed was a metal-free system that constitutes the essence of the extended use of asymmetric organocatalysis [235, 236].

As their starting materials the authors prepared by conventional means the diene **458** (E. Dane's diene [237]) and the α-diketone **459** (Fig. 85). For the Diels-Alder reaction one equivalent of **459** and 0.15 equivalents of the axially chiral bisamidinium catalyst **460** were dissolved in dichloromethane cooled to −30°C in a polypropylene tube, and then treated with a precooled solution of 1.5 equivalents of **458** in dichloromethane. After two days at −30°C, the mixture was warmed overnight to room temperature. The resulting material **461** was then transformed conventionally into (+)-estrone (**457**).

A simplified approach to understand the specificity of the reaction is illustrated in Fig. 85 (bottom left) where the specific hydrogen bonds between the bisamidinium catalyst **460** and diketone **459** stabilize a situation leaving open the system for attack by the diene component **458** from one side only. This example vividly illustrates how far the Diels–Alder key step in steroid synthesis has advanced. A comprehensive review of the Diels–Alder steroid syntheses became available in 2015 [238].

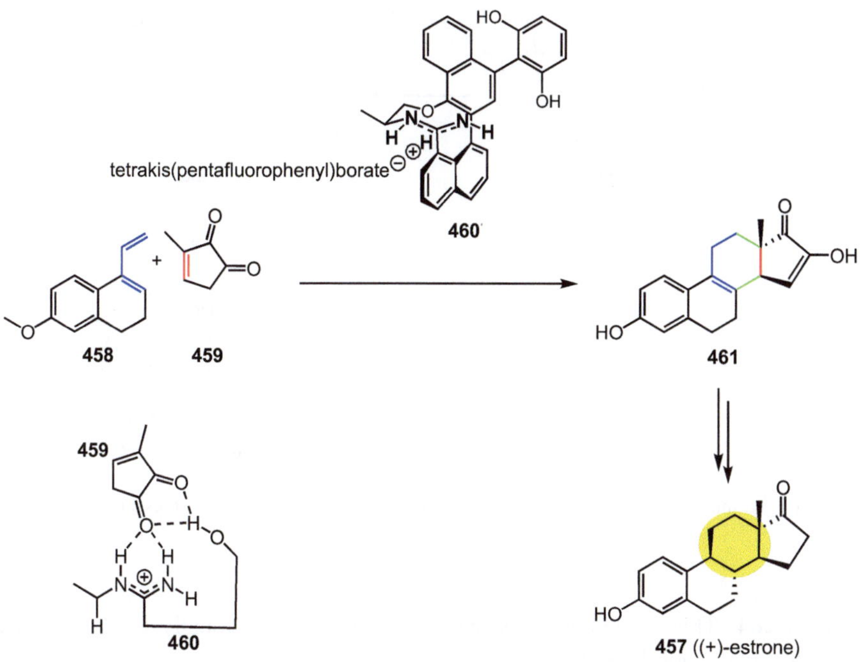


**Fig. 85** The enantioselective hydrogen bond-promoted Diels–Alder reaction in the synthesis of (+)-estrone (**457**)

## *7.2 Caribenol A*

Caribenol A (**462**) (Fig. 86) was isolated in 2007 by A. D. Rodríguez and co-workers from the West Indian gorgonian octocoral *Pseudopterogorgia elisabethae* (Plate 25), which was encountered already in Sect. 6.10 as a source of the diterpene colombiasin A (**200**). It represents a family of novel *nor*-diterpenes, possessing a unique structure and prominent biological activities [239]. Of particular interest is its strong inhibitory activity against *Mycobacterium tuberculosis* (H37Rv) that causes tuberculosis, a disease leading to over three million deaths worldwide each year [240].

Caribenol A (**462**) has an unprecedented tricarbocyclic ring system. It carries three *cis*-configured methyl groups at the C-1, C-4, and C-8 positions and a 5-hydroxyfuran-2(5*H*)-one motif, posing a particular synthesis challenge, which was met by Z. Yang and his group in 2010 with an asymmetric synthesis featuring an intramolecular Diels–Alder key step (Fig. 86) [241].

In the course of this synthesis the conventionally prepared dien-in arrangement **463** underwent the intramolecular Diels–Alder cyclization in toluene at 120°C with a trace of di-*tert*-butyl-4-hydroxytoluene as antioxidant added, to yield the tricyclic

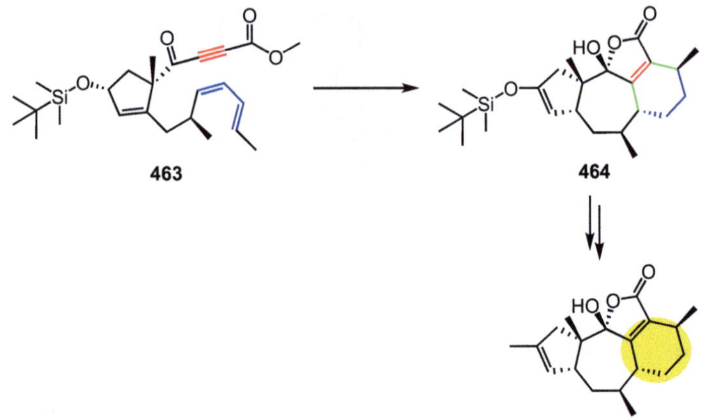


**Fig. 86** The intramolecular Diels–Alder cyclization key step in the total synthesis of caribenol A (**462**)

system **464**. This compound was then elaborated conventionally into the target molecule caribenol A (**462**).

## *7.3 (−)-Pycnanthuquinone C and (±)-Rossinone B*

In 2010, D. Trauner and coworkers proposed a biosynthesis mechanism in which a vinyl quinone Diels-Alder reaction was suggested as the key transformation [242]. Notably, this hypothesis was validated by their elegant work on the biomimetic total synthesis of pycnanthuquinone C (**465**), isolated from the brown alga *Cystophora harveyi* Womersley from the Sargassaceae family (Plate 46) [243]. In comparison to the classical Diels–Alder reaction involving quinones and electron-rich dienes, a vinyl quinone Diels-Alder reaction involves the inverse electron demand cycloaddition of a suitable dienophile to a vinyl quinone, generating an isoquinone methide intermediate. The latter could then tautomerize to yield quinone methides, bicyclic quinones, or hydroquinones. If the isoquinone methide is intercepted by water as a nucleophile, a functionalized tetralin hydroquinone will result, which can be easily oxidized into the corresponding tetralin quinone.

Accordingly, vinyl quinone **466** was prepared conventionally and then heated in a biphasic toluene/water mixture at 60°C to yield mainly (−)-pycnanthuquinone C (**465**), via its intramolecular cyclization intermediates **467** and **468** (Fig. 87).

From an Antarctic collection of the ascidian *Aplidium* species rossinone B (**469**) was isolated by B. R. Copp and coworkers in 2009 [244]. Biological investigations

**Plate 46** *Cystophora harveyi*. Photograph courtesy A. Pessarrodona, CC BY-NC

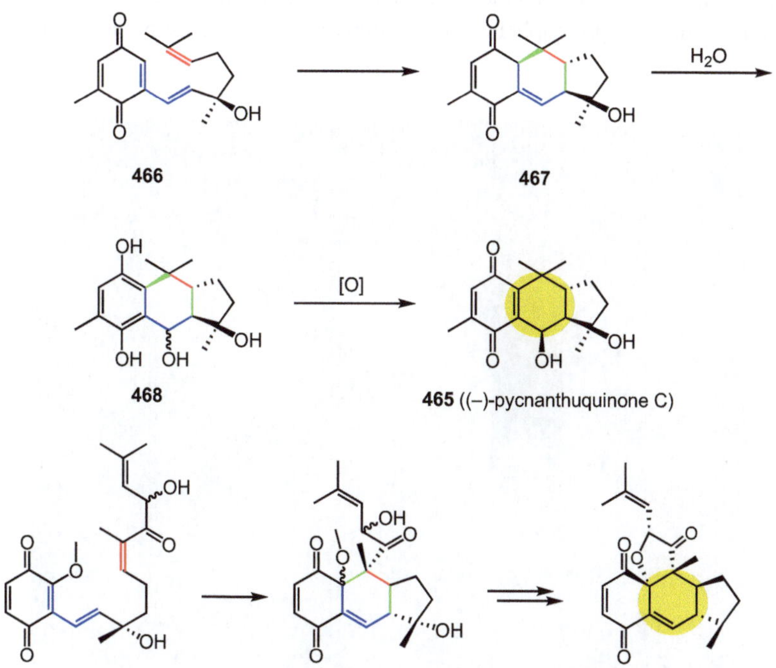


**466**

**467**

**468**

**465** ((–)-pycnanthuquinone C)

**470**

**471**

**469** ((±)-rossinone B)

**Fig. 87** The vinyl quinone Diels-Alder reaction key steps in the syntheses of (±)-(−)-pycnanthuquinone C (**465**) and rossinone B (**469**)

have shown that **469** exhibited a significant biological profile, including potent antiviral, anti-inflammatory, and antileukemic activities. For instance, it was found to show sub-micromolecular antiproliferative activity toward the P388 murine leukemia cell line.

The synthesis of (±)-rossinone B (**469**) was conducted by Y. Tang and coworkers along the lines investigated above using an intramolecular vinyl quinone Diels-Alder reaction as the key step (Fig. 87) [245]. Thus, when vinyl quinone **470** was heated in a sealed tube in toluene to 150°C it reacted to form the tricyclic compound **471**, which then was reacted further by conventional means into the target compound, (±)-rossinone B (**469**).

## *7.4 Chalcomoracin Heptamethyl Ether*

*Morus alba* L. (Plate 47) from the Moraceae family, known as the silkworm mulberry, is an abundant source of isoprenoid-substituted phenolic compounds. In particular, the rather large class of the so called mulberry Diels–Alder type adducts has attracted considerable interest [246]. Naturally occurring cyclohexenes, like chalcomoracin (**472**) [247], were the first examples of natural products biosynthesized by an enzyme-controlled intermolecular Diels-Alder reaction between a dehydroprenylphenol diene derived from an isoprenoid-substituted

**Plate 47** *Morus alba*. Photograph courtesy GerardM, Creative Commons Attribution-Share Alike 2.1 Spain

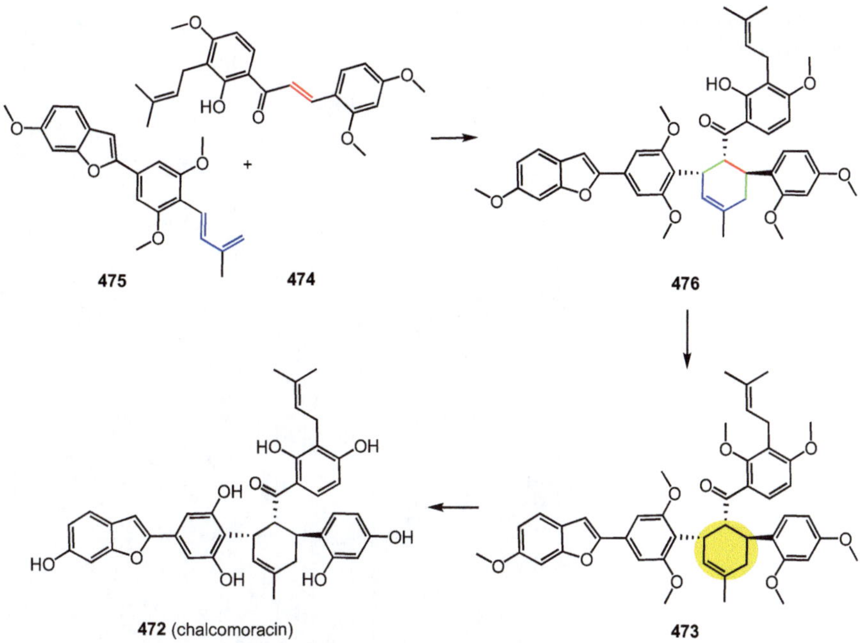


**Fig. 88** The intermolecular Diels–Alder cycloaddition key step in the synthesis of chalcomoracin (**472**)

phenolic compound and an alkene of a chalcone as the dienophile. Chalcomoracin (**472**) exhibited considerable antibacterial activity against MRSA strains K3 and ST [248] and moderate cytotoxicity against five human cancer cell lines [249].

According to biogenetic considerations, an intermolecular Diels–Alder cycloaddition between the 2′-hydroxychalcone dienophile **474** and the benzofurandehydroprenyl diene **475** proceeded at 180°C in a sealed tube to yield the *endo*/*exo* (2/1) mixture from which the *endo* derivative **476** could be separated easily (Fig. 88) [250]. The latter compound provided the targeted chalcomoracin heptamethyl ether (**473**) after methylation. It is interesting to note, that if the fully methylated version of the chalcone was used, no reaction occurred, pointing to the catalytic function of the hydroxy group. Finally, the heptamethylether **473** was demethylated into the natural product chalcomoracin (**472**).

## *7.5 (±)-Galanthamine*

Galanthamine (**477**) (Fig. 89) is an Amaryllidaceae alkaloid that was isolated from the bulbs of the snowdrop, *Galanthus nivalis* L. (Plate 48) [251], and is now more

**Plate 48** *Galanthus nivalis*. Photograph H. F. (author)

efficiently obtained, on an industrial scale, from related species such as *Leucojum aestivum* L. and certain *Narcissus* cultivars via acid-base solvent extraction followed by chromatographic purification to achieve >98% enantiopurity [252, 253]. Although plant extraction remains the commercial mainstay for **477**, more than a dozen total syntheses have been reported, from catalytic asymmetric approaches, to recent concise, enantioselective sequences that exploit biomimetic oxidative phenolic coupling or chiral pool starting materials such as (−)-shikimic acid. Also, semi-synthesis strategies that elaborate the abundant precursor lycoramine also contribute to the supply security of galanthamine (**477**) [251, 254].

Pharmacologically, galanthamine (**477**) is a reversible, competitive acetylcholinesterase inhibitor that penetrates the blood–brain barrier and allosterically modulates nicotinic receptors; it is approved worldwide (as Reminyl®/Razadyne®, 8–24 mg/day) for mild-to-moderate Alzheimer's disease, where it slows cognitive decline and improves memory, and it continues to be explored for postoperative ileus, organophosphate poisoning, and other cholinergic deficits [255].

A particular synthesis of (±)-galanthamine (**477**) to serve as an advanced example involves a delicately balanced cascade of Stille coupling and intramolecular Diels–Alder cyclization starting from 3,5-dibromo-2-pyrone (**478**) (Fig. 89) [256]. When

479

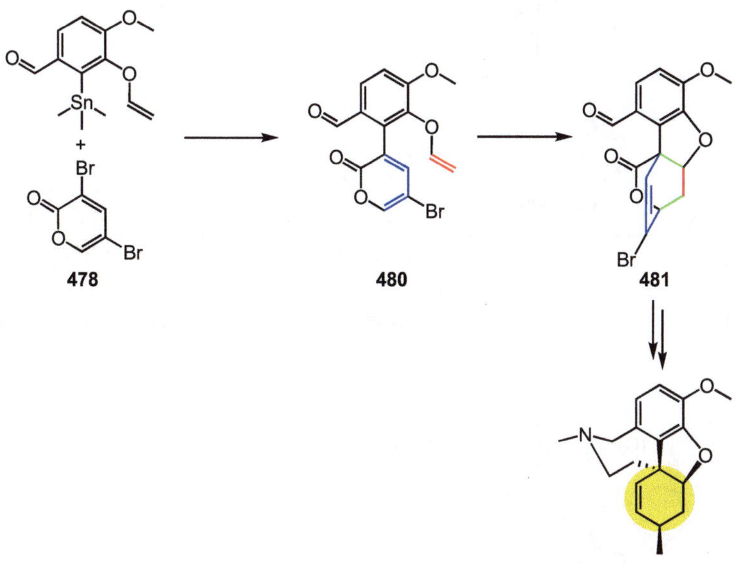


478 480 481

477 ((±)-galanthamine)

**Fig. 89** The cascade of Stille coupling and intramolecular Diels–Alder cyclization in the synthesis of (±)-galanthamine (**477**)

**478** was submitted to the optimum conditions of Stille coupling (5 mol% $Pd(PPh)_3$, 10 mol% copper(I) iodide in dimethylformamide at 95°C) with the tin reagent **479**, a C-3-selective coupling into intermediate **480** took place in tandem with an intramolecular Diels–Alder reaction to yield the cyclized product **481**. The latter could then be transformed by classical means into the targeted (±)-galanthamine (**477**).

## 7.6 *ent-Ledol*

The sesquiterpene alcohol *ent*-ledol (**482**) (Fig. 90), in the class of aromadendrenes, is contained in several taxa, as for instance, in angiosperms like *Calophyllum teysmanii* Miq. (Clusiaceae), a tree growing mainly in the wet tropical biome [257]. It was first isolated from *Ledum palustre* L. (= *Rhododendron tomentosum* Harmaja, family Ericaceae), also commonly known as Labrador tea or wild rosemary, a plant occurring in sphagnum meadows (Plate 49), although as the dextrorotatory enantiomer [258]. Its biological interest comes from its antifungal, anti-inflammatory, and antitussive effects, but it also displays considerable toxicity. However, synthesis chemists value **482** mainly as a stereochemical benchmark. It

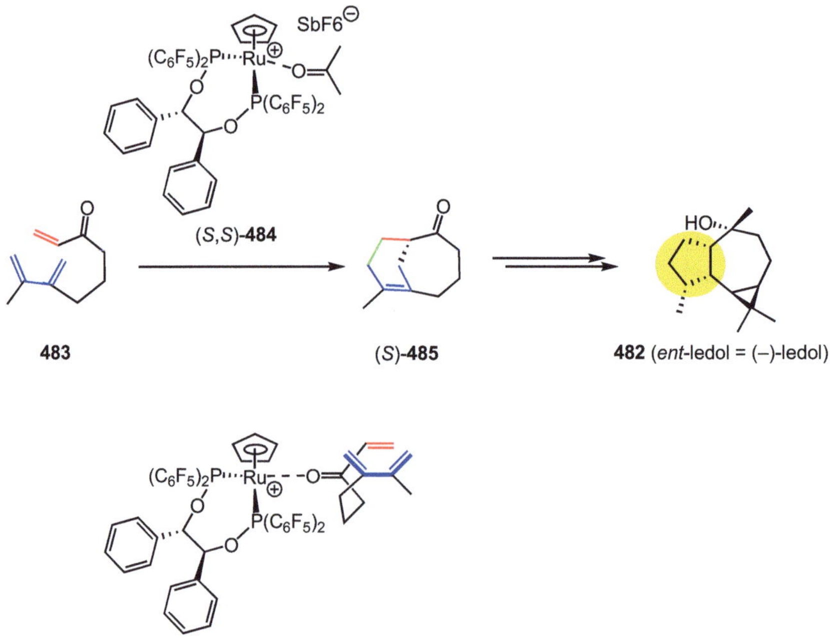


**Fig. 90** The intramolecular Diels–Alder reaction using a chiral ruthenium Lewis acid catalyst as the key step in the synthesis of *ent*-ledol (**482**) and the conformational structure of the transition state (bottom)

**Plate 49** *Rhododendron tomentosum*. Photograph courtesy H. Zell, Creative Commons Attribution-Share Alike 3.0 Unported

might be mentioned that *ent*-ledol (**482**) was assigned to (−)-ledol, and the "*ent*" was used to discriminate it from the (+)-ledol enantiomer, which is also produced by some organisms (see above).

Accordingly, *ent*-ledol (**482**) was used as the target to develop a novel intramolecular Diels–Alder reaction using a chiral ruthenium Lewis acid catalyst inducing asymmetric induction with up to 96% *ee* [259]. Thus, the rather unreactive vinylketone-diene **483** was reacted in the presence of 5 mol% of the ruthenium catalyst (*S*,*S*)-**484** in dichloromethane and 2,6-lutidine at room temperature for five days, which produced 70% yield of bicyclic (*S*)-**485** with 92% *ee* (Fig. 90). As studied by X-ray analysis of the complexed form of **484** the conformation is that having a *syn-s-trans* arrangement necessary for the observed stereoselectivity, as shown in Fig. 90 (bottom). The intermediate (*S*)-**485** could then be transformed into *ent*-ledol (**482**) by conventional steps.

## 7.7 *(+)-Pseudodeflectusin*

Pseudodeflectusin (**486**) was isolated from the culture broth of *Aspergillus pseudodeflectus* Samson & Mouchacca from the Aspergillaceae family, by Y. Mizushina and co-workers [260]. It exhibited cytotoxicity against several human cancer cell lines derived from the stomach (NUGC-3), cervix (HeLa-S3), and peripheral blood (HL-60), but did not display any inhibition of DNA metabolic enzymes. Thereby, it was considered to be a promising candidate as a new type of antitumor agent.

The synthesis of (+)-pseudodeflectusin (**486**) was based on an intermolecular Diels-Alder step with the following cascade of lactonization and decarboxylation in situ (Fig. 91) [261]. Thus, pyrone **487**, activated by deprotonation with a 10% excess of sodium hydride, was reacted in an intermolecular Diels–Alder reaction with the (*R*)-configured alkyne **488** in the dipolar solvent dioxane in a sealed tube at 200°C for 5 h. The resulting cycloadduct lactonized and decarboxylated in situ to yield (−)-(*R*)-7-bromo-8-hydroxy-3-methylisochroman-1-one (**489**), which was then elaborated into the target compound (+)-pseudodeflectusin (**486**).

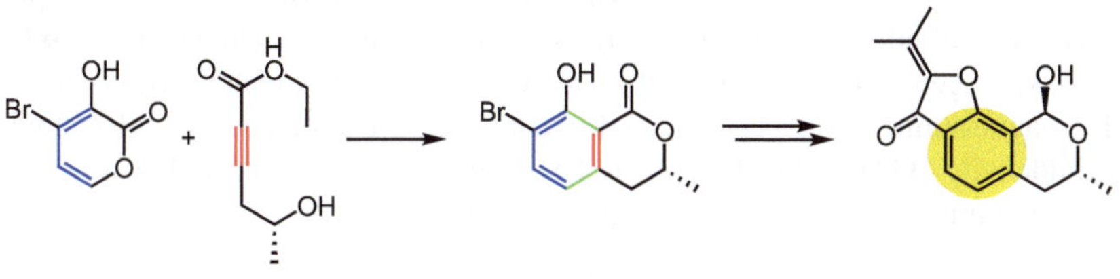


**487** **488** **489** **486** ((+)-pseudodeflectusin)

**Fig. 91** The intermolecular Diels-Alder key step with a following cascade of lactonization and decarboxylation in the synthesis of (+)-pseudodeflectusin (**486**)

## 7.8 (+)-Curcuphenol

From the marine sponge *Didiscus flavus* van Soest (= *D. oxeatus*) of the Raspailiidae family (Plate 50), P. J. McCarthy and his group isolated (+)-curcuphenol (**490**) [262]. As there has been considerable uncertainty later on with the structures of a series of compounds, called helirananes [263], also isolated from marine sponges, a synthesis study revealed the erroneously assigned eight-membered rings in these compounds: they were shown to be alkenylphenols instead [264].

**Plate 50** *Didiscus flavus*. Photograph courtesy S. Zea, spongeguide.org

The key step in the syntheses of the proposed structures including (+)-curcuphenol (**490**) (Fig. 92) and the confirmation of their actual structures was a diastereoselective inverse electron demand Diels–Alder reaction between the optically active enol ether **491** and the short-lived *ortho*-quinone methide species **492**, which was generated in situ at −40°C by the sequential addition of methylmagnesium bromide and di-*tert*-butyl dicarbonate to 4-methylsalicylaldehyde (**493**) in dimethyl ether. The resulting chroman ketal **494** was then transformed into the targeted (+)-curcuphenol (**490**).

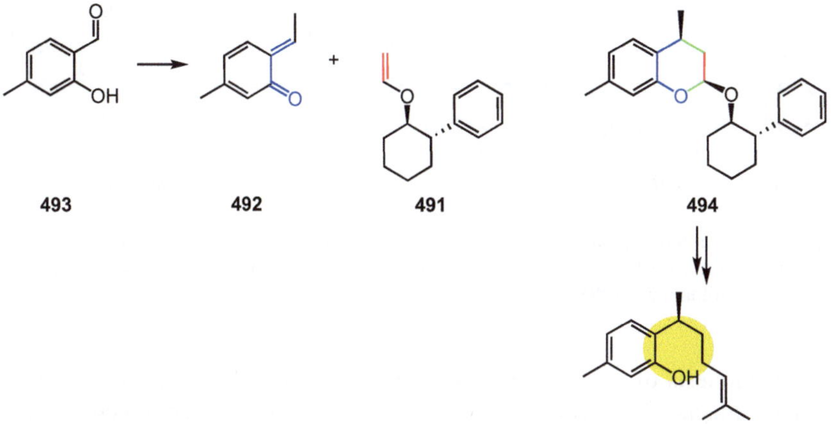


493 492 491 494

490 ((+)-curcuphenol)

**Fig. 92** The intermolecular Diels-Alder key step in the synthesis of (+)-curcuphenol (**490**)

## *7.9 Urolithin M7*

The urolithins, which are metabolites of ellagitannins, have a common tricyclic aromatic 6*H*-dibenzo[*b,d*]pyran-6-one core and differ in the number and position of hydroxy substituents. Kim and co-workers reported the isolation of urolithin M7 (3,8,10-trihydroxy-6*H*-benzo[*c*]chromen-6-one) (**495**) from the dried feces of the squirrel, *Trogopterus xanthipes* A. Milne-Edwards, of the family Sciuridae (Plate 51) [265]. These feces (excrementum pteropi) have use in Chinese traditional medicine for the treatment of angiostasis and dysmenorrhea [266].

**Plate 51** *Trogopterus xanthipes*. Photograph courtesy tong-tong, Attribution-NonCommercial 4.0 International

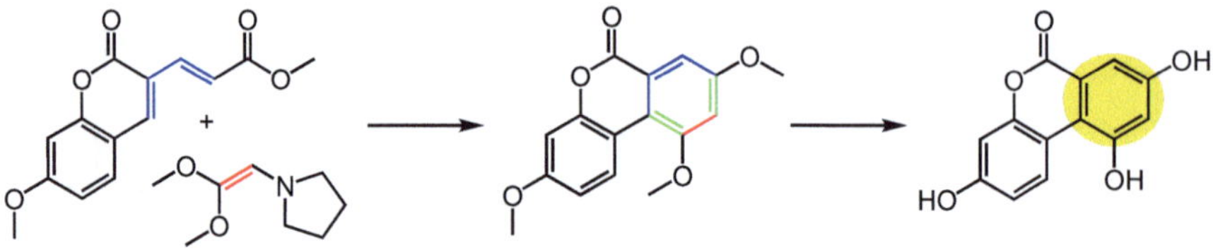


**Fig. 93** The domino sequence involving an inverse electron demand intermolecular Diels–Alder reaction followed by a 1,2-elimination of pyrrolidine and a dehydrogenation as the key step in the synthesis of urolithin M7 (**495**)

The synthesis of urolithin M7 (**495**) of G. J. Bodwell and co-authors had a domino sequence involving an inverse electron demand intermolecular Diels–Alder reaction followed by a 1,2-elimination of pyrrolidine and a dehydrogenation as the key step (Fig. 93) [267]. When the electrondeficient diene **496** was heated with the dienophilic enamine **497** (generated in situ from the reaction of dimethoxyacetaldehyde and pyrrolidine) in benzene at 80°C for seven days, the product of the cascade, Diels–Alder reaction, elimination, and dehydrogenation, the O-methylated dibenzopyranone **498** was obtained in quantitative yield. Demethylation of **498** in 57% hydroiodic acid at 127°C for 30 min resulted in a nearly quantitative yield of the target compound, urolithin M7 (**495**).

## 7.10 *Echinopines A and B*

In 2008, a Chinese group reported the isolation and structural elucidation of two sesquiterpenoids named echinopines A (**499**) and B (**500**) from the roots of the perennial spiny globe thistle *Echinops spinosus* L., of the Asteraceae family (Plate 52) [268]. They possess a unique carbocyclic framework characterized by a [3,5,5,7] ring system, which has been suggested to originate from a guaiane-type structure. The compounds have garnered attention from the synthesis community for their novel structure and synthesis challenges rather than for their biological activity [269].

One of the approaches to echinopine synthesis was the synthesis of echinopines A (**499**) and B (**500**) with a nicely constructed cascade based on a cycloisomerization followed by an intramolecular Diels–Alder reaction as the key step (Fig. 94) [270].

Upon treatment of the vinyl-acetylenic precursor **501** with palladium diacetate/triphenylphosphine at 80°C in toluene for 2 h, a cycloisomerization took place smoothly to provide the enoate **502**. The temperature was then elevated to 160°C for 6 h. This action provided the intramolecular Diels–Alder [5,6,7]-tricyclic derivative **503**, which could then be further elaborated into the targeted [3,5,5,7] ring system of echinopines A (**499**) and B (**500**).

**Plate 52** *Echinops spinosus*. Photograph courtesy KPFC, Creative Commons Attribution-Share Alike 4.0 Unported

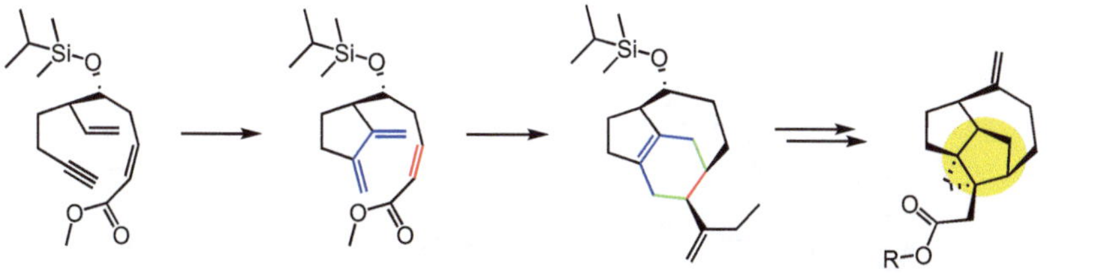


**501** **502** **503** **499** (echinopine A; R = H) **500** (echinopine B; R = $CH_3$)

**Fig. 94** Cascade of a cycloisomerization followed by an intramolecular Diels–Alder reaction as the key step in the synthesis of echinopines A (**499**) and B (**500**)

## *7.11 (±)-Kuwanone V and (±)-Dorsterone Methyl Ethers*

Among the mulberry Diels–Alder compounds (see Sect. 7.4 on chalcomarin (**472**)) isolated from *Morus alba* (Plate 47), (±)-kuwanone V (**504**) and (±)-dorsterone (**505**) were deemed to be of high interest with respect to their biological activities. Again, a biomimetic intermolecular Diels–Alder reaction between a highly electron-rich dienophile and a Lewis acid-sensitive diene derived from a chalcone was the key step to success (Fig. 95) [271].

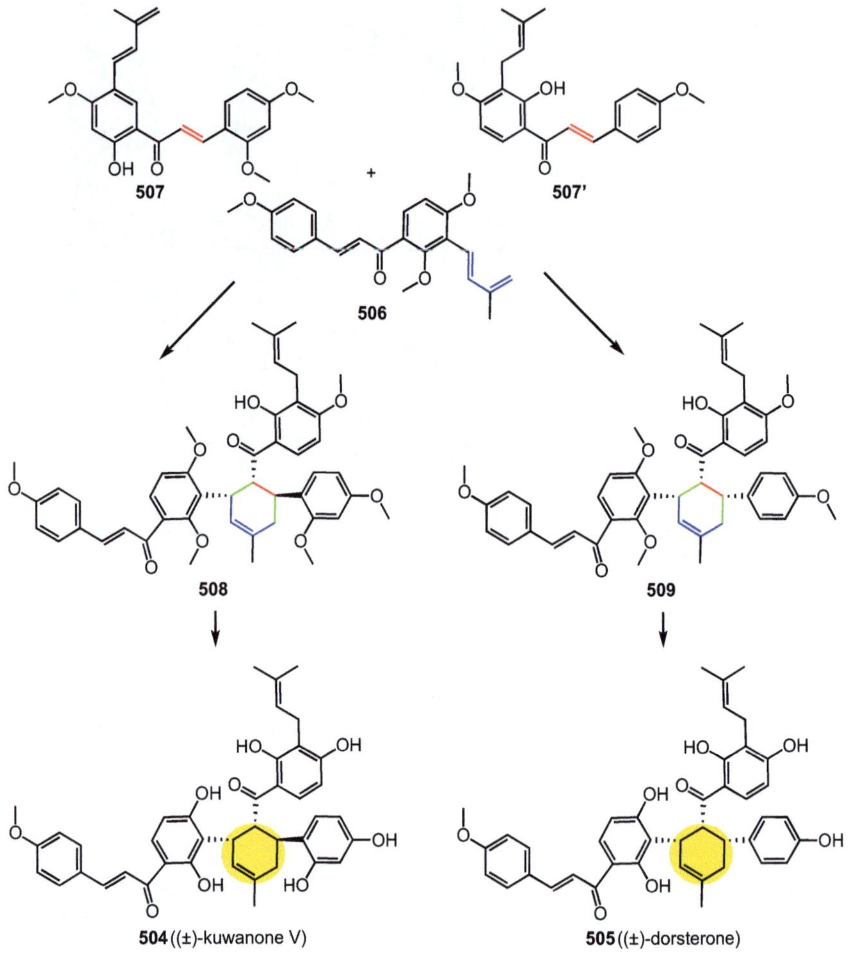


**Fig. 95** The intermolecular Diels–Alder cycloaddition key step in the synthesis of (±)-kuwanone V (**504**) and (±)-dorsterone (**505**)

The chalcones **506** and **507** or **507′** were subjected to a [4 + 2] Diels–Alder cycloaddition reaction. Heating these two compounds in toluene in a pressure tube at 160°C for 18 h afforded the *endo* and *exo* adducts **508** and **509**, in a 3:2 ratio and 55% yield. These could then be demethylated into the natural products (±)-kuwanone V (**504**) and (±)-dorsterone (**505**).

## 7.12 (±)-Sorocenol B

From the root bark of the woody plant *Sorocea bonplandii* Baill. (Plate 53), in the family Moraceae, another biogenic Diels–Alder cycloadduct, sorocenol B (**510**) was isolated in a very small (3 mg/kg) amount [272].

In the synthesis of (±)-sorocenol B (**510**), which displayed low micromolar-level cytotoxic activities, the key step consisted of a novel silver nanoparticle-catalyzed Diels-Alder cycloaddition of a chalcone and a diene (Fig. 96) [273]. Thus, chalcone

**Plate 53** *Sorocea bonplandii.* Photograph courtesy M. S. Rother, 

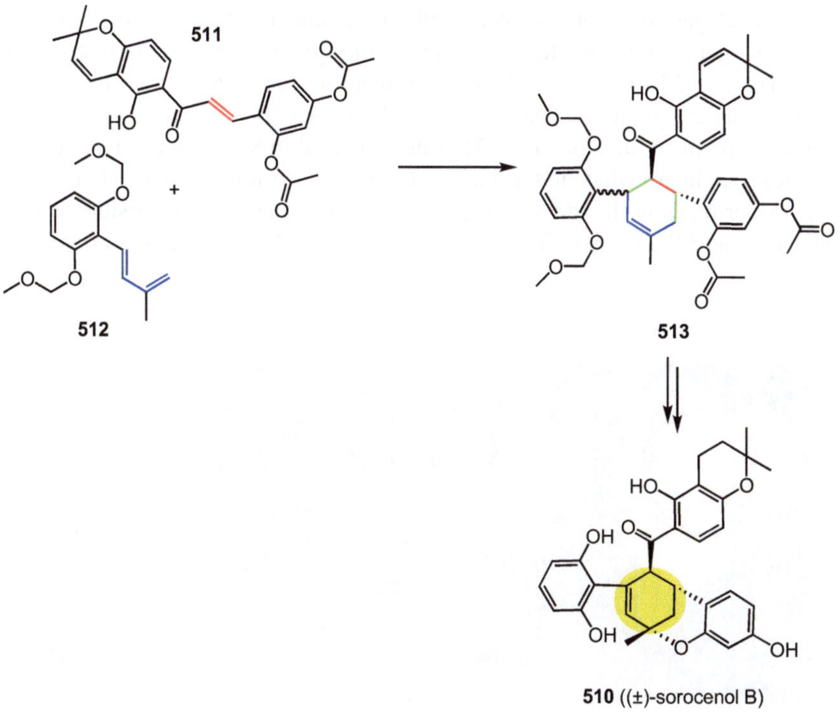


**Fig. 96** The intermolecular silver nanoparticle catalyzed Diels–Alder cycloaddition key step in the synthesis of (±)-sorocenol (**510**)

**511** and diene **512** were dissolved in dichloromethane and treated with silica-supported silver nanoparticles with 0.1% silver loading at 50°C for two days. The resulting *endo*/*exo* (2/1) mixture **513** (90% yield) could then be transformed into the target bicyclo[3.3.1] framework of (±)-sorocenol B (**510**).

## *7.13 (±)-Schefflone*

The trimeric monoterpene and weak mosquito larvicidal agent, (±)-schefflone (**514**) (Fig. 97), which is an apparent derivative of the antiparasitic aromatic monoterpene espintanol (**515**), was isolated from the antimalarial extracts of the root bark of the African shrub *Uvaria scheffleri* Diels, family Annonaceae (Plate 54), together with **515**. The structure determination was confirmed by single-crystal X-ray diffraction analysis. (±)-Schefflone (**514**) can be considered a product of a non-enzymatic Diels-Alder-type cycloaddition reaction of the quinone methide derivative of **515** as the diene and dienophile [274].

(±)-Schefflone (**514**) has a unique spiro-chroman-type trimeric skeleton, which could be derived from a tandem hetero-Diels–Alder trimerization of *ortho*-quinodimethane precursors (Fig. 97) [275].

For the elegant one-pot, one-step synthesis, espintanol (**515**) was chosen as the educt for generation of the necessary *ortho*-quinodimethane derivative **516** [275]. Treating **516** with silver(I) oxide in benzene for 16 h at room temperature resulted in a 72% yield of the targeted (±)-schefflone (**514**). It was also proven by a trapping experiment that the dimer **517** (also obtained in 8% yield from the reaction mixture) is an intermediate in this tandem oxidation two-fold intermolecular hetero-Diels–Alder reaction. In addition, the methide **516** undergoes a reversible reaction with the dimer **517**, whereas the reaction from dimer **517** to trimer **514** is irreversible.

**Plate 54** *Uvaria scheffleri*. Photographs courtesy K. Wambua and P. M. Kirika, The East African Herbarium, National Museums of Kenya, Nairobi, Kenya

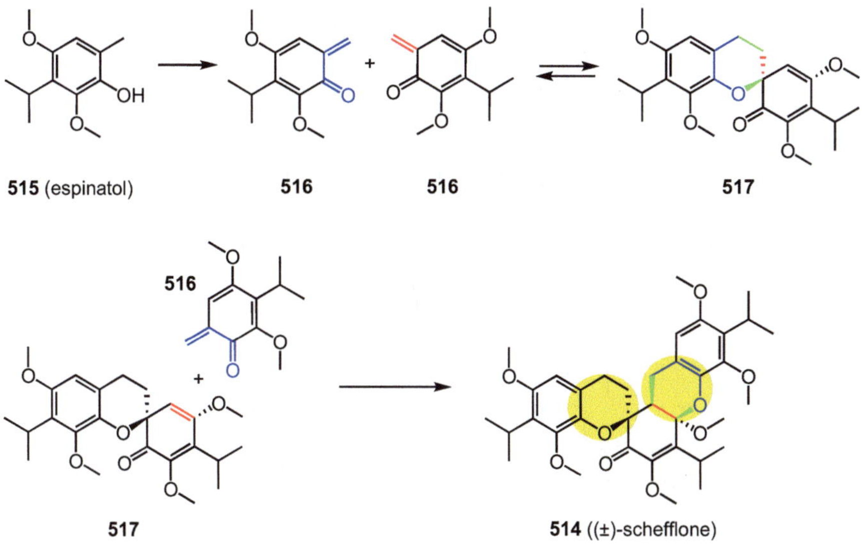


**Fig. 97** The tandem oxidation twofold intermolecular hetero-Diels–Alder reaction key step in the synthesis of (±)-schefflone (**514**) from espinatol (**515**)

## *7.14 Cannabinol*

From the cannabinoids forming a class of over 100 natural products, cannabinol (**518**) (Fig. 98) is certainly a most prominent member. Isolated from the plant *Cannabis sativa* L., family Cannabaceae (Plate 55), it is known for its mild psychoactivity that acts as a low affinity partial agonist at the $CB_1$ and $CB_2$ receptors in the endocannabinoid system [276, 277].

Of course, there have been many synthesis approaches to the cannabinoids, but the present focus will be on the application of an inverse electron demand Diels–Alder reaction as the key step (Fig. 98) [278]. In addition, the G. J. Bodwell group developed a multicomponent system, which allowed the assembling of all necessary components in a single step.

Thus, a solution of the salicylaldehyde derivative **519**, pyrrolidine, dimethyl (*E*)-pentenedioate (**520**), and acetone in 1,4-dioxane was reacted at room temperature for 2 h and then heated to 90–100°C for 24 h and resulted in 48% yield of the desired 6*H*-dibenzo[*b*,*d*]pyranone **523**. The latter could then be transformed by classical procedures into the target compound cannabinol (**518**). In the course of this reaction, both the diene **521** and the dienophile **522** for the inverse electron demand Diels–Alder reaction step, are generated in situ by the secondary amine-mediated process,

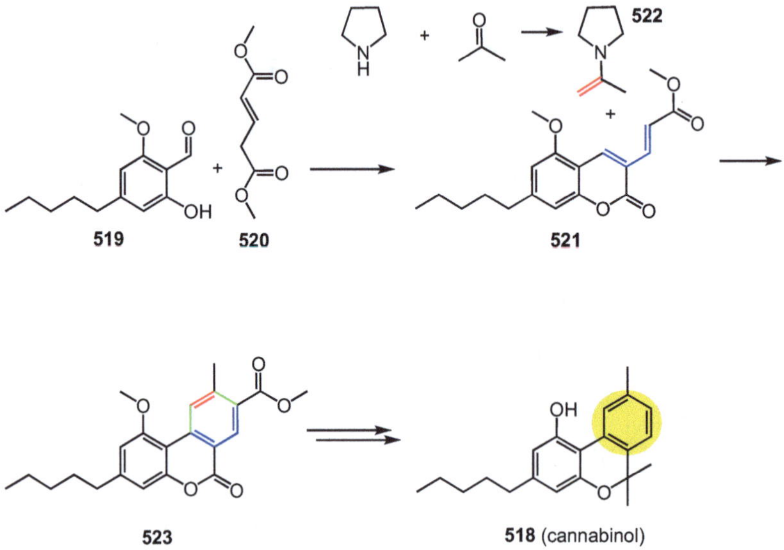


**Fig. 98** The domino sequence of inverse electron demand Diels–Alder reaction, 1,2-elimination, and dehydration in the synthesis of cannabinol (**518**)

**Plate 55** *Cannabis sativa*. Photograph courtesy T. Tuason, Creative Commons Attribution-Share Alike 4.0 International

as already encountered in the synthesis of urolithin M7 (**495**) (Sect. 7.9). The domino reaction is finalized by 1,2-elimination and dehydration.

## 7.15 Defucogilvocarcin V

Defucogilvocarcin V (**524**) was isolated from the fermentation broth of the spore-forming mesophilic bacterium *Streptomyces arenae* 2064 (family Streptomycetaceae) by the group of R. C. Pandey and displays antibiotic activity on activation by light [279]. The same group as that synthesizing cannabinol (**518**) in the previous section applied their strategy also to this synthesis challenge to construct the 6*H*-benzo[*d*]naphtho[1,2-*b*]pyran-6-one skeleton of **524** (Fig. 99) [280].

An inverse electron demand Diels–Alder reaction between diene **525** and enamine **526** with domino 1,2-elimination and dehydration by heating in benzene at 80°C for 48 h resulted in the properly substituted 6*H*-benzo[*d*]naphtho[1,2-*b*]

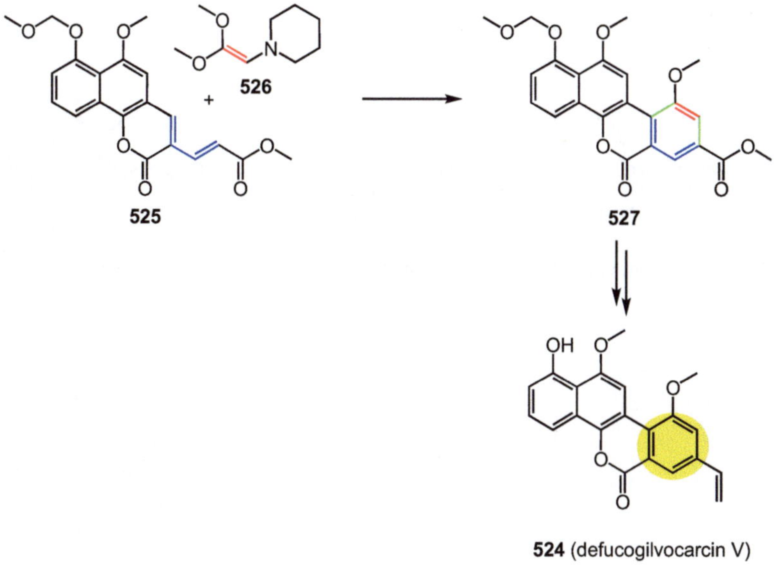


**Fig. 99** The domino sequence of inverse electron demand Diels–Alder reaction, 1,2-elimination, and dehydration in the synthesis of defucogilvocarcin (**524**)

pyran-6-one skeleton **527**. This compound was then elaborated conventionally into the targeted compound, defucogilvocarcin V (**524**).

## *7.16 Carpanone*

Carpanone (**528**) (Fig. 100) is the most well-known member of a small group of naturally occurring benzoxanthenones. They belong to the lignans and are characterized by an oxygen-containing tetracyclic basic skeleton. The natural product with its highly oxygenated hexacyclic ring system and five contiguous stereogenic centers was isolated from the bark of the carpano tree (a previously unknown species of *Cinnamomum*, family Lauraceae) as a racemate [281].

Besides many other syntheses of carpanone (**528**), a most elegant approach was that reported by U. Beifuss and his team in 2012 (Fig. 100) [282]. As the starting material for the dimerization, (*E*)-2-propenylsesamol (**529**) was chosen. It underwent in acetate buffer/acetone (8/2) at room temperature and in the presence of oxygen for 2 h a laccase-catalyzed domino phenol oxidation/*anti*-β,β-radical coupling/intramolecular hetero-Diels–Alder reaction via the radical **530**, and its coupling product **531**, to yield carpanone (**528**) and small amounts of its diastereomer.

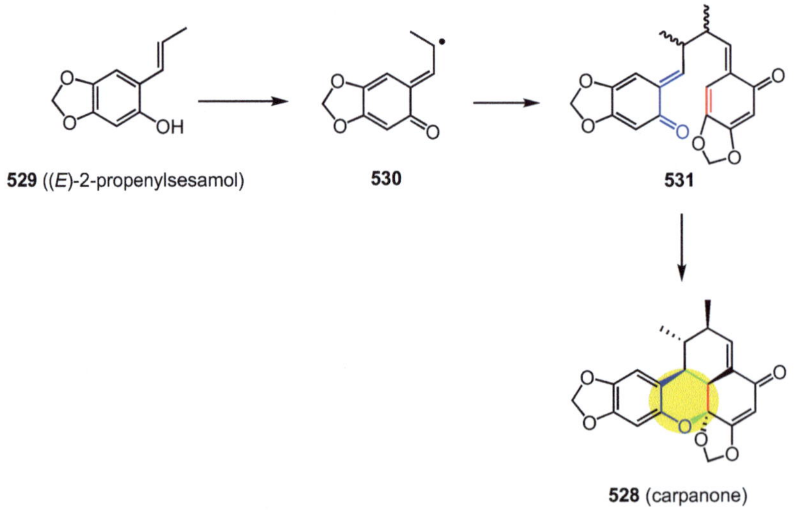

**529** ((*E*)-2-propenylsesamol) **530** **531**

**528** (carpanone)

**Fig. 100** The laccase-catalyzed domino phenol oxidation/*anti*-β,β-radical coupling/intramolecular hetero-Diels–Alder reaction in the synthesis of carpanone (**528**)

## 7.17 (±)-α-Lycorane

Species of the Amaryllidaceae family are rich in alkaloids. One such alkaloid, (±)-α-lycorane (**532**) (Fig. 101), occurring in the red spider lily, *Lycoris radiata* (L'Hér.) Herb. (Plate 56), has proved to be notable for its biological activities. It is of

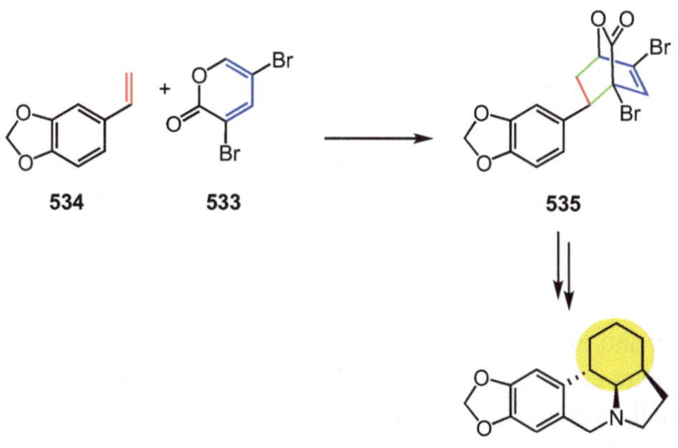


**Fig. 101** The intermolecular Diels–Alder reaction key step in the total synthesis of (±)-α-lycorane (**532**)

**Plate 56** *Lycoris radiata*. Photograph courtesy Blue Lotus, Creative Commons Attribution 2.0 Generic

even more interest from a synthesis standpoint, because of its pyrrolo[*de*] phenanthridine skeleton [283, 284].

An intermolecular Diels–Alder reaction by heating dibromopyrone **533** with the styrene dienophile **534** in benzene yielded the *endo*-adduct **535**, which was easily separated from a small amount of the *exo*-adduct (Fig. 101) [284]. This product could then be elaborated into (±)-α-lycorane (**532**) using conventional steps.

## 7.18 (−)-Scabronine A

The cyathane diterpenoid, (−)-scabronine A (**536**) (Fig. 102) was isolated among other compounds of this type from the bitter hedgehog mushroom, *Sarcodon* (= *Hydnellum*) *scabrosum* (Fr.) E. Larss., family Bankeraceae (Plate 57), by a Japanese group [285]. It contains six contiguous stereogenic centers in its seven-membered ring. (−)-Scabronine A (**536**) demonstrates a potent induction effect upon the production and excretion of neurotrophic factors, including nerve growth factor in 1321N1human astroglial cells [286]. The nerve growth factor is a member of a family of small, secreted proteins known as neurotrophins, which are important signaling molecules for the growth, maintenance, and survival of neural cells [287]. It has been suggested that administration of exogenous nerve growth factor could result in cytoprotection and stimulate the outgrowth of neuritic projections [288]. Consequently, direct administration of nerve growth factor to the brain has

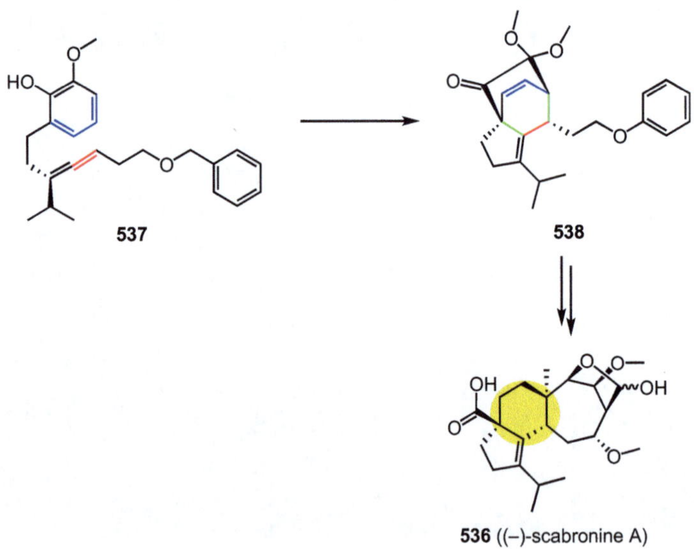


**Fig. 102** The cascade of oxidative dearomatization and intramolecular inverse electron demand Diels–Alder reaction in the synthesis of (−)-scabronine A (**536**)

**Plate 57** *Sarcodon* (= *Hydnellum*) *scabrosum*. Photograph courtesy caspar s, Creative Commons Attribution 2.0 Generic

potential applicability in the treatment of neurodegenerative disorders, such as Alzheimer's, Parkinson's, and Huntington's diseases [289]. However, this approach is problematic clinically, because the nerve growth factor does not cross the blood-brain barrier and is rapidly metabolized in vivo [288]. Hence, the search for compounds able to stimulate the endogenous production of the nerve growth factor has led to the isolation of structurally unique natural products with this property [290]. Owing to their lipophilic nature and small size, cyathane diterpenoids like (−)-scabronine A (**536**) exhibit promising structural characteristics amenable to the crossing of the blood-brain barrier. Accordingly, their therapeutic relevance and synthesis challenges have evoked intense investigation by several research groups since their initial description in the literature.

A synthesis involving a cascade of oxidative dearomatization and intramolecular inverse electron demand Diels–Alder reaction was devised by Y. Kobayakawa and M. Nakada in 2013 (Fig. 102) [291]. Upon oxidative dearomatization of allene **537** by means of (diacetoxyiodo)benzene, the intramolecular cyclization to **538** took place in a one-pot reaction in methanol at room temperature standing for seven days with 97% yield. Compound **538** was then elaborated by conventional approaches into (−)-scabronine A (**536**) as well as a few more natural products of this type.

## 7.19 Gelsemiol

Gelsemiol (**539**) is a C-6,C-11-oxidized iridoid, which was isolated from the twining vine, *Gelsemium sempervirens* (L.) J. St.-Hil., family Gelsemiaceae (Plate 58)

**Plate 58** *Gelsemium sempervirens*. Photograph courtesy J. Evans, Creative Commons Attribution-Share Alike 4.0 International

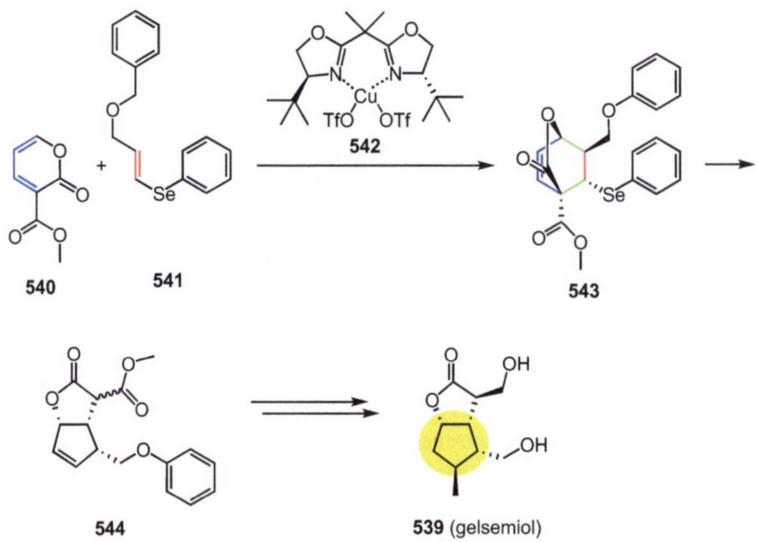


**Fig. 103** The inverse electron demand Diels–Alder reaction on the way to gelsemiol (**539**)

[292]. Later, it was also isolated from the seashore vervain, *Verbena litoralis* Kunth (family Verbenaceae) and shown that it displays neuritotrophic properties in the presence of the nerve-growth factor [293]. Hence, this is another compound of interest in this same biological area as (–)-scabronine A (**536**) (Fig. 103) as

discussed in Sect. 7.18. In 2013, the Swiss group of K. Gademann set out for a rational total synthesis involving a stereoselective inverse electron-demand Diels–Alder reaction as the key step (Fig. 103) [294].

After several failed attempts by varying the reagents and conditions, the inverse electron demand Diels–Alder reaction with diene **540** and the dienophilic vinylselenide **541** in the presence of the chiral copper catalyst **542** under neat conditions and reduced pressure was conducted overnight. This led to the bicyclic compound **543** in 91% yield and with a high enantioselectivity of 92% *ee*. This bicyclic compound then underwent a radical reaction/skeletal rearrangement sequence driven by tributyltin hydride/azobisisobutyronitrile in benzene under reflux and then silica for 60 h to yield **544**, which was then transformed into gelsemiol (**539**) using conventional chemistry. The study authors also confirmed that **539** induced neuritic outgrowth in a cellular model [294].

## 7.20 Brosimones A and B

Brosimones A (**545**) and B (**546**) (Fig. 104) were both isolated from the Brazilian plant *Brosimopsis oblongifolia* Ducke (syn. *Brosimum lactescens*; family Moraceae) (Plate 59) [295, 296], and feature dimeric structures derived from a prenyl chalcone, as we have already encountered in the mulberry Diels–Alder natural compounds in Sects. 7.11 and 7.12 ((±)-kuwanone V (**504**), (±)-9-dorsterone (**505**), and (±)-sorocenol B (**510**)). The intriguing biosynthesis relationships and structures of the brosimones, and, in particular the [3.3]metacyclophane core of **545**, underscores these compounds as attractive synthesis targets.

This challenge was met by the group of J. A. Porco Jr. in 2013 by employing platinum on charcoal-cyclopentene or dicyanodichlorobenzoquinone to effect the dehydrogenation of prenylchalcone substrates in combination with silver nanoparticles to promote the subsequent intermolecular Diels–Alder reactions [297]. The synthesis of brosimone A (**545**) started with the chalcone **547** (Fig. 104), which was treated with 10 mol% platinum on charcoal, 0.2 mol% silver nanoparticles, and cyclopentene in dichloroethane at 110°C for 48 h, to yield mainly the *exo*-adduct **548**. When **548** was reacted with 1.5 equivalents of dicyanodichlorobenzoquinone and 0.3 mol% silver nanoparticles in chlorobenzene at 130°C for 72 h, a 60% yield of the *exo-exo*-cyclophane **549** resulted, which upon reaction with 60 mol% palladium on charcoal and two equivalents of ammonium acetate in 1,4-cyclohexadiene/acetone at 40°C for 24 h, was converted into brosimone A (**545**).

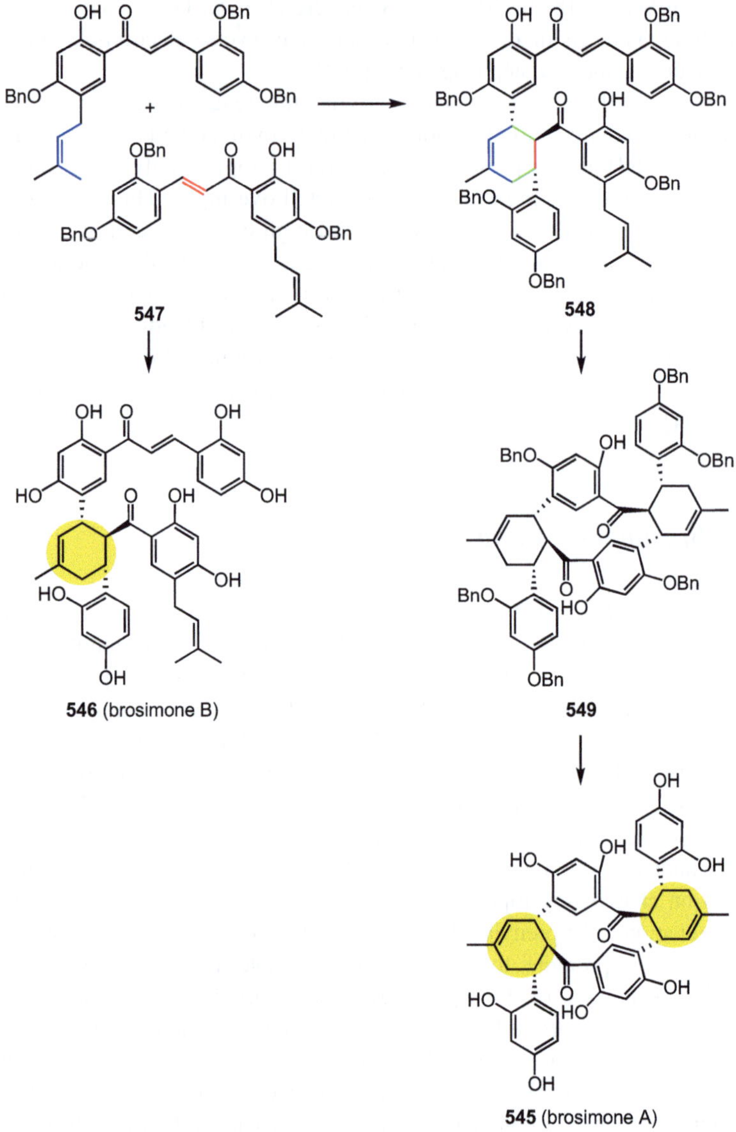


**Fig. 104** The oxidative intermolecular silver nanoparticle catalyzed Diels–Alder cycloaddition steps in the syntheses of brosimone A (**545**) and B (**546**)

On reacting the chalcone **547** with 40 mol% palladium on charcoal, two equivalents of ammonium acetate, and 1,4-hexadiene in acetone at 40°C for 24 h, an 85% yield of brosimone B (**546**) was obtained.

**Plate 59** *Brosimopsis oblongifolia* (syn. *Brosimum lactescens*). Photograph courtesy R. B. Foster, Field Museum of Natural History, Chicago — 

## *7.21 (±)-Pancratistatin*

Pancratistatin (**550**), along with its related Amaryllidaceae natural products, has been pursued vividly by synthesis-oriented chemists because of its potent antineoplastic activities [298]. Pancratistatin (**550**) was isolated from the bulbs of the spider lily *Lycoris radiata* (L'Hér.) Herb., family Amaryllidaceae (Plate 56) [299]. Structurally, **550** is built on a heavily fused tricyclic framework entailing six contiguous stereogenic centers on ring C, five substituents on the aromatic A ring, and a highly strained B-ring lactam.

For the synthesis of (±)-pancratistatin (**550**), the (*E*)-β-borylstyrene **551** was used as a novel dienophile for the key Diels–Alder reaction with the dibromopyrone **533** (Fig. 105) [300], which was encountered earlier as a suited diene. When the two components were heated in toluene at 80°C the cycloaddition took place smoothly to afford the *endo*-bicyclolactone **552** in nearly 90% yield. Following conventional methodology, compound **552** was as then transformed to the targeted (±)-pancratistatin (**550**).

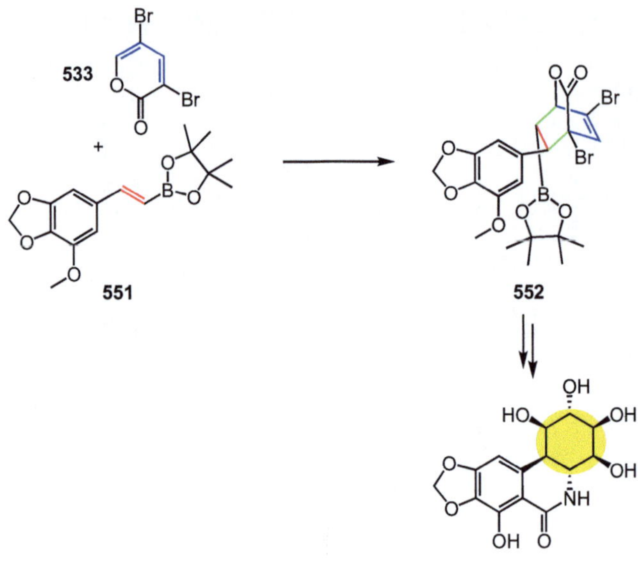


**Fig. 105** The intermolecular Diels–Alder reaction key step in the total synthesis of (±)-pancratistatin (**550**)

## 7.22 *(+)-Cavicularin*

The cyclophane natural product cavicularin (**553**) (Fig. 106), isolated from the liverwort *Cavicularia densa* Steph., family Blasiaceae (Plate 60) [301], which is endemic to Japan, displays a strained molecular architecture and thus exhibits conformational chirality, thereby representing a challenge for the synthesis community.

To achieve a stereospecific synthesis of (+)-cavicularin (**553**), P. Zhao and C. M. Beaudry used an intramolecular regioselective pyrone Diels–Alder reaction of a vinyl sulfone as the key step (Fig. 106) [302]. The vinyl sulfone pyrone synthon **554** was synthesized in a conventional manner, and, when it was kept in ethyl acetate solution with 3 Å molecular sieve and the chiral *Cinchona*-based thiourea catalyst **555** at 45°C it underwent first the intramolecular Diels–Alder reaction between the vinyl dienophile and the pyrone diene, leading to intermediate **556**. The latter lost in situ benzenesulfinic acid and then carbon dioxide to give **557**, which could then be transformed readily in a few additional steps into (+)-cavicularin (**553**).

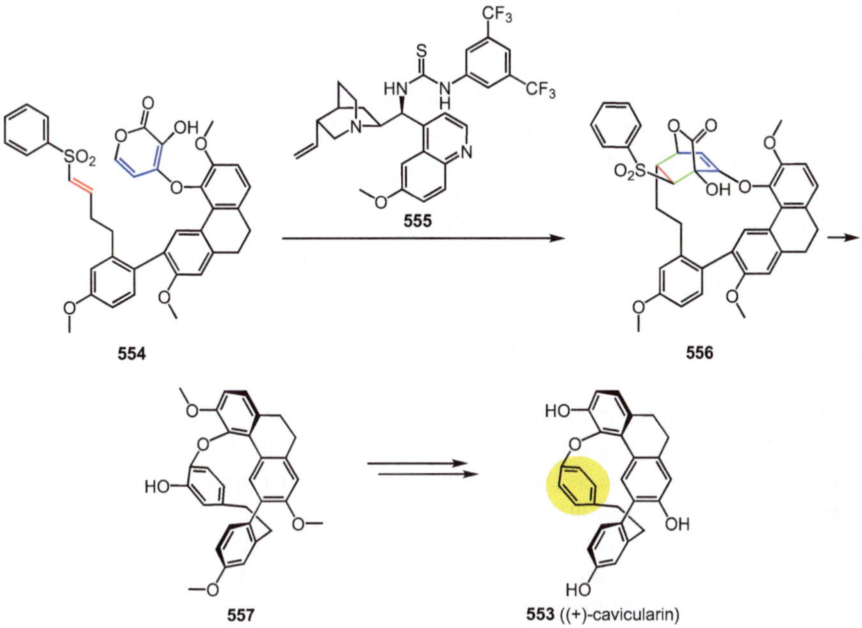


**Fig. 106** The intramolecular chirally catalyzed regioselective pyrone Diels–Alder reaction of a vinyl sulfone as the key step in the stereoselective synthesis of (+)-cavicularin (**553**)

**Plate 60** *Cavicularia densa*. Photograph courtesy Daderot, Creative Commons CCO 1.0 Universal Public Domain

## *7.23 Kopsinine*

As a second prominent example, kopsinine (**558**) represents an *Aspidosperma* alkaloid of the *Kopsia* group, first isolated from *Kopsia longiflora* Merr. (syn. *K. arborea* Blume) from the family Apocynaceae (Plate 61) [303]. It is characterized mainly by a [2.2.2]octane ring central to the core of the structure.

The synthesis of (−)-kopsinine (**558**) by D. L. Boger and his group [304, 305] introduced a novel tandem cycloaddition strategy. The route hinged on an intramolecular [4 + 2]/[3 + 2] cycloaddition cascade using a 1,3,4-oxadiazole as the key dipolarophilic moiety (Fig. 107). In practice, the 1,3,4-oxadiazole-containing precursor **559** underwent in *o*-dichlorobenzene at 180°C an initial inverse electron demand Diels–Alder (forming a new six-membered ring) to yield the intermediate **560**, which after the loss of nitrogen became **561**, and immediately triggered a 1,3-dipolar [3 + 2] cycloaddition, yielding the pentacyclic *Aspidosperma* skeleton **562** in 70% yield in one step. This cascade Diels–Alder approach created all six stereocenters of the kopsinine core with complete control. The authors demonstrated the power of such a unique diene/dienophile combination, here an oxadiazole acting as a masked aza-diene and dipole to rapidly construct complexity. This methodological advance also featured strategic flexibility: the same cascade intermediate was used to synthesize several *Aspidosperma* alkaloids, like (+)-fendleridine, (−)-deoxoapodine, and (−)-kopsifoline D, highlighting a divergent synthesis design enabled by the Diels–Alder key step. Compound **562** could then be elaborated into the targeted natural product, kopsinine (**558**).

**Plate 61** *Kopsia longiflora* (syn. *K. arborea*). Photograph courtesy T. Gerus, Creative Commons Attribution 2.0 Generic

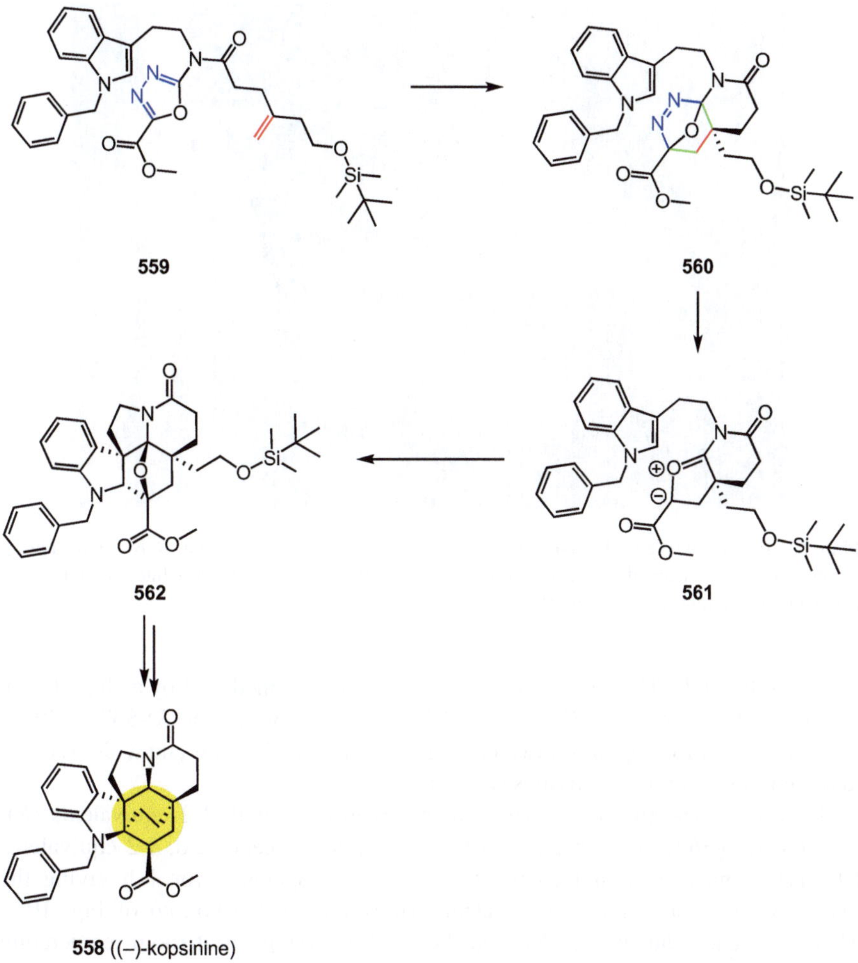


**Fig. 107** An intramolecular [4 + 2]/[3 + 2] cycloaddition cascade using a 1,3,4-oxadiazole as the key dipolarophilic moiety in synthesis of (−)-kopsinine (**558**)

## *7.24 (−)-Nicolaioidesin C and (−)-Panduratin A*

(−)-Nicolaioidesin C (**563**) and (−)-panduratin A (**564**) showing antiausterity activities against the PANC-1 human pancreatic cancer line, can be isolated from the finger-root plant, *Boesenbergia pandurata* (Roxb.) Schltr. (syn. *B. rotunda* (L.) Mansf.) (Plate 62), of the ginger family Zingiberaceae, which is native to Southeast Asia [306].

To synthesize these two natural products, an asymmetric Diels–Alder cycloaddition of the 2′-hydroxychalcone **565** with an acyclic diene has been devised

**Plate 62** Flower (left) and rhizomes (right) of *Boesenbergia pandurata* (syn. *B. rotunda*). Photographs courtesy Ariearsad, Creative Commons Attribution-Share Alike 4.0 International (left); Kembangraps, public domain (right)

successfully [307]. This Diels–Alder cycloaddition was mediated by a chiral boron complex with (*R*)- or (*S*)-3,3′-diphenyl-2,2′-bi-1-naphthol ((*R*)- or (*S*)-**567** = (*R*)- or (*S*)-VANOL), affording the corresponding products in high yields and with excellent diastereo- and enantioselectivities (Fig. 108).

Thus, chalcone **565** was reacted in tetrahydrofuran with 1.2 equivalents (*R*)-VANOL ((*R*)-**567**) in the presence of 1.2 equivalents acetic acid, 1.2 equivalents $BH_3$.THF, and a 5 Å molecular sieve at room temperature for 1 h giving the intermediate boron complex of chalcone (illustrated at the bottom of Fig. 108). Then, myrcene (**566**) was added and the reaction continued at room temperature for 68 h yielding (−)-nicolaioidesin C (**563**) in 78% with 96% *ee*.

Alternatively, when chalcone **565** was reacted in tetrahydrofuran with two equivalents of (*S*)-VANOL ((*S*)-**567**) in the presence of two equivalents acetic acid, two equivalents of $BH_3$.THF, and a 5 Å molecular sieve at room temperature for 1 h, it gave the intermediate boron complex of chalcone (illustrated at the bottom of Fig. 108). Then, (*E*/*Z*)-ocimene (1/3) (**568**) was added and the reaction continued at room temperature for 68 h, finally yielding (−)-panduratin (**564**) in 58% with 87% *ee*.

These two syntheses present a nice illustration of the progress made with tailoring intermolecular asymmetric Diels-Alder reactions as needed for a specific synthesis purpose.

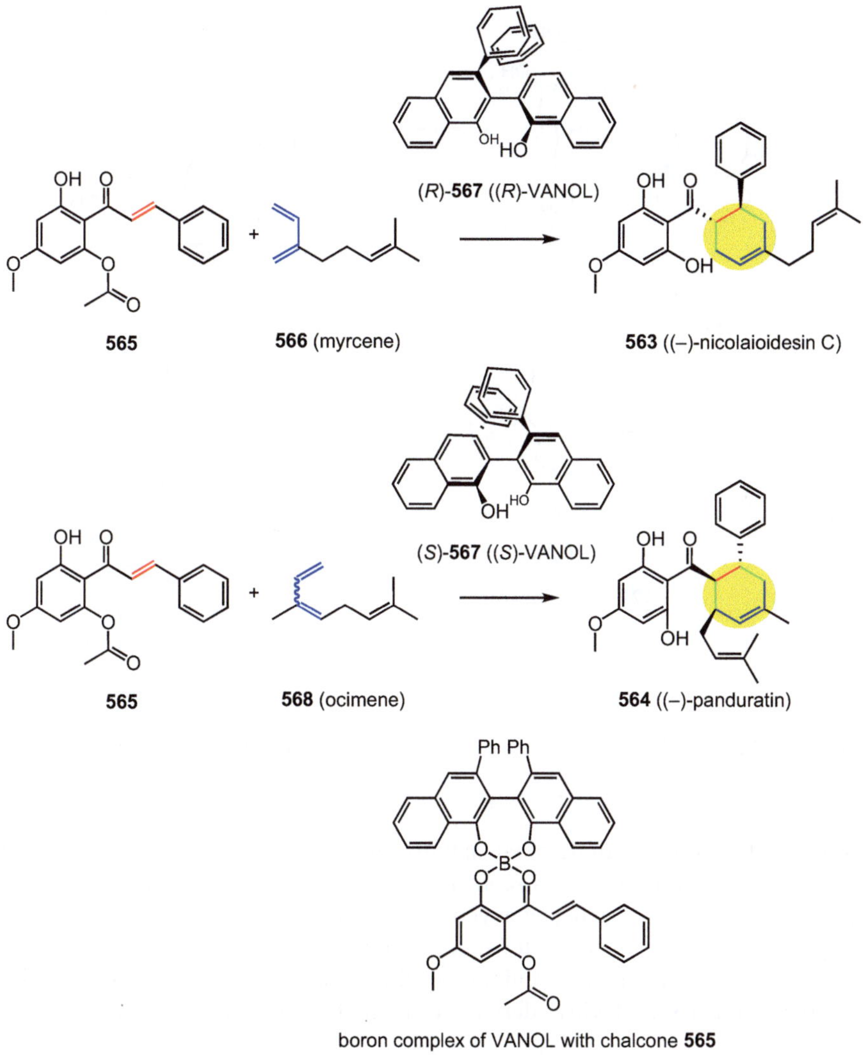


**Fig. 108** A Diels–Alder cycloaddition is mediated by a chiral boron complex in the syntheses of (–)-nicolaioidesin C (**563**) and (–)-panduratin (**564**)

## *7.25 (+)-Cytosporolide A*

The meroterpenoid (+)-cytosporolide A (**569**) (Fig. 109) was isolated in 2010 by Y. Che and co-workers from the fungus *Cytospora* sp., which was found in a soil sample collected on the Qinghai-Tibetan plateau at high altitude [308]. The

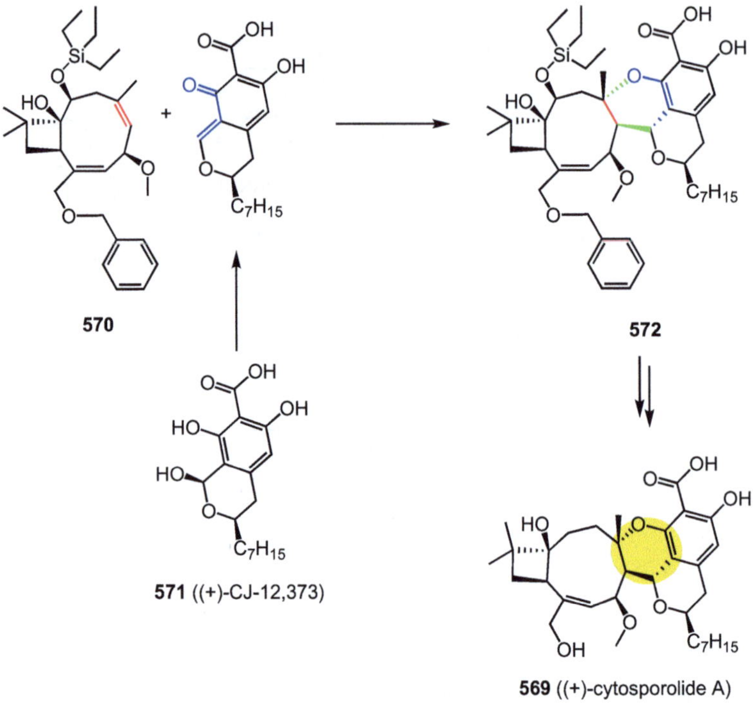


**Fig. 109** The hetero-Diels–Alder reaction in the synthesis of (+)-cytosporolide A (**569**)

compound showed antimicrobial activity against the Gram-positive bacteria *Staphylococcus aureus* and *Streptococcus pneumoniae*. The structure of **569** consists of a complicated pentacyclic ring system containing an oxygenated caryophyllene skeleton connected to a substituted isochroman ring. A biogenetic study indicated that an *ortho*-quinone methide reacting in a hetero-Diels–Alder reaction with a suited vinyl derivative might be responsible for this type of skeleton.

Following this concept, K. I. Takao and his group synthesized in 2015 the necessary educts, the protected (−)-fuscoatrol (**570**) and the *ortho*-quinone methide precursor (+)-CJ-12,373 (**571**), using common methodology (Fig. 109) [309]. Their hetero-Diels–Alder reaction proceeded in toluene at 100°C for 5 h giving the protected pentacyclic derivative **572** with complete chemo-, regio-, and stereoselectivity. Deprotection in conventional ways provided the target compound (+)-cytosporolide A (**569**).

## 7.26 *Δ^7^-Mesembrenone, Mesembrine, and Gracilamine*

Of the various hydroindoline alkaloids, three have attracted an intense synthesis interest by the group of S. A. Snyder, who used an intramolecular pyrone Diels–Alder key step for the construction of their hydroindoline skeleton [310]. These were $\Delta^7$-mesembrenone (**573**), mesembrine (**574**), and gracilamine (**575**). The first two were isolated from the medicinal Kanna plant *Mesembryanthemum tortuosum* L. (syn. *Sceletium tortuosum* (L.) N.E. Br.) from the family Aizoaceae (Plate 63) [311]. This has been used traditionally as a mood elevator, stress reducer, analgesic, anxiolytic, and even a narcotic. The main activity is attributed to the mesembrine-type alkaloids present, which have shown promise in providing pharmacologically and clinically relevant antidepressant and anxiolytic effects. These alkaloids are serotonin reuptake inhibitors, which provides a rationale for the plant's traditional use as an antidepressant, but other biological activities are beginning to appear in the literature. For example, mesembrenone has reasonably potent PDE4 inhibitory activity [312].

Gracilamine (**575**), a cytotoxic pentacyclic Amaryllidaceae alkaloid with an unprecedented hydroindoline structure, was obtained from *Galanthus gracilis* Čelak (Plate 63) [313]. It seems that it displays some cytotoxicity [313, 314].

For the synthesis of $\Delta^7$-mesembrenone (**573**) and mesembrine (**574**) the substrate **576** was prepared conventionally (Fig. 110). Upon heating **576** in toluene for 10 h at 160°C using microwave irradiation, an intramolecular pyrone Diels–Alder reaction occurred, and the resulting intermediate underwent in situ a retro Diels–Alder loss of carbon dioxide when standing for 15 h at 23°C in chloroform solution with some added silica gel, to yield the indoline derivative **577**. Conventional elaboration of

**Plate 63** Left: *Mesembryanthemum tortuosum* (syn. *Sceletium tortuosum*). Photograph courtesy T. Nummelin, Creative Commons Attribution-Share Alike 3.0 Unported. Right: *Galanthus gracilis*. Photograph courtesy SiGarb, Creative Commons Attribution-Share Alike 4.0 Unported

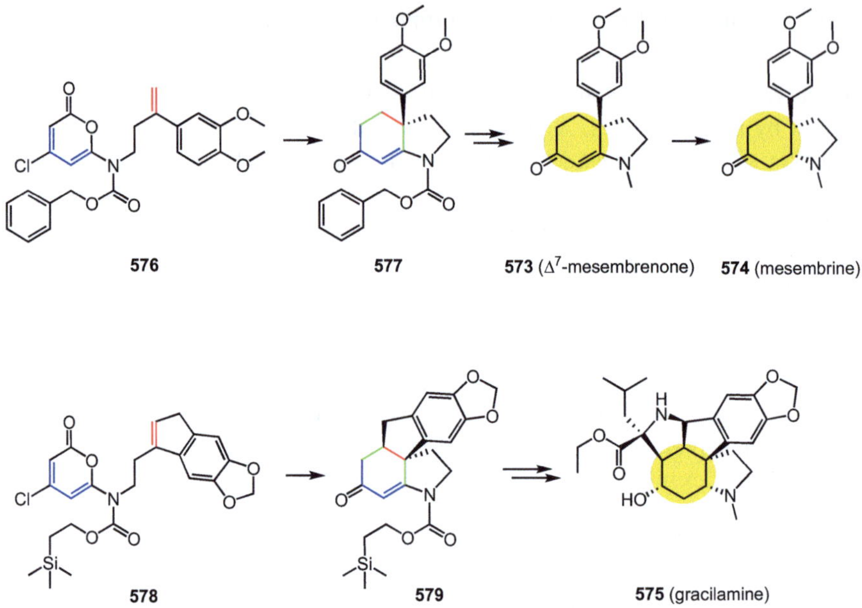


**Fig. 110** The intramolecular pyrone Diels–Alder and in situ retro Diels–Alder loss of $CO_2$ in the synthesis of $\Delta^7$-mesembrenone (**573**), mesembrine (**574**), and gracilamine (**575**)

this compound resulted in $\Delta^7$-mesembrenone (**573**), which could then be transformed into mesembrine (**574**) by a single-step procedure [314].

Along the same lines, substrate **578** was heated in toluene at 170°C under microwave irradiation for 12 h and then treated with silica gel in chloroform at 23°C for 15 h. The resulting cyclization product **579** gave gracilamine (**575**) after conventional preparative steps.

## *7.27 (−)-Neocosmosin A*

(−)-Neocosmosin A (**580**) is a resorcylic acid macrolide, isolated from the fungus *Neocosmospora* sp. (UM-031509) [315]. It was shown to have a strong binding affinity for human opioid and cannabinoid receptors [316]. The chiral C-2 carbinyl center, proved to have (*S*)-configuration [317].

For the synthesis of (−)-neocosmosin A (**580**) an intramolecular pyrone Diels–Alder reaction with accompanying cycloreversion of carbon dioxide was devised starting from the alkyne (+)-(*S*)-substrate **581** as the key step (Fig. 111) [318]. The reaction proceeded smoothly in refluxing mesitylene (0.1 *M*) for 46 h in the presence of 0.5 equivalents 2,6-di-*tert*-butyl-4-methylphenol as antioxidants, to give the macrolide **582** in 65% yield. Conventional elaboration of **582** then provided the natural product, (−)-neocosmosin A (**580**).

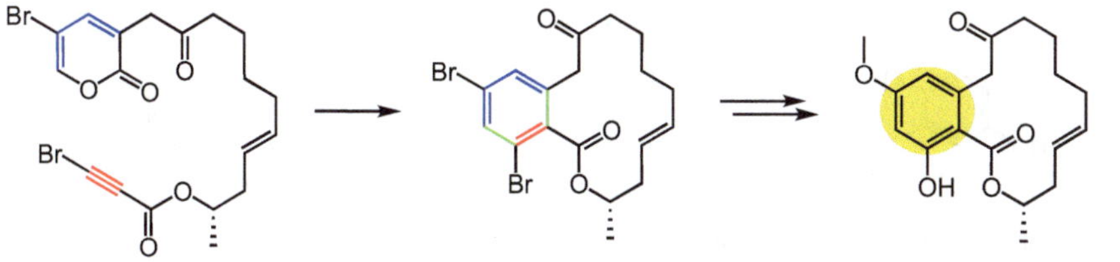


(+)-(*S*)-**581** **582** **580** ((−)-neocosmosin A)

**Fig. 111** The intramolecular pyrone Diels–Alder reaction with accompanying cycloreversion of carbon dioxide as key step in the synthesis of (−)-neocosmosin A (**580**)

## *7.28 Paeoveitol*

The roots of the herbaceous perennial plant, *Paeonia veitchii* Lynch of the family Paeoniaceae (Plate 64), are used as a treatment for depression-like disorders in traditional Chinese medicine ("chuan shi shao") and is also recognized for activities related to immunomodulation, anti-inflammation, hypoglycemia, and circadian regulation [319]. One of its chemical constituents is paeoveitol (**583**), which is interesting for its unprecedented [3,2-*b*] fused 2,3-dihydrobenzofuran and chromane units in a tetracyclic framework with three contiguous stereogenic centers [320].

For the synthesis of paeoveitol (**583**), the group of Z. Xie devised a bioinspired hetero Diels–Alder reaction between paeoveitol D (**584**), prepared by classical means, and the readily accessible substrate **585** (Fig. 112) [321]. Heating the two compounds in toluene solution in a sealed tube at 130°C resulted first in an elimination reaction of **585** to give the *ortho*-quinone methide **586**, which then underwent the intermolecular hetero Diels–Alder reaction with **584** to yield the tetracyclic compound **587** as a single diastereomer in 55% yield. Saponification of

**Plate 64** *Paeonia veitchii*. Photograph courtesy Vladaig, public domain

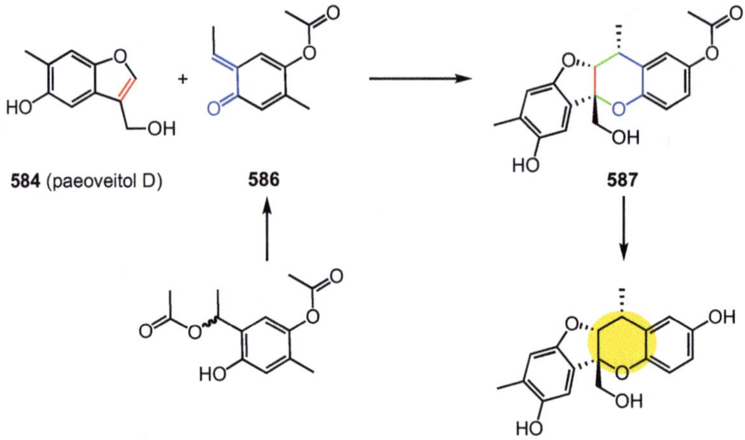


**584** (paeoveitol D) **586** **587** **585** **583** (paeoveitol)

**Fig. 112** The intermolecular hetero Diels–Alder reaction key step in the synthesis of paeoveitol (**583**)

**587** with potassium carbonate in methanol yielded 80% of the target compound, paeoveitol (**583**).

## 7.29 *Lancifodilactone G Acetate*

Lancifodilactone G (**588**) (Fig. 113) was isolated from the medicinal plant *Schisandra lancifolia* Michx., family Schisandraceae (Plate 65), an endemic forest vine from Sichuan and Yunnan, China, which is rarely cultivated; its structure was determined by X-ray crystallographic analysis [322, 323]. Unlike the other members of the terpenoid family, **588** contains a C/D ring system that bears a rare nonresonance-stabilized aliphatic enol, a highly congested tricyclic FGH ring system containing six contiguous stereogenic centers, and an unusual two-fold anomerically stabilized bis-spiro system. Compounds related structurally to lancifodilactone (**588**) have been reported to exhibit antihepatitic, antitumor, and anti-HIV activities [324].

An early asymmetric intermolecular Diels–Alder reaction step paved the way to a total synthesis of lancifodilactone G acetate (**589**) (Fig. 113) [325]. The easily accessible dienophile and diene substrates **590** and **591** were reacted in dichloromethane solution at −78°C in the presence of the chiral oxazaborolidine catalyst **592**, to yield the cyclohexene derivative **593** in 95% yield with nearly 90% *ee*. This reaction worked well on a 100 g scale, which was of high importance to guarantee a sufficiently available chiral educt for the further synthesis, which required a total of 28 steps to provide ultimately lancifodilactone G acetate (**589**). Saponification could then also lead to lancifodilactone (**588**).

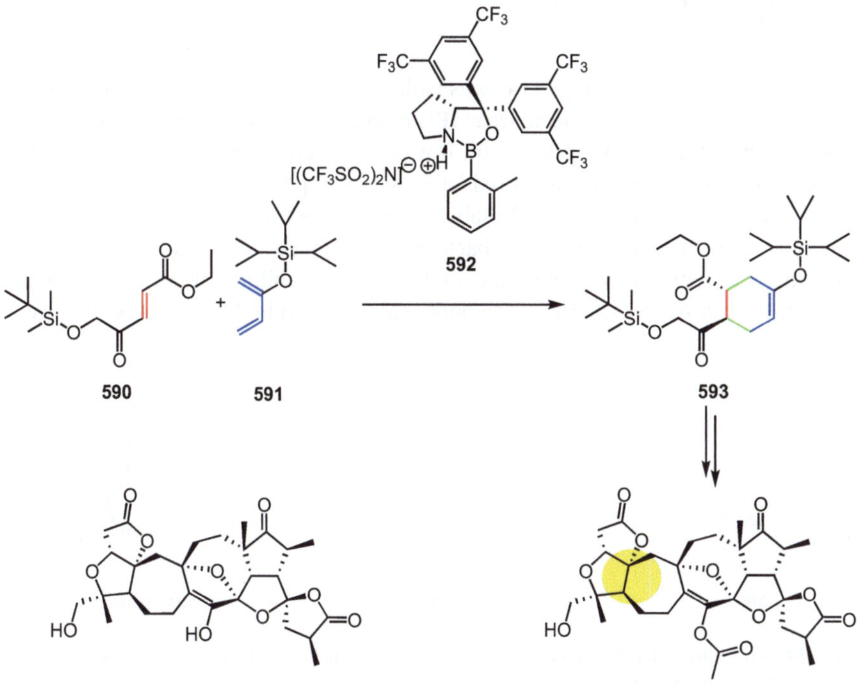


**Fig. 113** The asymmetric intermolecular Diels–Alder reaction key step in the synthesis of lancifodilactone acetate (**589**)

**Plate 65** *Schisandra lancifolia*. Photograph courtesy Board of Trustees, RBG Kew, Creative Commons BY 3.0

## 7.30 Homodimericin A

Homodimericin A (**594**) (Fig. 114) was isolated from the soil fungus *Trichoderma harzianum* Rifai, family Hypocreaceae (Plate 66), an organism with potential for the suppression of fungal pathogens [326]. It was produced by the fungus as a chemical defense to counter the exogenous oxidative stress imposed by a co-cultured bacterium *Streptomyces* sp. 4231. Notably, the production of **594** was strongly upregulated by bafilomycin C1, a bacterially produced antifungal agent, thereby implying that **594** and related metabolites may exhibit antibacterial activity. Homodimericin A (**594**) possesses an unprecedented and highly complex hexacyclic

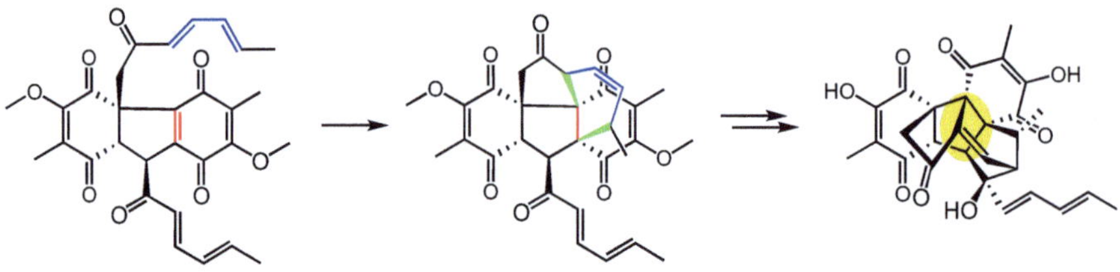


**595** **596** **594** (homodimericin A)

**Fig. 114** The intramolecular tandem Diels–Alder reaction key step in the construction of the fundamental skeleton of homodimericin A (**594**)

**Plate 66** *Trichoderma harzianum*. Photograph, public domain of USA

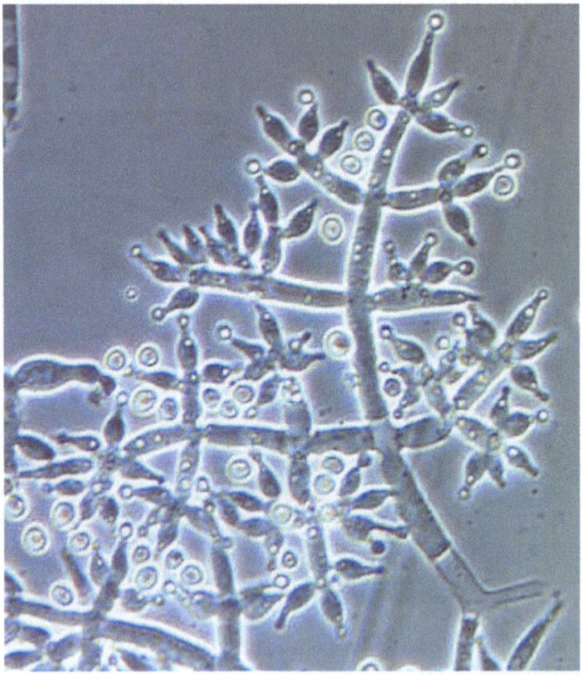

framework. In particular, its central B/D/E/F tetracyclic core contains eight consecutive stereogenic centers, three of which are adjacent quaternary centers [327]. The highly congested nature of **594** made it a formidable and challenging synthesis project.

An intramolecular tandem Diels–Alder reaction/carbonyl–ene cyclization to forge the congested D/E/F tricyclic cage motif was chosen and optimized as a key step (Fig. 114) [328]. Thus, the conventionally prepared 1*H*-fluorene-1,4,5,8(4a*H*)-tetraone **595** was heated in 1,2-dichloroethane at 180°C for 5 h, whereby mainly the 10*H*-10,5a-prop[1]enocyclopenta[*e*]fluorene-1,4,6,9,11(12*H*)-pentaone **596** was obtained. The latter could then be elaborated via a carbonyl-ene cyclization by classical means into the natural product homodimericin A (**594**). This is a wonderful example of the progress the Diels-Alder reaction has made in the construction of highly complex compounds.

## 7.31 *Myrtucommulones K, N, and O*

The evergreen shrub *Myrtus communis* L., family Myrtaceae (Plate 67), has a long history as an antiseptic, disinfectant, and hypoglycemic agent used in traditional medicine [329, 330]. The plant has raised the interests of both the medicinal chemistry and drug discovery communities. In their isolation report, L. Wang and co-workers postulated that myrtucommulones K (**597**), N (**598**), and O (**599**) (Fig. 115) might be derived biosynthetically from flavesone (a β-triketone) and (−)-β-caryophyllene, which had been isolated previously from the same plant [331].

Based on this biosynthesis hypothesis a biomimetic intermolecular hetero-Diels–Alder reaction of a neat mixture of the easily available isobutylidene syncarpic acid (**600**) as hetero-diene and (−)-β-caryophyllene (**601**) as dienophile at 80°C resulted in a mixture of myrtucommulones K (**597**) (42%), N (**598**) (10%), and O (**599**) (25%), which could be separated readily by flash chromatography (Fig. 115) [331]. Myrtucommulones K (**597**) and N (**598**) resulted from the endocyclic double

**Plate 67** *Myrtus communis*. Photograph courtesy LIGURIAN VASCULAR FLORA, Creative Commons Attribution 2.0 Generic

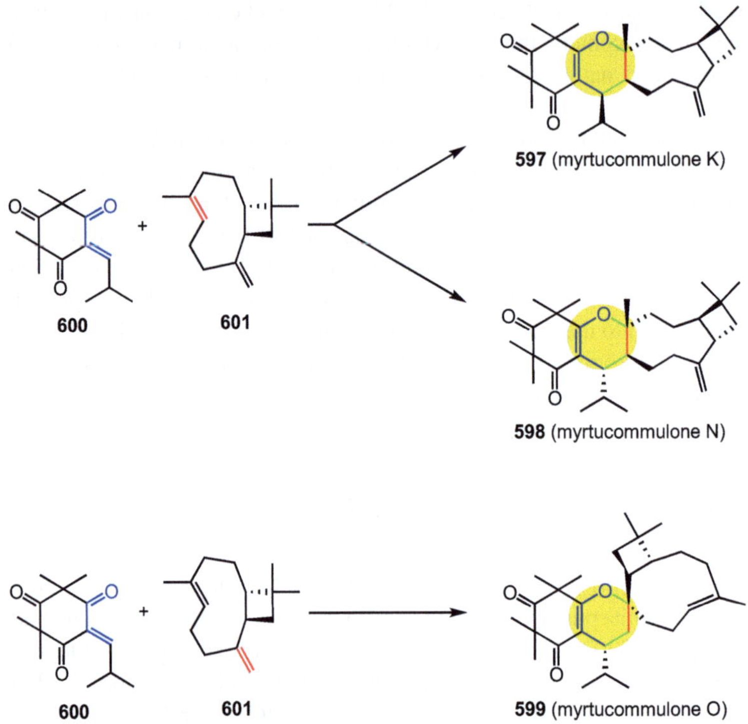


**Fig. 115** The biomimetic intermolecular hetero-Diels–Alder reaction key step in the syntheses of myrtucommulones K (**597**), N (**598**), and O (**599**)

bond as the dienophile, whereas production of myrtucommulone O (**599**) was due to an exocyclic reaction approach.

## *7.32 (+)- and (−)-Paeoveitol*

In the year previously, the synthesis of racemic paeoveitol (**583**) was described (Sect. 7.28), and, along the same lines in 2017, using an intermolecular hetero-Diels–Alder reaction catalyzed by both enantiomers of the chiral Brønsted acid **602**, another Chinese group succeeded with excellent diastereo- and enantioselectivity (>99% *dr* and 90% *ee*) to obtain (+)-**583** and (−)-**583** (Fig. 116) [332].

The reaction proceeded from the educts paeoveitol D (**584**) and the phenol **585** in halogenated solvents with 5 mol% of the (*S*)- and (*R*)-configured enantiomers of the catalyst **602** at room temperature in about 90% yield of the cyclization product **587** (see Fig. 112), which was then transformed into the target enantiomers, (+)-paeoveitol (**583**) and (−)-paeoveitol (**583**).

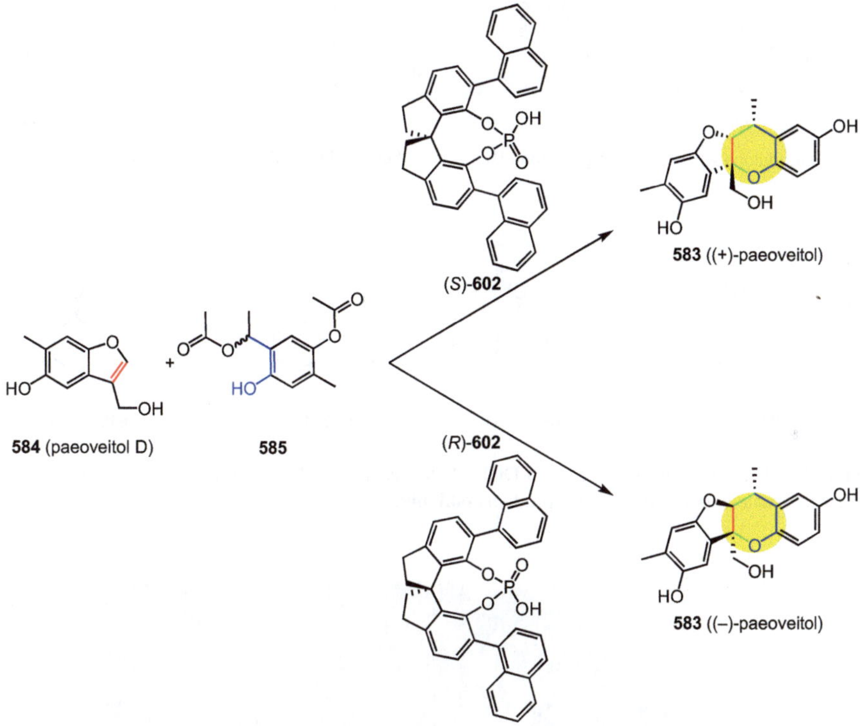


**Fig. 116** The intermolecular hetero-Diels–Alder reaction catalyzed by the enantiomers of the chiral Brønsted acid in the synthesis of (+)- and (−)-paeoveitol (**583**)

## *7.33 Nodulones A–C and Xestolactone A Frameworks*

Nodulones A–C (**603**–**605**) are small, highly oxygenated polyketides that belong to the dihydronaphthalen-1-one (tetralone) family (Fig. 117). They are produced by fungi like *Daldinia eschscholtzii* (Ehrenb.) Rehm (Plate 68), of the family Hypoxylaceae, a filamentous wood-inhabiting endophyte, and they are thought to arise from the 1,8-dihydroxynaphthalene polyketide pathway [333]. Nodulone A (**603**) displays potent immunosuppressive activity (inhibition of ConA-induced mouse splenocyte proliferation) and cytotoxicity toward SF-268 glioma cells [334]. Nodulone B (**604**) exhibits moderate antibacterial and antifungal activities against *Bacillus subtilis*, *Candida albicans*, and other organisms. Nodulone C (**605**) shows broad-spectrum antimicrobial action comparable to **604** and weak antiproliferative effects for the K-562 myelogenous leukemia cell line [335].

Xestolactone A (**606**) (Fig. 117) was the first, and so far only compound, obtained from in the demosponge *Xestospongia vansoesti* Bakus & Nishiyama in the family Petrosiidae (Plate 68), collected off Palawan Island, Philippines [336]. It belongs to the sponge-specific xestoquinone merosesquiterpenoid family. A bicyclic

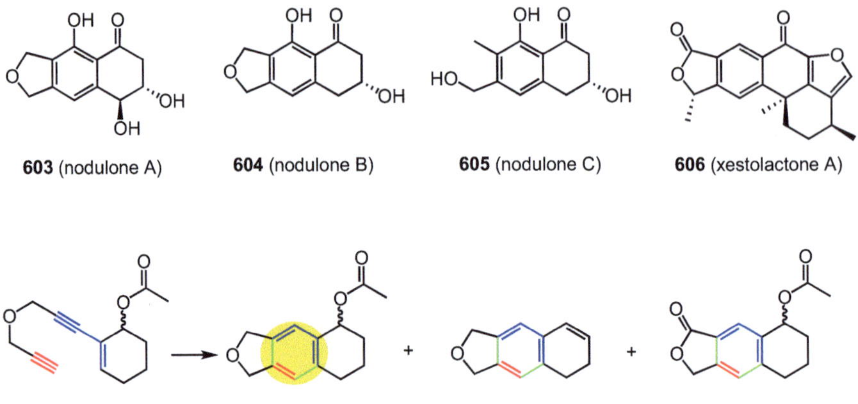


**Fig. 117** The intramolecular dehydro-Diels–Alder reaction as key step in the construction of the tricyclic frameworks of the natural products **603–606**

**Plate 68** Left: *Daldinia eschscholtzii*. Photograph courtesy Byrain, Creative Commons Attribution-Share Alike 3.0 Unported. Right: *Xestospongia muta*, a species looking much the same as *X. vansoesti*. Photograph courtesy Twilight Zone Expedition Team 2007, NOAA-OE, Creative Commons Attribution 2.0 Generic

sesquiterpene core is fused to a highly oxygenated 2*H*-chromen-3-one (xestolactone) ring. The skeleton is unprecedented among marine natural products and rationalized from the standpoint of biogenesis as an oxidative cleavage product of xestoquinone. Xestolactone A (**606**) showed indoleamine-2,3-dioxygenase-1 inhibition with moderate activity in an enzyme-based kynurenine assay [336].

With respect to the synthesis of the above-mentioned four natural products, their tricyclic frameworks present in the isofuranonaphthalenone could be constructed [337] (Fig. 117) with an intramolecular dehydro-Diels–Alder reaction [338, 339]. Thus, upon heating the enyne acetate **607** in 1,2-dichlorobenzene at

140°C for 24 h, the cyclization product **608** was obtained together with its elimination product **609** and an oxidation product, the phthalide **610**. Accordingly, framework substrates for the further elaboration of the natural products were available by this ingenious methodology.

## 7.34 *Selaginpulvilin D*

From the spikemoss *Selaginella pulvinata* Hook. & Grev. (Plate 69), family Selaginellaceae, a species included in the Chinese Pharmacopoeia, which is used for the treatment of dysmenorrhea, asthma, and traumatic injury, several 9,9-diphenyl-1-(phenylethynyl)-9*H*-fluorene derivatives were isolated, and among them selaginpulvilin D (**611**) could be isolated [340]. It was found to be a potent natural phosphodiesterase-4 inhibitor [341].

**Plate 69** *Selaginella pulvinata.* Photograph courtesy K. Schablewski, RBG Kew, Creative Commons BY 4.0

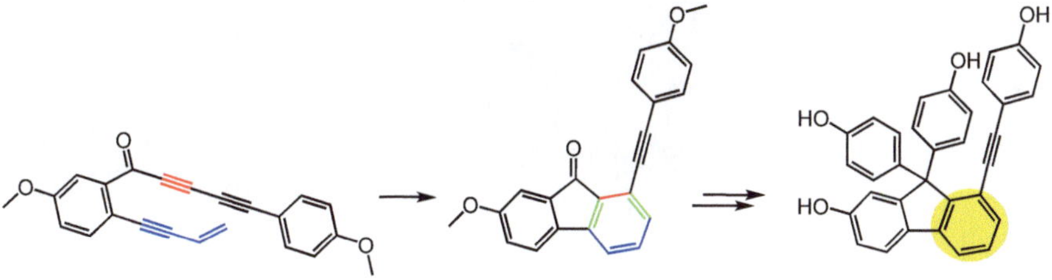


**Fig. 118** The chemoselective enyne–alkyne dehydro-Diels–Alder reaction key step in the synthesis of selaginpulvilin D (**611**)

A highly chemoselective enyne–alkyne dehydro-Diels–Alder reaction was employed for the construction of the tricyclic fluorene framework present in the natural product selaginpulvilin D (**611**) (Fig. 118) [342]. For this key step the enyne-alkyne precursor **612**, synthesized using previously known methodology, was heated in 1,2-dichlorobenzene for 2 h at 80°C, resulting exclusively in the formation of fluoreneone **613**. The latter could then be transformed by conventional procedures into the target product, selaginpulvilin D (**611**). Again, a dehydro-Diels–Alder reaction paved the way to an aromatic kernel of a natural product.

## *7.35 Magellanine*

Magellanine (**614**) is a member of the *Lycopodium* alkaloids isolated in 1976 from the club moss *Lycopodium magellanicum* (P. Beauv.) Holub, family Lycopodiaceae (Plate 70) [343]. It is mainly of interest to the synthesis community due to its tricyclic angular core.

The key step in the synthesis of magellanine (**614**) by L. Barriault and his group used as the substrate the ene-yne compound **615**, prepared along a conventional path (Fig. 119) [344]. In the presence of 1 mol% of the gold(I) catalyst **616** in toluene solution at 20°C and treatment with one equivalent of camphorsulfonic acid 91%, the intermediate **617** was obtained using a dehydro-Diels–Alder reaction in a highly regioselective and diastereoselective manner (>95%). The latter was then transformed in several classical steps into the desired (racemic) magellanine (**614**).

**Plate 70** *Lycopodium magellanicum*. Photograph courtesy franek2, Creative Commons Attribution-Share Alike 3.0 Unported

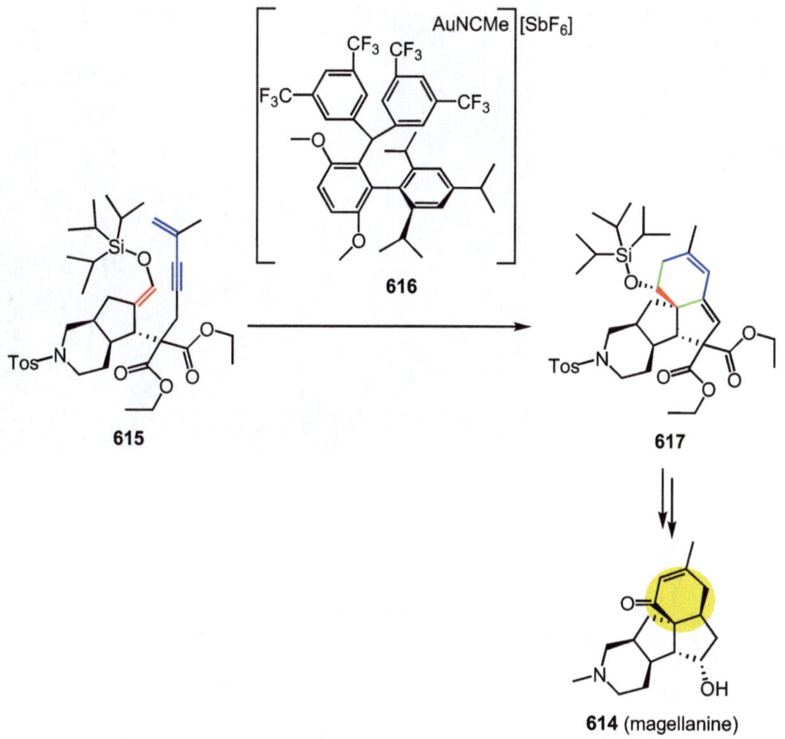


**Fig. 119** The gold(I)-catalyzed dehydro-Diels–Alder reaction key step in the synthesis of magellanine (**614**)

## *7.36 (±)-Verrubenzospirolactone*

The highly fused meroterpenoid (±)-verrubenzospirolactone (**618**) was isolated from the soft coral *Sinularia verruca* van Ofwegen, family Alcyoniidae (Plate 71), which is endemic to Palau in the Western Central Pacific [345, 346]. Again, as in the example above, it was mainly of interest to the synthesis community due to its stereochemical complexity.

In their bioinspired total synthesis, J. H. George and his team used a biomimetic, intramolecular Diels–Alder reaction of the conventionally synthesized 2*H*-chromene **619** to form two rings and four stereocenters in the final step. This reaction occurred spontaneously at 30°C in water leading to (±)-verrubenzospirolactone (**618**) (Fig. 120) [347]. This example very nicely illustrates the enormous potential of variants of the Diels–Alder reaction to construct highly complex stereostructures.

**Plate 71** *Sinularia verruca.* Photograph courtesy l. van Ofwegen, WoRMS, Creative Commons Attribution-Share Alike 4.0 International

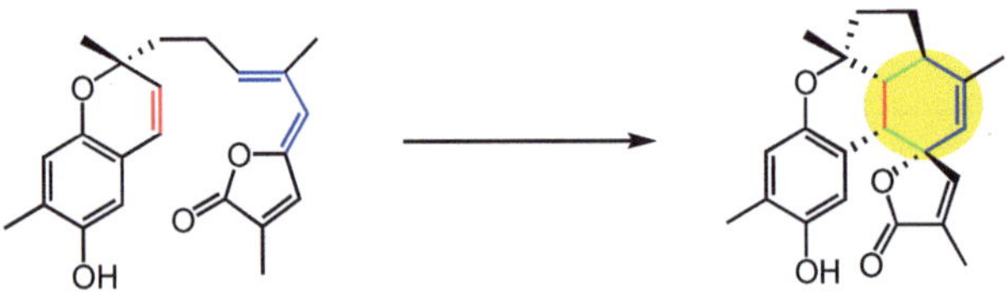


**Fig. 120** The biomimetic intramolecular Diels–Alder reaction key step in the synthesis of (±)-verrubenzospirolactone (**618**)

## *7.37 (−)-Cycloclavine*

In the final example from 2017, an intramolecular strain-promoted Diels–Alder methylenecyclopropane reaction was used as the spectacular key to the enantioselective synthesis of (−)-cycloclavine (**620**). This compound was isolated initially from the seeds of the African morning glory shrub *Ipomoea hildebrandtii* Vatke, of the Convolvulaceae family (Plate 72) [348], and later from the filamentous fungus *Aspergillus japonicus* [349]. In biological evaluations it was demonstrated to possess promising insecticidal and antiparasitic activities [350]. Cycloclavine (**620**) belongs to the clavine subset of ergot alkaloids, a prominent group of natural products that share a fused indole core and are favored synthesis challenges.

In the synthesis of (−)-cycloclavine (**620**) described by S. R. McCabe and P. Wipf, the diene **621** underwent in tetrahydrofuran solution an *anti*-selective intramolecular Diels–Alder reaction with the pendant methylenecyclopropane. Upon heating by microwave irradiation to 95°C for 30 min the procedure yielded

**Plate 72** *Ipomoea hildebrandtii*. Photograph courtesy L. Comte, Arbres et flore

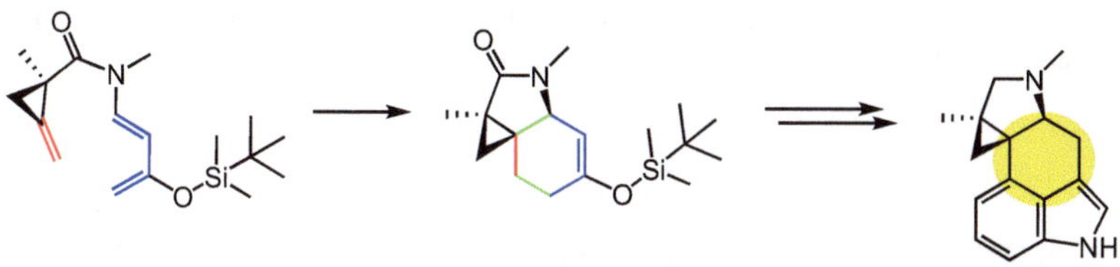


**621** **622** **620** ((–)-cycloclavine)

**Fig. 121** The intramolecular strain-promoted Diels–Alder methylenecyclopropane reaction in the enantioselective synthesis of (−)-cycloclavine (**620**)

stereoselectively the enolether **622** (Fig. 121) [351]. Further elaboration along conventional lines led to the targeted (−)-cycloclavine (**620**).

## *7.38 (+)-Aplykurodinone-1*

To start the year 2018, (+)-aplykurodinone-1 (**623**) (Fig. 122), a degraded steroidal tricyclic lactone isolated from the sea hare *Syphonota geographica* A. Adams & Reeve (Plate 73) [352] belonging to the family Aplysiidae, can be illustrated as a further example of the developing synthesis methodology. This compound, like other congeners, is characterized by an unusual *cis*-fused hydrindane skeleton, containing six contiguous stereogenic centers, including an all-carbon-bonded quaternary carbon center and a methyl-substituted side chain. While no studies aimed at

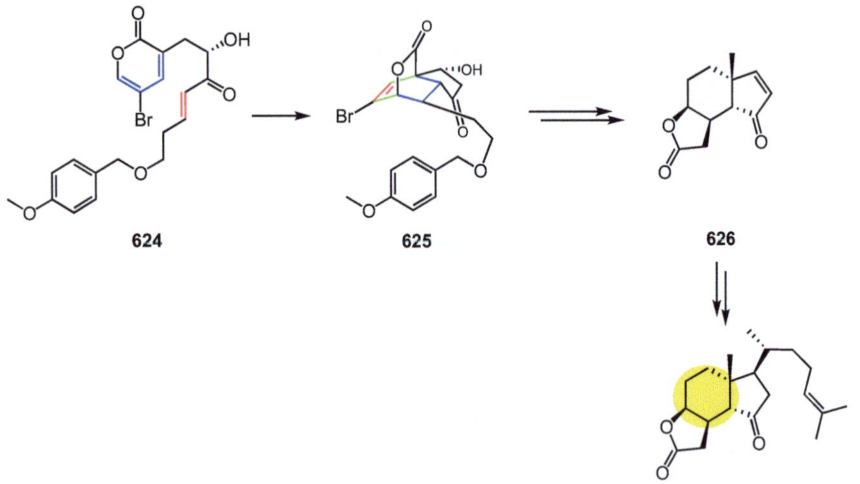


624 625 626

623 ((+)-aplykurodinone-1)

**Fig. 122** The stereoselective intramolecular Diels–Alder reaction key step in the synthesis of (+)-aplykurodinone-1 (**623**)

**Plate 73** *Syphonota geographica*. Photograph courtesy R. Zerpe, Creative Commons Attribution 2.0 Generic

assessing the biological activities of **623** have been reported, several analogous aplykurodins have been shown to possess cytotoxic activities against various human cancer cell lines [353].

An asymmetric total synthesis of (+)-aplykurodinone-1 (**623**) was achieved using a route in which hydrogen bonding serves as the stereochemical control element governing the π-facial selectivity of the intramolecular Diels–Alder reaction in refluxing toluene of the enone-tethered 5-bromo-2-pyrone (**624**) (Fig. 122) [354]. In this process, the configuration at the stereogenic hydroxy group in the enone dienophile was conveyed in a highly effective manner through intramolecular hydrogen bonding with the enone carbonyl oxygen, yielding the intermediate **625**. The tricyclic lactone **626** resulting after decarboxylation and several further steps from **625**, was then converted to the target (+)-aplykurodinone-1 (**623**) by conventional steps.

## *7.39 (−)-Artemisinin*

Artemisinin (**627**) (qinghaosu) is a sesquiterpene lactone endoperoxide isolated from the leaves of the annual short-day plant, sweet wormwood *Artemisia annua* L., family Asteraceae (Plate 74). It is native to temperate Asia and was investigated under China's then secret antimalarial research program [355, 356]. The team of Tu Youyou abandoned traditional hot-water decoctions in favor of a low-temperature ether wash, obtaining a neutral crystalline compound that cleared *Plasmodium-*

**Plate 74** *Artemisia annua*. Photograph courtesy Pixabay

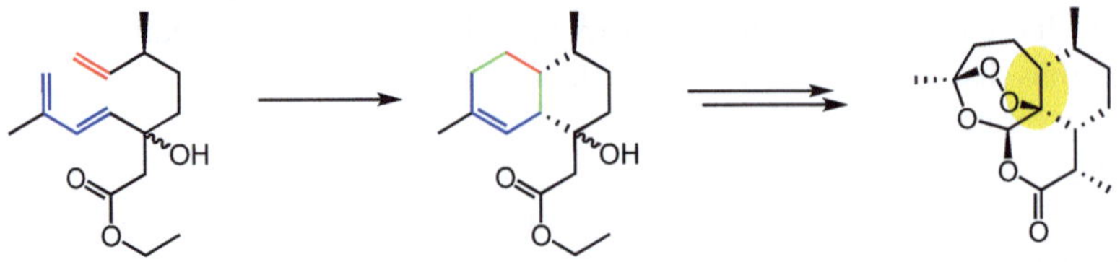


**Fig. 123** The intramolecular Diels–Alder key step in the synthesis of (−)-artemisinin (**627**)

infected mice and monkeys of this parasite, a study that earned her the 2015 Nobel Prize in Physiology or Medicine [357]. The unique 1,2,4-trioxane bridge in its structure seems to be responsible for its antiplasmodial potency. Artemisinin and derivatives have shown also promising antineoplastic, antiviral, and immunomodulatory activities in pre-clinical studies [358].

Hence, this sesquiterpene endoperoxide has been a challenging target for syntheses studies. With respect to obtaining its enantiomers for biological studies the group of A. Giannis approached the generation of (−)-artemisinin (**627**) and its enantiomer by undertaking an intramolecular Diels–Alder reaction of the triene **628** (derived from chiral pool member (+)-(*S*)-citronellene) in toluene at 190°C for 30 h to yield mainly the cyclization product **629** (Fig. 123) [359]. The latter could then be elaborated into the targeted (−)-artemisinin (**627**).

## 7.40 Delitschiapyrone A

In a search for bioactive small-molecule natural products from plant-associated fungi, *Delitschia* sp. FL1581 (family Delitschiaceae), isolated from the fallen leaves of the saw palmetto *Serenoa repens* (Bartram) J.K. Small of the Arecaceae family (Plate 75), collected in south-central Florida, was investigated. Among other compounds, delitschiapyrone A (**630**) was isolated, which possesses an unprecedented pentacyclic ring system [360]. *Delitschia* is best known for occurring on decaying wood and fallen leaves [361]. Biosynthetically, the unique 6/6/5/7/6 pentacyclic core of **630** was hypothesized to be formed by an intermolecular Diels–Alder-type addition of the precursors derived from (1′*R*)-2′,3′-dihydropyrenocine C and 6-ethyl-2,7-dimethoxyjuglone, found to co-occur with **630** in this fungus. This compound was evaluated for cytotoxicity against a small panel of human tumor cell lines (MCF-7, H460, HepG2, and U2OS), and showed cytotoxic activity for all of these [360].

**Plate 75** *Serenoa repens*, which hosts the fungus *Delitschia* sp. FL1581 on its decaying leaves. Photograph courtesy CoolGuySlate, Creative Commons Attribution-Share Alike 4.0 International

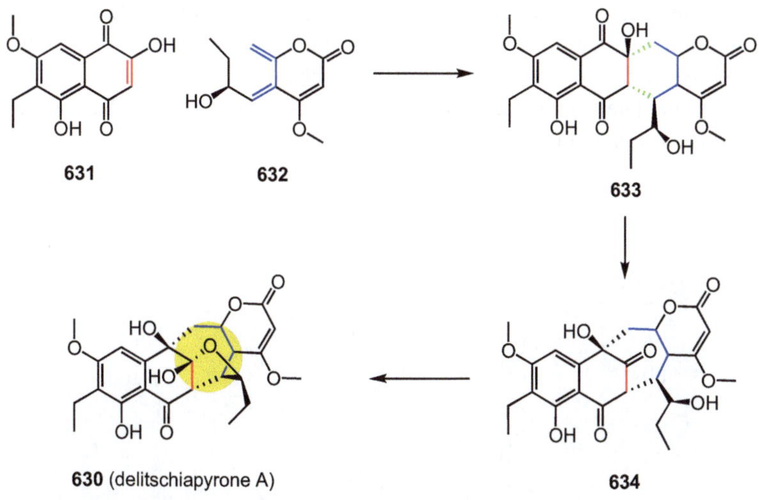


**Fig. 124** The exclusively regioselective and diastereoselective reaction cascade consisting of the Diels–Alder reaction yielding **633**, an α-ketol rearrangement into **634**, and a cyclic hemiacetalization to provide delitschiapyrone A (**630**)

Inspired by the biogenesis hypothesis mentioned above, the synthesis of delitschiapyrone A (**630**) was addressed (Fig. 124) [362]. The key step of this procedure was an exclusively regioselective and diastereoselective reaction cascade consisting of the Diels–Alder reaction to yield **633**, an α-ketol rearrangement into **634**, and a cyclic hemiacetalization, all achieved by simply stirring a heterogeneous mixture of the two Diels–Alder substrates **631** (one equivalent) and **632** (four equivalents, derived by dehydration from (*R*)-5-(1-hydroxybutyl)-4-methoxy-2*H*-pyran-2-one), the putative biosynthesis intermediates, in water at 35°C for 2.5 days. This furnished directly the pentacyclic delitschiapyrone A (**630**), in 75% yield.

## *7.41 (+)-Sarcophytin*

*Sarcophyton* soft corals have been the subject of extensive research due to the variety of cembranoid diterpenes produced by these organisms, which have challenged the synthesis community for decades [363]. In 1998, A. S. R. Anjaneyulu and co-workers reported the isolation of (+)-sarcophytin (**635**) from *Sarcophyton elegans* Moser, family Sarcophytidae (Plate 76), collected from the Indian Ocean at the Andaman and Nicobar islands [364]. The primary biological function of these compounds is believed to be a chemical defense against predators [365].

E. M. Carreira ventured into a synthesis of (+)-sarcophytin (**635**) in 2018 (the last example for this year), relying on an intramolecular Diels–Alder reaction (Fig. 125) [366]. In this key step, the ester **636** was heated in a 0.05 *M* dimethylsulfoxide solution at 100°C to yield the cyclization product **637** in 76% yield. In the latter part of the synthesis, a reaction cascade involving lactone opening, alcohol oxidation, and ketone epimerization then provided the targeted (+)-sarcophytin (**635**). This synthesis also led to the reassignment of its absolute configuration, because the synthesis of the Diels–Alder substrate **636** was based on (*S*)-citronellal and (*R*)- and (*S*)-carvone from the chiral pool.

**Plate 76** The yellow Fiji leather soft coral *Sarcophyton elegans*. Photograph courtesy P. Asman and J. Lenoble, Creative Commons Attribution 2.0 Generic

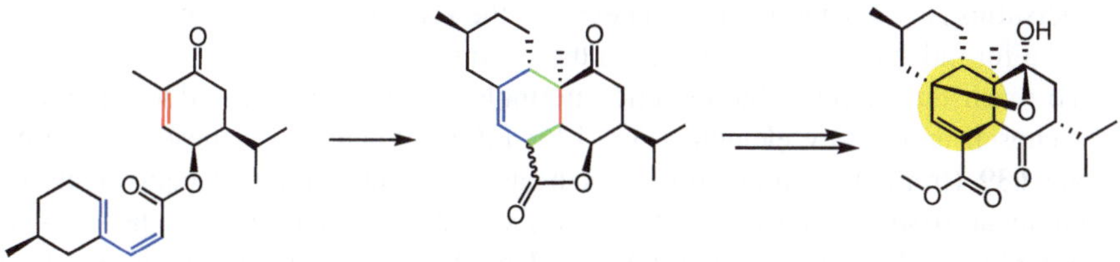


**Fig. 125** The intramolecular Diels–Alder reaction key step in the synthesis of (+)-sarcophytin (**635**)

## *7.42 Core Skeleton of Xenovulene A*

Xenovulene A (**638**), an oxygenated humulene-containing meroterpene was isolated from the environmentally widespread saprotroph, *Acremonium strictum* (W. Gams) Sumerb., family Sarcocladiaceae (Plate 77), found in the soil, rotting mushrooms, and plant debris, and is also a pathogen in immunosuppressed persons and an agent of hyalohyphomycosis [367]. In addition, it was found to be an inhibitor of benzodiazepine binding to the human γ-aminobutyric acid receptor, and antidepressant effects were reported [368]. The compound has a 5,6,11-tricyclic core structure, which is also present in the related *Stereum hirsutum*-derived compounds

**Plate 77** *Acremonium strictum*. Photograph H. F. (author)

sterhersutins A (**643**) and B (**644**). Therefore, the synthesis of this core structure has posed a formidable problem for the synthesis community.

To reach this target, a biomimetic intermolecular hetero-Diels–Alder approach was chosen as the key element (Fig. 126) [369]. When the D-ribose-originating enone **639** (from the chiral pool) was heated together with five equivalents of α-humulene (**640**) in toluene at 150°C in a sealed tube for up to five days, about 50% yields of the diastereomers **641a** and **641b** were obtained, which could be

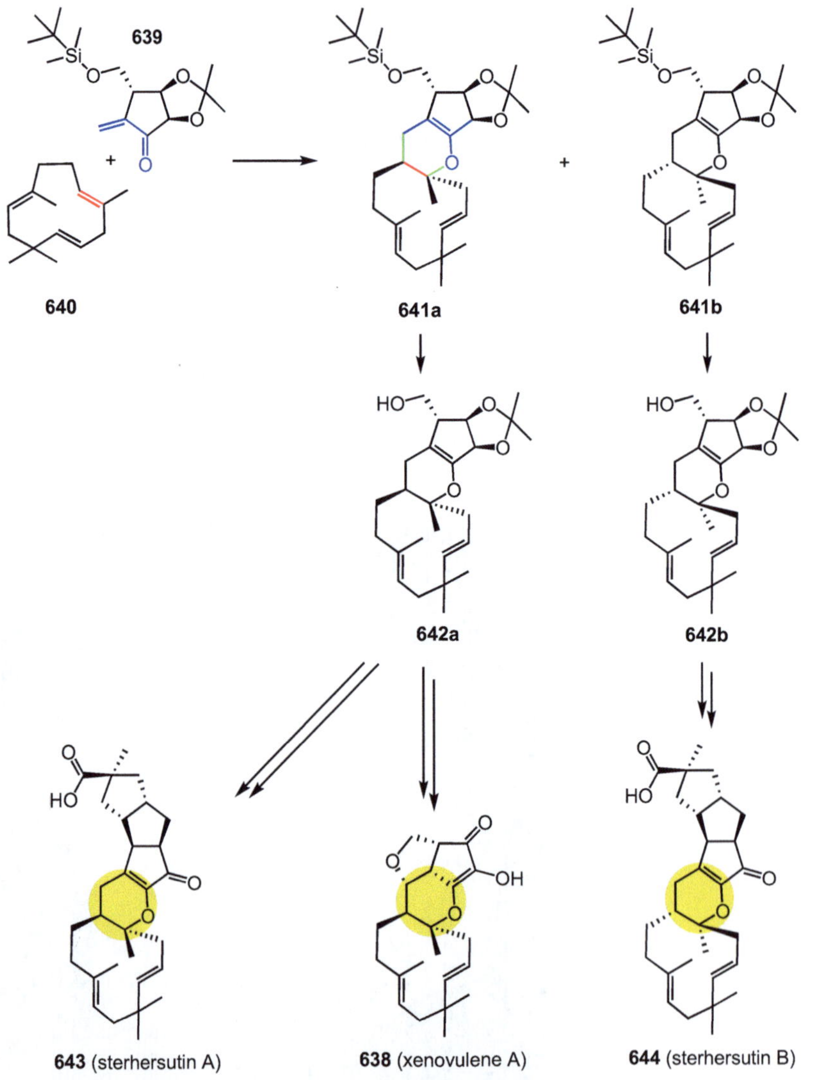


**Fig. 126** The biomimetic intermolecular hetero-Diels–Alder key step leading to the core skeletons **642a** and **642b** to synthesize xenovulene A (**638**) and sterhersutins A (**634**) and B (**644**)

separated easily. From the **641a** and **641b** reaction with two equivalents of tetrabutylammonium fluoride in tetrahydrofuran for one day at room temperature, this led to the desired skeletons **642a** and **642b**, templates that could be derivatized into the above-mentioned natural products, xenovulene A (**638**) and sterhersutins A (**643**) and B (**644**).

## *7.43 Inulavosin*

Racemic inulavosin (**645**) was isolated in 1995 from the roots of the veined-leaf Inula, *Inula nervosa* Wall. ex DC., syn. *Duhaldea nervosa* (Wall. ex DC.) Anderb. (family Asteraceae) (Plate 78) [370]. This racemic natural product displays a range of biological activities, such as antibacterial, piscicidal, and antifungal properties [371].

**Plate 78** *Inula nervosa.* Photograph courtesy California Botanic Garden Herbarium, CC BY-NC-SA

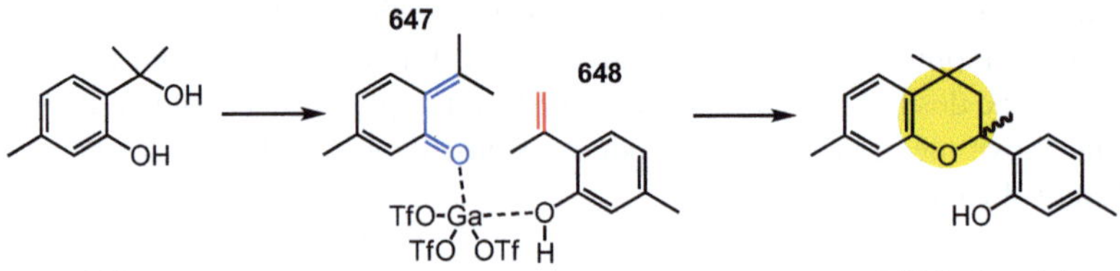


**Fig. 127** The gallium(III) trifluoromethanesulfonate-catalyzed hetero-Diels–Alder reaction key step to synthesize inulavosin (**645**)

To synthesize inulavosin (**645**), a novel gallium(III) trifluoromethanesulfonate-catalyzed hetero-Diels–Alder reaction was devised [372]. When optimized, the thymol derivative **646** was dissolved in dichloromethane and treated for 2 h at room temperature with the catalyst yielding 88% of the target **645** (Fig. 127). This dimerization was based on the catalyzed development of a complex equilibrium between the quinone methide **647** and the alkene **648**, which can react in the hetero-Diels–Alder reaction.

## *7.44 (−)-Daphenylline*

Daphenylline (**649**), an alkaloid isolated in 2009 from the fruits of *D. longeracemosum* Rosenth., family Daphniphyllaceae (Plate 79), by X.-J. Hao and co-workers [373], was the first member to be obtained of the numerous *Daphniphyllum* alkaloids containing a benzene ring in the hexacyclic core structure. The *Daphniphyllum* alkaloids display a range of biological activities, such as antitumor and antioxidant properties, as well as the promotion of nerve growth factors and vasorelaxation. A main challenge for the synthesis community has been their highly complex structures.

This challenge was met by the group of F. G. Qiu, who reported a concise, 19-step synthesis of (−)-daphenylline (**649**), featuring a stereoselective intramolecular Diels–Alder reaction to build a large portion of this hexacyclic framework (Fig. 128) [374]. Starting from the chiral pool member (*S*)-carvone, they assembled the tetracyclic intermediate **650** poised for cycloaddition. An intramolecular Diels–Alder reaction in xylene as the solvent at 200°C in the presence of a trace amount of 2,6-di-*tert*-butyl-4-methylphenol as antioxidant and then constructed the A/B/C/D-ring system **651** in one step, forging a bridged bicyclic structure with excellent control of five new stereocenters. This Diels–Alder step was crucial for rapidly generating the core connectivity and chirality of the *Daphniphyllum* alkaloid skeleton. The synthesis further utilized a Robinson annulation and Friedel–Crafts closure to complete the formation of daphenylline (**649**). The Diels–Alder methodology described herein exemplified stereocontrol via Lewis-acid catalysis in a complex

**Plate 79** *Daphniphyllum longeracemosum.* Photograph courtesy Daderot, public domain

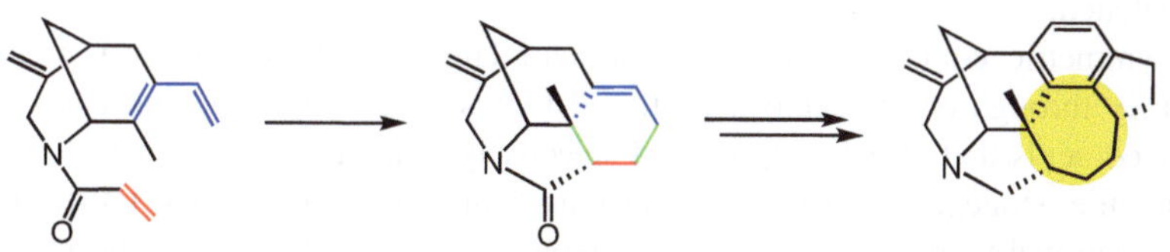


650 651 649 (daphenylline)

**Fig. 128** The intramolecular Diels–Alder reaction key step in the synthesis of (−)-daphenylline (**649**)

setting and demonstrated a cascade tactic for assembling a hallmark polycycle. Overall, this work highlighted a strategic use of an intramolecular Diels–Alder reaction to tackle the architectural complexity of *Daphniphyllum* alkaloids in a streamlined fashion.

At the end of this decade, it should be mentioned briefly that considerable progress has also been made in the theoretical understanding and consolidation of the Diels–Alder methodology. It is widely regarded that the Diels–Alder reaction belongs to the class of pericyclic transformation, which usually proceeds in a concerted fashion via a six-membered aromatic transition state [375]. However, in certain cases, stepwise reaction mechanisms involving diradicals or zwitterionic intermediates are also feasible [376]. The most prominent theoretical tools, which are still presently used, are Fukui's frontier MO theory [377] and the Woodward–Hoffmann rules [8, 42], which played a significant role in the use of cycloaddition reactions in synthesis [378]. Other approaches, such as the conceptual density functional theory and the Hard and Soft Acid and Base theories [379] have also contributed to the current understanding of pericyclic reactions. Even calculated activation barriers for the Diels–Alder reaction between substituted butadienes and ethenes with density functional theory at the M06–2X level using a basis set with TZ2P quality are possible currently [380].

# 8 Recent Applications

The Diels–Alder reaction, once celebrated simply as the most reliable way to knit two π-systems into a six-membered ring, has over the last two decades grown into a richly diversified platform for molecular construction. Advances in catalyst design, reaction engineering, and mechanistic understanding have transformed what was historically a thermal, *endo*-selective transformation into a family of finely tunable cycloadditions that can be coaxed to forge multiple rings, install densely packed stereochemistry, and unfold in concert with retro, cascade, or electrocyclic events. Modern variants now accommodate features from photo-generated transient dienes to fleeting cyclic allenes, and from water-accelerated, biomimetic unions to hetero-, aza- and inverse electron-demand manifolds that unite partners previously thought mismatched.

Asymmetric control, once the domain of chiral auxiliaries, is now achieved routinely through organocatalysis, tailored Lewis-acid complexes, and even chiral Brønsted acids that choreograph facial selectivity with sub-mole-percent loadings. Meanwhile, strategic intramolecular designs, which are transannular, tethered, and conformationally pre-organized, deliver stereochemical fidelity that rivals enzymatic precision. The compatibility of this reaction with photochemical activation, mild Lewis-acid mediation, and aqueous or solvent-free conditions further underscores its continuing relevance to sustainable synthesis.

The pages that follow illustrate how these methodological leaps have enabled concise access to some of the most intricate natural-product frameworks disclosed in the recent literature. Besides offering a chronological history, the compilation highlights recurring themes: chiral-pool fragments primed for intramolecular closure; catalytic, facially selective unions embedded in multi-step cascades; inverse electron demand variants that fuse heavily functionalized rings in a single stroke; and bio-inspired sequences that mirror putative biosynthetic logic. Collectively, they reveal a discipline in which mechanistic creativity and strategic minimalism go hand-in-hand, where a single pericyclic event can orchestrate half a molecule's complexity, set a suite of stereocenters, and launch the remainder of a synthesis on a downhill trajectory.

In surveying these recent accomplishments, the following Section aims not only to celebrate individual synthetic tours de force but also to trace the evolving mindset around the Diels–Alder reaction itself — from a dependable step within a linear plan to a design principle that dictates retrosynthetic vision. It is this shift, from using the cycloaddition to close a ring, to leveraging it as a springboard for architectural invention, that defines the current era of Diels–Alder methodology and sets the stage for the examples that follow.

## *8.1 (−)-Scabrolide*

Scabrolide A (**652**) is a flagship member of the natural product family of polycyclic furanobutenolide-derived norcembranoid diterpenoids. It was first isolated in 2002 from the soft coral *Sinularia scabra* Tixier-Durivault of the family Alcyoniidae (Plate 80), along with several other norcembranoids [381]. Since its initial isolation, **652** has been demonstrated to inhibit IL-6 and IL-12 production in vitro, suggesting its potential as an anti-inflammatory agent [382]. Structurally, scabrolide A (**652**) is characterized by a fused [5/6/7] carbocyclic framework featuring six stereogenic centers, of which five are situated contiguously about the compact and densely functionalized western region of the molecule.

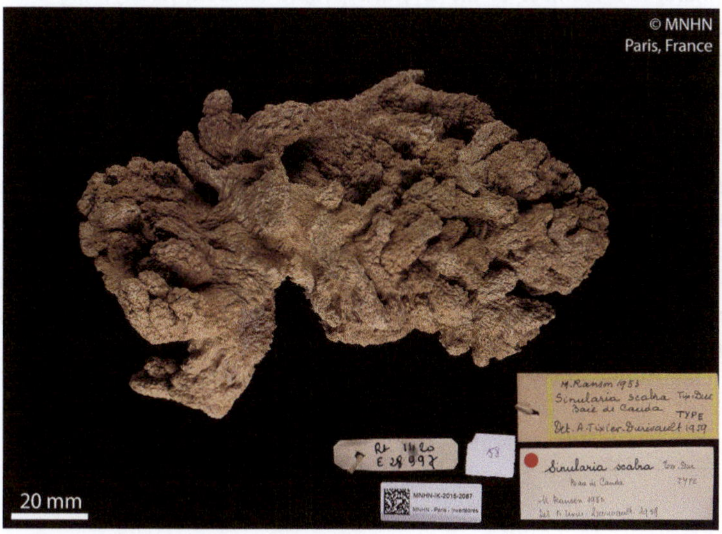


**Plate 80** *Sinularia scabra.* Photograph courtesy M. HENNION, MNHN - Paris, Creative Commons Attribution 4.0 International

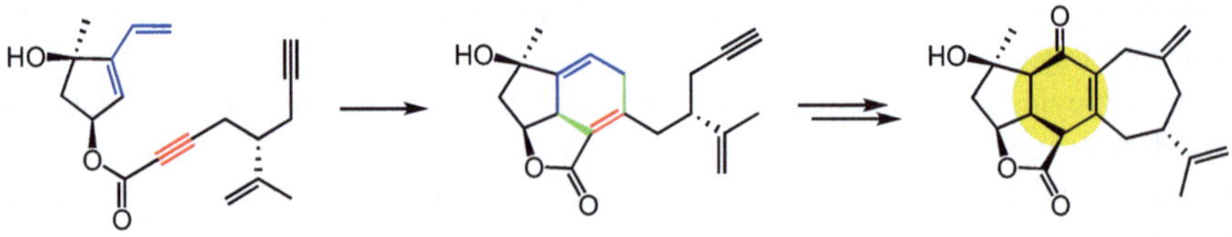


**Fig. 129** The intramolecular Diels–Alder reaction key step in the synthesis of (−)-scabrolide A (**652**)

An intramolecular Diels–Alder reaction was chosen as the key step in the construction of the western tricycle (Fig. 129) [383]. The left part of the cyclization substrate **653** was derived from the chiral pool member (*R*)-linalool, and the acetylenic side chain from (*R*)-carvone, both members of the chiral pool. The transformation of **653** proceeded smoothly under thermal conditions, by simply heating in a xylene mixture at 149°, which produced the tricycle **654** in high yield and as a single diastereomer. Further common transformations then led to the target compound, (−)-scabrolide A (**652**).

## 8.2 *Talatisamine*

Talatisamine (**655**) (Fig. 130), isolated from an *Aconitum* species [384], an herbaceous perennial plant of the buttercup family (Ranunculaceae) found in the Northern Hemisphere, is a member of the $C_{19}$-diterpenoid alkaloid family, and exhibits $K^+$ channel inhibitory and antiarrhythmic activities [385]. The formidable synthesis challenge that **655** presents is due to its highly oxidized and intricately fused hexacyclic 6/7/5/6/6/5-membered-ring structure with twelve contiguous stereocenters. One of the key steps in the total synthesis of **655** was an intramolecular Diels–Alder reaction (Fig. 130) [386].

The key substrate was first chosen by the group of M. Inoue with an unsubstituted cyclohexadienone, but the yields of the cyclization were found to be much too low due to intermolecular dimerization. Substitution with bromine, thus reducing this intermolecular process by steric hindrance had the desired result, and thus, when **656** was heated in refluxing toluene, it underwent an intramolecular Diels–Alder reaction to yield 61% **657** together with 28% of its diastereomer. Further elaboration of **657** finally yielded the highly complex natural product talatisamine (**655**).

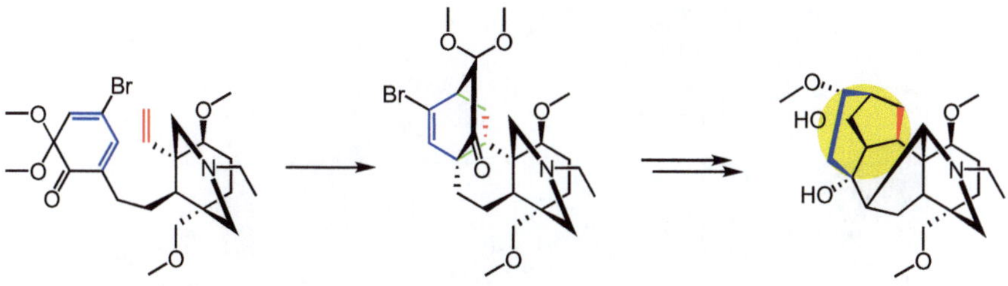


656 657 **655** (talatisamine)

**Fig. 130** The intramolecular Diels–Alder reaction key step in the synthesis of talatisamine (**655**)

## *8.3 Tetrodotoxin*

Tetrodotoxin (**658**) is one of the most well-investigated natural products [387]. This toxin was found originally in the puffer fish, *Takifugu rubripes* Temminck & Schlegel of the Tetraodontidae family (Plate 81) [388]. Later, it was shown to be identical to the toxin isolated from the eggs of the California newt, *Taricha torosa* Rathke, family Salamandridae (Plate 81) [389].

Tetrodotoxin (**658**) (Fig. 131) became highly notable since it blocks voltage-gated sodium channels [390]. The unique structure of **658**, composed of a tetracyclic skeleton with a tricyclic orthoester (a dioxaadamantane core) and a cyclic guanidine moiety, is another attractive feature of the molecule. The skeleton is highly oxidized and is substituted with five hydroxy groups. In addition, a variety of natural congeners have been isolated. Based on these structural relationships, biosynthesis pathways of **658** have been proposed [391]. Its structural complexity has made **658** a fascinating synthesis target. Although several total syntheses have been reported, an outstanding synthesis based on an intermolecular Diels–Alder reaction determining the stereochemistry of the primary ring system of **658** was described by the group of S. Yokoshima (Fig. 131) [392].

The dienophile **659** was prepared from the chiral pool member α-methyl-D-mannoside, and it reacted in an intermolecular Diels–Alder reaction in toluene at 40°C quantitatively with the siloxydiene **660** to give the hydroindene derivative **661**. Further elaboration of this compound finally provided the targeted toxin **658**.

**Plate 81** Left: puffer fish, *Takifugu rubripes*. Photograph courtesy Wayback Machine, Creative Commons Attribution-Share Alike 2.5 Generic. Right: California newt, *Taricha torosa*. Photograph courtesy C. Long, Creative Commons Attribution-Share Alike 3.0

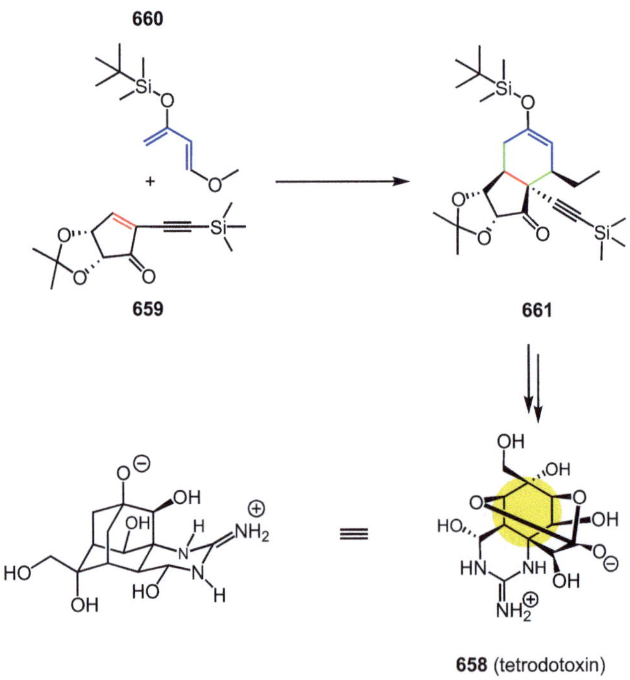


**Fig. 131** The intermolecular Diels–Alder reaction key step in the synthesis of tetrodotoxin (**658**)

## *8.4 (+)-Brevianamide A*

Brevianamide A (**662**) was isolated from the fungus *Penicillium brevicompactum* Dierckx of the Aspergillaceae family (Plate 82), and is one of the first known bicyclo [2.2.2]diazaoctane alkaloids [393]. It is interesting for agricultural purposes because it exhibits potent antifeedant activity against the larvae of the insect pests *Spodoptera frugiperda* J. E. Smith (fall armyworm) and the tobacco budworm *Chloridea virescens* Fabricius (syn. *Heliothis virescens*) (Plate 82), both from the family Noctuidae [394].

The critical intramolecular Diels–Alder reaction in the synthesis of (+)-brevianamide A (**662**) was a bioinspired part of a reaction cascade, in which the stereochemistry of the substrate **663** originated from the chiral pool member L-tryptophan methyl ester (Fig. 132) [395]. The reaction proceeded in water at ambient temperature for 30 min, resulting in a mixture of the two diastereomers (+)-brevianamide A (**662**) and (+)-brevianamide B (**664**) in a ratio of 93/7. Interestingly, this is also the ratio closely matching those of the two isolated natural products from *Penicillium brevicompactum* Dierckx.

**Plate 82** *Penicillium brevicompactum*. Photograph courtesy A. Rockefeller, Creative Commons Attribution-Share Alike 3.0 Unported (upper left). *Spodoptera frugiperda*. Photograph courtesy FW. Peairs, Creative Commons Attribution Alike 3.0 USA (lower left). *Heliothis virescens*. Photograph courtesy W. Cranshaw, Creative Commons Attribution Alike 3.0 USA (right)

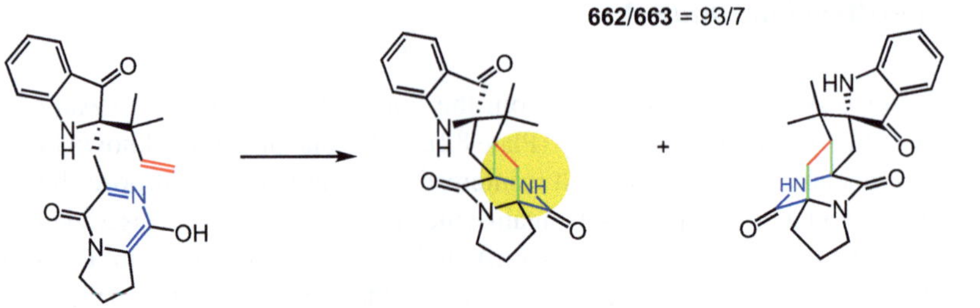


663 662 ((+)-brevianamide A) 664 ((+)-brevianamide B)

**Fig. 132** The intramolecular Diels–Alder reaction as key step in the synthesis of (+)-brevianamide A (**662**)

## *8.5 Canataxpropellane*

Taxane diterpenes [396] are a family of natural products exhibiting potent anticancer activity [397] that were originally isolated from slow-growing evergreen shrubs in the genus *Taxus*, commonly known as yews. In the 1990s, major synthetic efforts culminated in the first total syntheses of the most prominent anticancer drug, taxol (paclitaxel) (**127**) among which one involving use of the Diels–Alder reaction was discussed in Sect. 5.2. This drug turned out to be one of the top-selling anticancer drugs (with peak sales in 1999 of 1.5 billion USD) over the last three decades [398]. Subsequently, different *Taxus* species have been screened for their constituents and >500 taxanes have been isolated to date [396]. Among them, (−)-canataxpropellane (**665**) was isolated from *Taxus canadensis* Marsh from the Taxaceae family [399] (for the similar species *T. brevifolia*, see Plate 14), and represents a member of the complex taxane group. This highly oxygenated diterpene is one of the most intricate and complex natural products that has ever been isolated. As in the case of its analogue taxol (**127**), it suffers from extremely inefficient sourcing from its natural producing organisms. (−)-Canataxpropellane (**665**) comprises a heptacyclic [5,5,5,4,6,6,6] carbon framework (Fig. 133). It is densely functionalized and highly oxidized (five hydroxy, one ketone groups), containing only two methylene groups. Among the features distinguishing the structural complexity of this compound are the following: it is the only natural product harboring a [3.3.2] and [4.4.2] propellane, and simultaneously it contains 12 contiguous stereocenters including five quaternary centers, of which four reside in a cyclobutane ring, and its backbone consists exclusively of neopentylic motifs.

The challenge was met by T. Gaich and her group at the University of Konstanz in Germany with a total stereoselective synthesis of canataxpropellane (**665**) involving

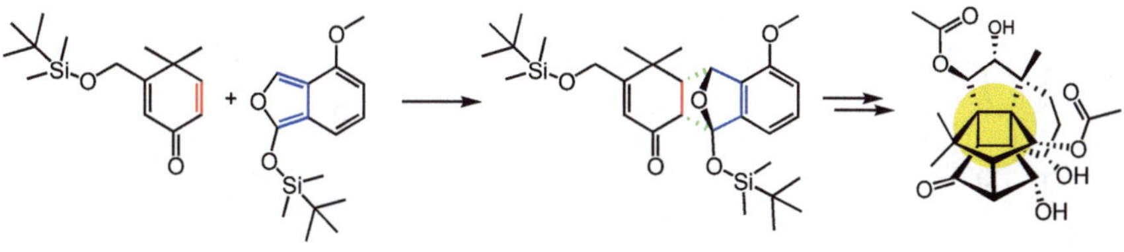

666 667 668 665 (canataxpropellane)

**Fig. 133** The intermolecular Diels–Alder reaction key step at the beginning of the synthesis of racemic canataxpropellane (**665**) and its levorotatory enantiomer

26 steps, with the key step of an initial intermolecular Diels–Alder reaction between the dienophile **666** and the diene **667** proceeding in tetrahydrofuran at room temperature, providing almost exclusively the *endo* cyclization product **668** [400]. The latter could then be forged into the racemic target compound canataxpropellane (**665**). For the enantioselective synthesis, a chiral isopropyl-chlorosilyl-α,α,α',α'-tetraphenyl-2,2-dimethyl-1,3-dioxolane-4,5-dimethanol catalyst was used in this key step.

## *8.6 (−)-PF-1018*

The unusual anionic lipid (−)-PF-1018 (**669**) (Fig. 134) was isolated from the fungal strain *Humicola* sp. 1018 (family Chaetomiaceae) in a screening procedure against the diamond back moth, *Plutella xylostella* L. from the Plutellidae family (Plate 83) [401], which is responsible for large agricultural losses worldwide [402].

The synthesis of (−)-PF-1018 (**669**) with a rare proline-derived tetramic acid linked to a rather complex tricyclic hydrocarbon core with six stereocenters and having two of them quaternary, was based by D. Trauner and his group on the elegant bioinspired combination of an 8π-electrocyclization with an intramolecular Diels–Alder reaction [403]. Thus, the pentaene **670** produced in situ by a Stille–Liebeskind cross-coupling reaction between an alkenyl stannane and an unsaturated ester, underwent in dimethylformamide solution at 125°C for 20 h the 8π-electrocyclization into **671** and its interception by an intramolecular Diels–Alder reaction to provide the tricycle **672** in 42% yield. The latter could then be elaborated into the target product (−)-PF-1018 (**669**), which was shown to exist in a thermal equilibrium of its inseparable (*Z*)- and (*E*)-diastereomers, occurring in a ratio of 3.5/1.

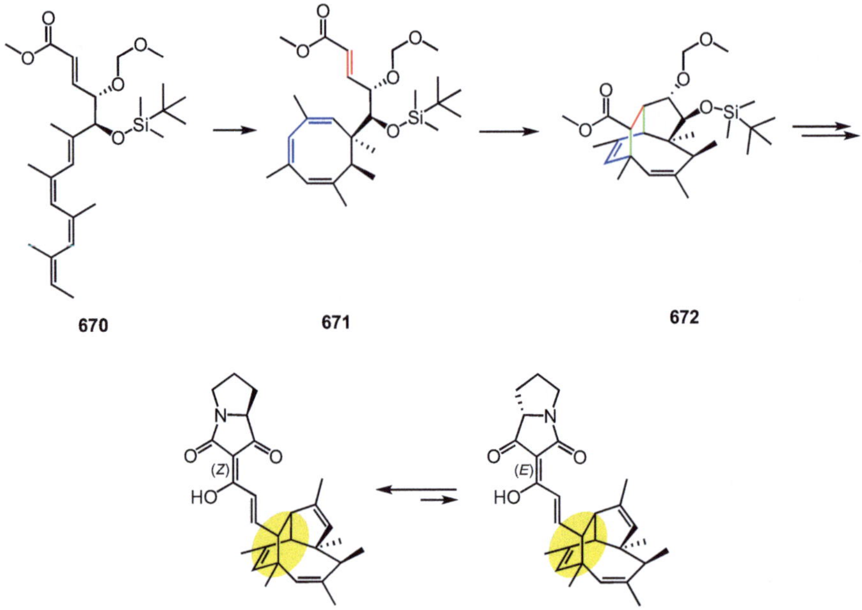


**Fig. 134** An 8π-electrocyclization with an intramolecular Diels–Alder reaction as key steps in the synthesis of (–)-PF-1018 (**669**)

**Plate 83** *Plutella xylostella*. Photograph courtesy O. Leillinger, Creative Commons Attribution-Share Alike 2.5 Generic

## *8.7 Marinoquinoline A*

Marinoquinoline A (**673**) (Fig. 135) was isolated from the aerobe mesophilic gliding bacterium *Ohtaekwangia kribbensis* Yoon together with other compounds of this

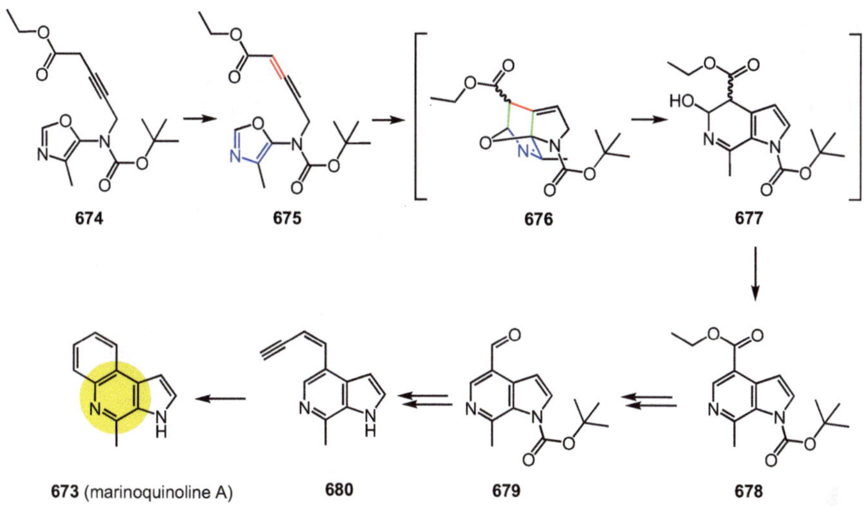


**Fig. 135** The intramolecular Diels–Alder oxazole cycloaddition reaction in the synthesis of marinoquinoline A (**673**)

type. It has a pyrroloquinoline structure and shows weak antibacterial and antifungal activities and moderate cytotoxicity against mammalian cancer cell lines and also inhibitory activity against *Plasmodium falciparum* K1 [404].

The key reaction for the synthesis of the azaindole core of this natural product was established as a new variant of the intramolecular Diels–Alder oxazole cycloaddition (Fig. 135) [405]. When the oxazole-propargylester **674** was heated in triethylamine at 60°C for 19 h a tandem alkyne–allene isomerization to the allene **675** took place, which was followed by an intramolecular Diels–Alder oxazole cycloaddition, and via the rearrangement intermediates **676** and **677**, afforded the azaindole **678** in 86% yield. A series of further transformations via **679** and **680** to build the additional benzene ring finally provided the targeted natural product, marinoquinoline A (**673**).

## *8.8 cis-Crotonin*

The 19-*nor*-clerodane diterpenoid *cis*-crotonin (**681**) was isolated from the shrub *Croton lucidus* L. of the Euphorbiaceae family (Plate 84) [406], and its structure was confirmed subsequently by further studies [407, 408]. The crotonins display a broad range of biological activities, but such data for **681** seem to be sparse.

As shown in Fig. 136, the spectacular enantioselective total synthesis of the *cis*-decalin derivative *cis*-crotonin (**681**) started with the stereodivergent inverse electron demand Diels–Alder reaction of 4-methyl-substituted 2-pyrone **682** and (*R*)-configured enantiopure (>99% *ee*) cyclohexadiene derivative **683** [409]. In the presence of

**Plate 84** *Croton lucidus*. Photograph public domain, CC0 1.0 Universal; Herbarium sheet NEON

10 mol% ytterbium tris-trifluoroacetate and 12 mol% (*S*)-configured catalyst **684**, the Diels–Alder reaction proceeded smoothly with good diastereoselectivity, providing after in situ removal of the silyl group a pair of diastereoisomers with the desired **685** in excess. Remarkably, this diastereoisomer had five contiguous chiral centers, of which three shared the same stereochemistry as *cis*-crotonin (**681**). The retro Diels–Alder extrusion of carbon dioxide from **685** (chlorobenzene, reflux), followed by the subsequent protection with ethylene glycol, led to the dienyl ester **686** in 95% yield. This intermediate was then elaborated into the desired *cis*-crotonin (**681**) by conventional steps. All in all, this is an exciting example of the developing art in harnessing the Diels–Alder reaction.

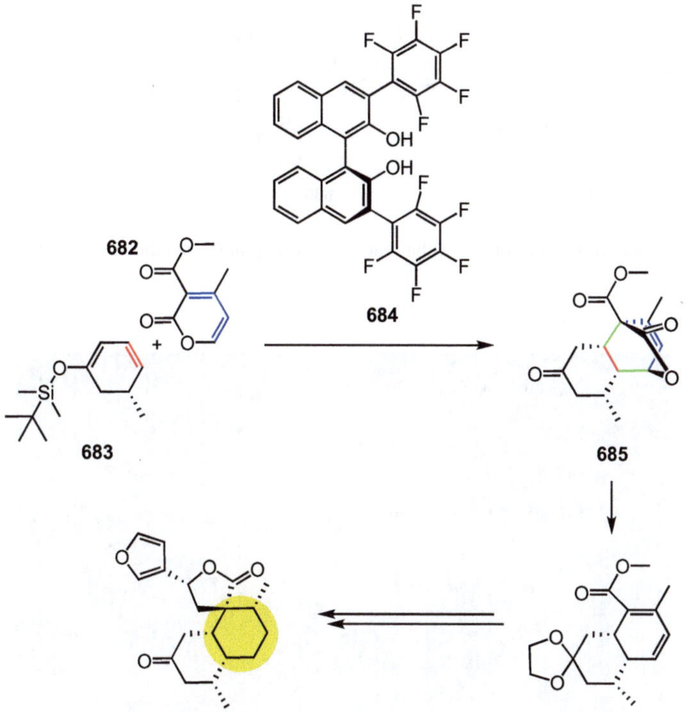


**Fig. 136** The stereodivergent inverse electron demand Diels–Alder pyrone reaction in the synthesis of *cis*-crotonin (**681**)

## *8.9 (±)-Clivonine*

The lycorenine-type Amaryllidaceae alkaloid clivonine (**687**) (Fig. 137) was isolated in 1965 from *Clivia miniata* (Lindl.) Verschaff. of the Amaryllidaceae family (Plate 85), and its structure finally fully resolved in 1971 [410, 411]. This compound has served as a synthesis methodology benchmark due to its intriguing homolycorine-type alkaloid structure.

Thus, an intermolecular Diels–Alder cycloaddition of 3,5-dibromo-2-pyrone (**478**) (which was encountered already in the synthesis of another Amaryllidaceae alkaloid, galanthamine (**477**), in Sect. 7.5) with the styrene pinacol boronate dienophile **688**, allowed to set up the required stereochemistry to be installed (Fig. 137) [412]. These two components reacted in toluene solution under reflux for three days to provide a 70% yield of the *endo*-addition product **689**. Its further elaboration by different procedures afforded in the target molecule (±)-clivonine (**687**).

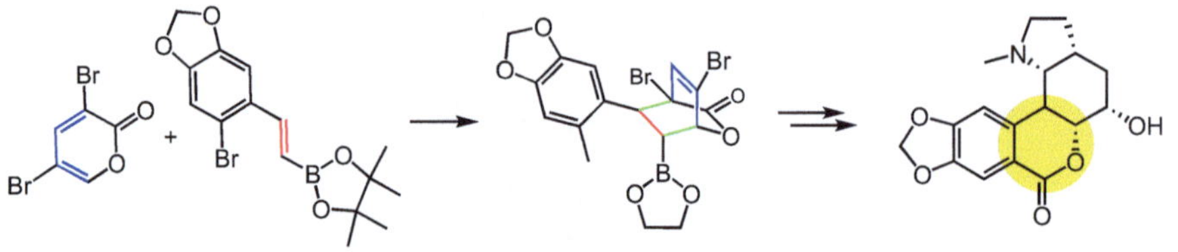


**Fig. 137** The intermolecular Diels–Alder reaction key step in the synthesis of (±)-clivonine (**687**)

**Plate 85** *Clivia miniata*. Photograph courtesy JMK, Creative Commons Attribution-Share Alike 3.0 Unported

## *8.10 (±)-Indanostatin*

Indanostatin (**690**) (Fig. 138) was isolated from the soil actinomycete *Streptomyces* sp. RAI-20 collected near Nagoya in Japan [413], and is one of the most potent known members of the small glutamate-toxicity blockers, with nanomolecular activity in the C6 model to mimic excitonic neuronal death, and thereby displays a neuroprotective promise [414].

Indanostatin (**690**) was selected to serve as an example of the versatility of a variation of Diels–Alder syntheses, namely, the use of a bisketene equivalent as diene component [415]. In this investigation, 2,5-bis(*tert*-butyldimethylsilyloxy) furans were established as vicinal bisketene equivalents for application as dienes in the Diels–Alder reaction. Cycloaddition with olefinic dienophiles, under exceptionally mild conditions, enabled convergent access to highly substituted *p*-

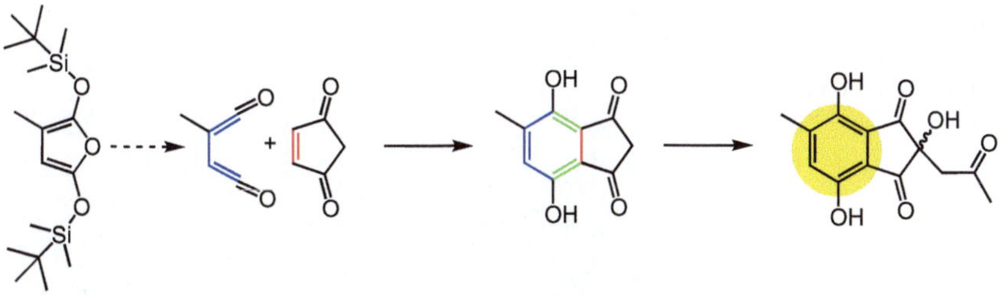


**Fig. 138** The one-pot Diels–Alder/ring-opening/tautomerization sequence in the synthesis of (±)-indanostatin (**690**)

hydroquinones in unprotected form via a one-pot Diels–Alder/ring-opening/tautomerization sequence. The synthesis of *p*-benzoquinones from acetylenic dienophiles, including benzynes, was also demonstrated, and 2,5-bis(*tert*-butyldimethylsilyloxy)pyrroles were established as competent dienes for the synthesis of *p*-iminoquinones. The application of this methodology then enabled access to the neuroprotective agent (±)-indanostatin (**690**) in a gram-scale.

As shown in Figs. 138, 3-methyl-2,5-bis(*tert*-butyldimethylsilyloxy)furan (**691**) prepared in one step from methylsuccinic anhydride, reacted smoothly in methanol at ambient temperature with 1,3-cyclopentenedione (**692**) to give 93% yield of the hydroquinone intermediate **693**, which underwent a subsequent one-pot oxidation/alkylation to give (±)-indanostatin (**690**) in 54% yield, without the need for chromatographic purification. It should be mentioned that this furan served as a bisketene equivalent, as indicated in Fig. 138. It is relevant to note that in 2022 an extensive review on the applications of the Diels–Alder reaction in natural product synthesis for the period of 2017–2020 was published [416].

## 8.11 Cephanolides A–D

The norditerpenoids cephanolides A–D (**694–697**) were isolated in 2017 from the coniferous shrub *Cephalotaxus sinensis* (Rehder & E.H. Wilson) H.L. Li from the Cephalotaxaceae family (syn. *C. harringtonii* (Knight ex J. Forbes) K. Koch)) (Plate 86), besides several other compounds of this type [417]. They are not only interesting for their bioactivity but are also notable for their synthesis aspects.

An intramolecular inverse electron demand Diels–Alder cycloaddition was chosen as the key reaction to forge the framework of these compounds (Fig. 139) [418]. The enolether Diels–Alder substrate **698** was produced in situ from the corresponding indanone, and it reacted smoothly in dichloromethane on warming from 0°C to room temperature to form exclusively the *endo* cycloadduct **699**. The structure of this product was established unequivocally by its X-ray crystallographic

**Plate 86** *Cephalotaxus sinensis*. Photograph courtesy Daderot, public domain

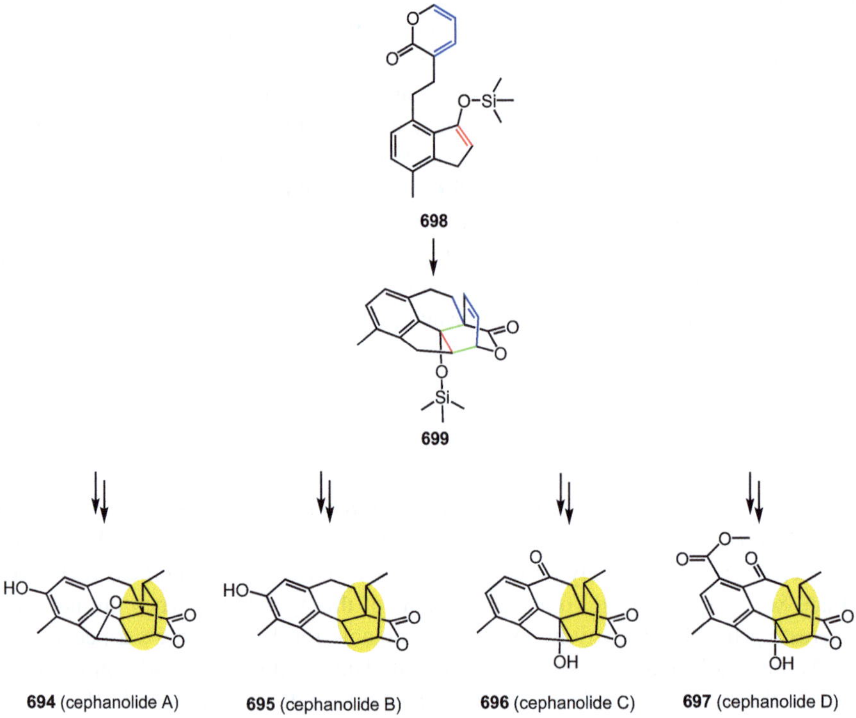


**Fig. 139** An intramolecular inverse electron demand Diels–Alder cycloaddition as key step in the synthesis of the cephanolides A–D (**694–697**)

analysis. The cycloadduct **699** could then be transformed by conventional procedures into the four targeted natural products, cephanolides A–D (**694**–**697**).

## 8.12 *Perovskatone D and Perovskone*

Perovskatone D (**700**), an icetexane diterpenoid, and perovskone (**701**), a triterpenoid, were isolated from the Russian sage *Perovskia atriplicifolia* Benth. from the Lamiaceae family (syn. *Salvia yangii* B.T. Drew) (Plate 87) and *P. abrotanoides* Kar., all from the *Salvia* genus [419, 420]. Like many constituents from *Salvia* medicinal species they display various biological properties [421]. Their unique structural parameters have made them attractive targets for synthesis.

Two Diels–Alder steps were envisaged on the synthesis path to perovskone (**701**) via perovskatone D (**700**), among other subroutes to different compounds of this type, by the group of S. Gao in 2021 (Fig. 140) [422].

First, an intermolecular asymmetric photoenolization/Diels–Alder reaction was developed to construct a tricyclic ring bearing three contiguous quaternary centers, which was used to build the core icetexane skeleton. Thus, the synthesis commenced with the photoenolization ($\lambda = 366$ nm) of the substituted benzaldehyde **703**. Upon addition of titanium tetraisopropoxide and the TADDOL ligand in dichloroethane, the titanium complex **704** of **702** and **703** formed, which underwent a Diels–Alder reaction and took 1.5 h to yield the *endo* cyclization product **705** in 94% yield. The latter could then be elaborated conventionally into perovskatone D (**700**).

Second, a bioinspired intermolecular Diels–Alder reaction of the thus obtained perovskatone D (**700**) with (*E*)-α-ocimene (**706**) in the presence of catalytic traces of europium tris(6,6,7,7,8,8,8-heptafluoro-2,2-dimethyl-3,5-octanedione) in toluene at 60°C in a sealed tube led the stereospecifically cyclized product **707**, which could then be converted into perovskone (**701**).

**Plate 87** *Perovskia atriplicifolia* (syn. *Salvia yangii*). Photograph courtesy A. Barra, Creative Commons Attribution-Share Alike 4.0 International

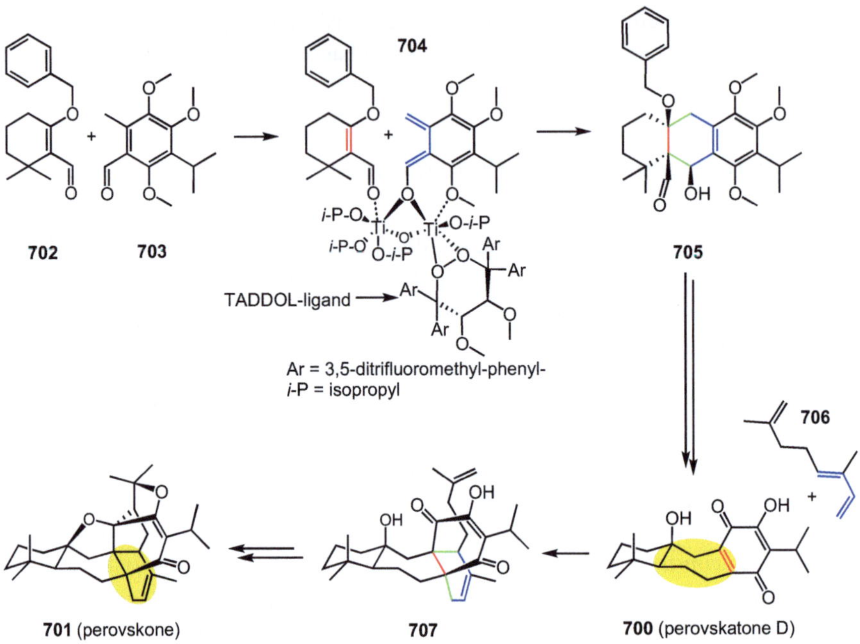


**Fig. 140** The intermolecular asymmetric photoenolization/Diels–Alder reaction key step in the synthesis of perovskatone D (**700**) and the bioinspired intermolecular Diels–Alder reaction to synthesize perovskone (**701**) from the latter

## *8.13 (−)-Strictosidine*

(−)-Strictosidine (**708**) was isolated by countercurrent distribution from the European bluestar, *Rhazya orientalis* Decne. (syn. *Amsonia orientalis*) of the Apocynaceae family (Plate 88) [423]. It turned out that **708** is the last common biosynthesis precursor to all monoterpene indole alkaloids, which today, amount to more than 3000 members with diverse structures and bioactivities. This compound contains nine stereocenters, a highly congested dihydropyran ring, a glucosyl moiety, and a bis-acetal linkage.

The challenge to synthesize (−)-strictosidine (**708**) was met by the group of Y. Tang, K. N. Houk, and N. K. Garg in 2021 [424]. The key step was a facially selective asymmetric inverse electron demand hetero-Diels–Alder reaction between the glucosyl-modified alkene **709** and the enal **710**, to set the precise stereotriad (Fig. 141).

Heating **709** and **710** in hexafluoroisopropanol at 50°C for 16 h gave rise to the desired cycloadduct **711** and some of its epimer. Thereby, this reaction provided mainly one of the 16 possible isomers dictated by the stereochemistry of the sugar moiety, which was also rationalized by DFT calculations. Product **711** was then transformed conventionally into the targeted (−)-strictosidine (**708**).

**Plate 88** *Rhazya orientalis* (syn. *Amsonia orientalis*). Photograph courtesy Linda, Creative Commons Attribution 2.0 Generic

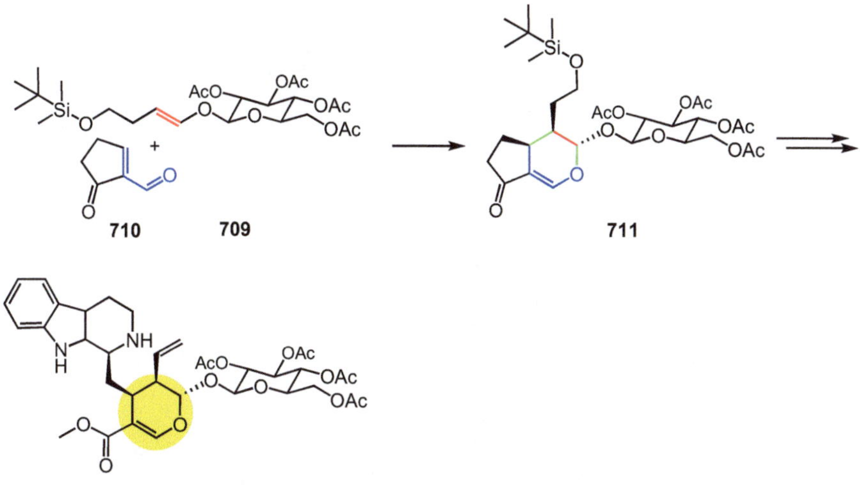


**Fig. 141** The facially selective asymmetric inverse electron demand hetero-Diels–Alder reaction key step in the enantioselective synthesis of (–)-strictosidine (**708**)

## *8.14 Spirooliganin*

Spirooliganin (**712**) was isolated from the stem bark of the toxic shrub *Illicium oligandrum* E.D. Merrill & W.Y. Chun, family Schisandraceae (Plate 89), which is distributed mainly in southern China and used in folk medicine for treating rheumatoid arthritis. Compound **712** features an unprecedented linear 6/6/6/7 tetracycle with six chiral centers and a spiro-pentacycle attached to the first ring. Biologically, this molecule exhibits potent activity against coxsackievirus B3 [425].

**Plate 89** *Illicium oligandrum*. Photograph courtesy Berlin-Dahlem Botanical Garden and Botanical Museum, Germany — 

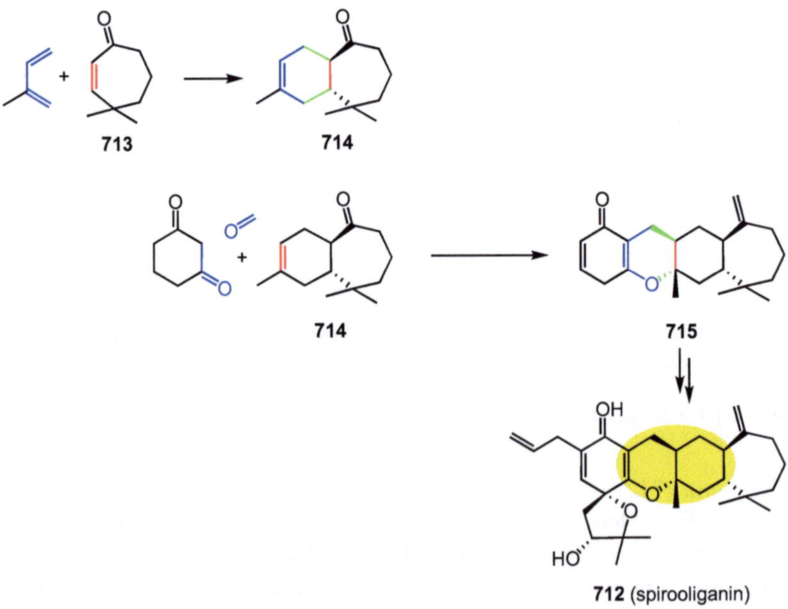


**Fig. 142** The Diels–Alder cycloaddition and three-component hetero-Diels–Alder cycloaddition key steps in the synthesis of spirooligman (**712**)

Diels–Alder cycloaddition of cycloheptenone **713** with isoprene, which was used also as the solvent, under irradiation ($\lambda = 365$ nm), generated a pair of *trans*-fused racemic regioisomers in a ratio of 47:53 that was separated (Fig. 142).

These compounds served as key intermediates in the synthesis of the stereoisomers and regioisomers of spirooliganin. A three-component hetero-Diels–Alder cycloaddition of **714**, formaldehyde, and 1,3-cyclohexanedione was used to construct the tetracyclic core. Two regioisomeric pairs of enantiomers were obtained. To ensure the complete conversion of **714**, 1,3-cyclohexanedione was strictly controlled by being added in portions after the reaction mixture was heated to 100°C. The hetero-Diels–Alder reaction afforded a *cis*-fused 6–6 ring system, resulting in the assembly of the key tetracyclic core **715**, for which the stereochemistry of which was confirmed by X-ray crystal structural analysis. From **715** to the target compound spirooliganin (**712**), conventional steps were pursued. The ensemble of all 32 stereoisomers could be synthesized this manner and then screened for their biological properties.

## 8.15 (+)-Pepluanol A

In 2016, the diterpene pepluanol A (**716**) was isolated among other related compounds from the invasive milkweed, *Euphorbia peplus* L., from the Euphorbiaceae family (Plate 90). Since the voltage-gated potassium channel Kv1.3 is a validated target for the treatment of autoimmune diseases, such as multiple sclerosis, type-1 diabetes, asthma, and psoriasis, Kv1.3 was studied in terms of its response to the new compounds, and they were found to be active in the μ*M* range [426]. Due to its unprecedented 3,5,6,7-tetracyclic angular carbon framework with a moderate oxidation pattern and a total of seven contiguous stereocenters, of which one represents a quaternary stereocenter, this molecule presented a formidable challenge for the synthesis community.

The asymmetric synthesis of (+)-pepluanol A (**716**), achieved by the group of T. Gaich in 2021, and was based mainly on a Curtin–Hammett (kinetics)-driven stereoconvergent intramolecular Diels–Alder reaction (Fig. 143) [427]. Thus, the (+)-3-carene-derived (chiral pool) compound **717** cyclized in *o*-xylene solution at 150°C in the presence of catalytic amounts of a basic catalyst for enolization, 1,8-diazabicyclo[5.4.0]undec-7-ene, into **718** with 88% yield on a gram scale. Further elaboration of this material led to the target compound (+)-pepluanol A (**716**).

**Plate 90** *Euphorbia peplus*. Photograph courtesy Rasbak, Creative Commons Attribution-Share Alike 3.0 Unported

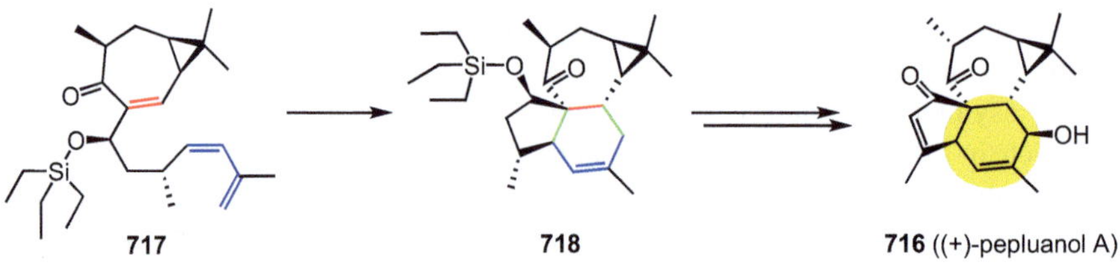


**717** **718** **716** ((+)-pepluanol A)

**Fig. 143** Curtin–Hammett (kinetics)-driven stereoconvergent intramolecular Diels–Alder reaction key step in the synthesis of (+)-pepluanol A (**716**)

## 8.16 *Nahuoic Acid A*

Nahuoic acid A (**719**) (Fig. 144) was isolated in 2013 from cultures of a *Streptomyces* sp. isolate RJA2928, collected in a marine sediment in Papua New Guinea. It was found to be the first known selective cofactor-competitive inhibitor of the epigenetic enzyme lysine methyl transferase SETD8, which is implicated, along with other

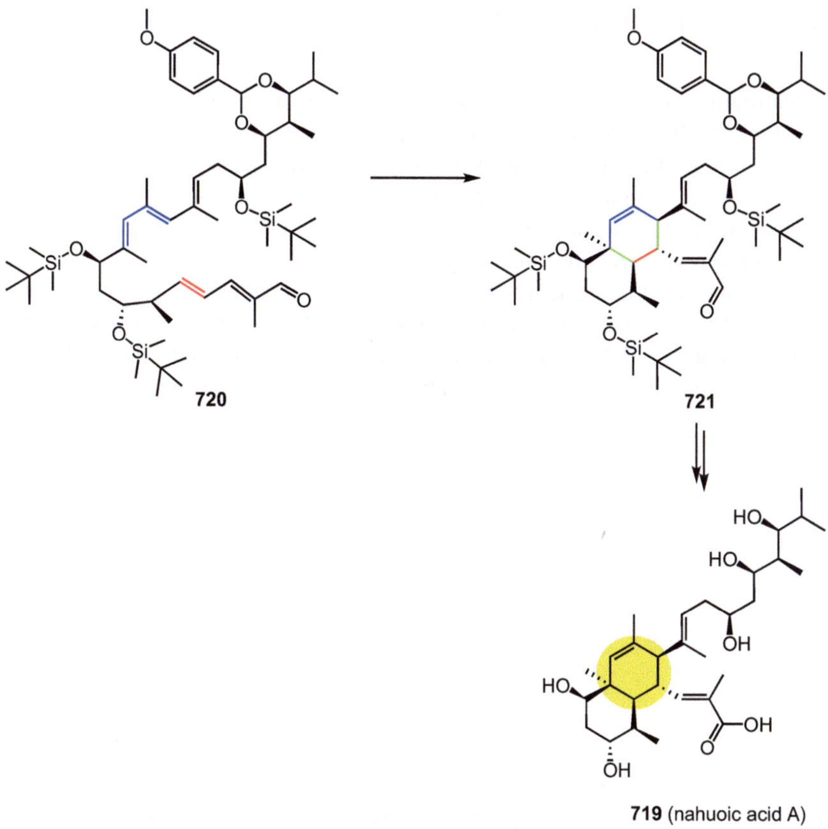

**Fig. 144** The biogenesis-inspired intramolecular Diels–Alder reaction key step in the synthesis of nahuoic acid A (**719**)

enzymes, in chemical modifications of histones, a group of highly basic proteins that pack the DNA into the nucleosomes in eukaryotic cells. Its structure was shown to contain an octahydronaphthalene core with seven contiguous stereocenters and unsaturated substituents of (*E*)-configuration at C-4 and C-13, namely, a 2-methylpropenoic acid, and a 2,4,10-trimethylundec-9-ene-3,5,7-triol side chain with four additional stereogenic centers (Fig. 144) [428].

The key step in the synthesis of nahuoic acid A (**719**) (Fig. 144) was a biogenesis-inspired intramolecular Diels–Alder reaction of the enantiopure substrate **720** [429]. Upon heating **720** in *o*-xylene at 200°C for 18 h, the reaction produced among two other diastereomers the octahydronaphthalene **721** stemming from a *Si-exo* attack. This product could then be transformed conventionally into the desired nahuoic acid A (**719**).

## 8.17 Aglacins A, B, and E

The natural lignans aglacins A, B, and E (**722–724**) were isolated by the group of P. Proksch in 2001 from the methanol extract of the stem bark of the evergreen tree *Aglaia cordata* Hiern of the Meliaceae family (Plate 91), which was collected in Indonesia when searching for insecticidal compounds [430, 431]. Their structures are characterized by an aryltetralin moiety with two highly oxygenated benzene rings, and a fused tetrahydrofuran ring.

To synthesize the aglacins, an intermolecular asymmetric photoenolization/Diels–Alder reaction, as encountered in Sect. 8.12 as a key step in the synthesis of perovskatone D (**700**), using the same chiral ligand system seen in Fig. 140, was established (Fig. 145) [432]. Thus, aldehyde **725** was irradiated at 366 nm with

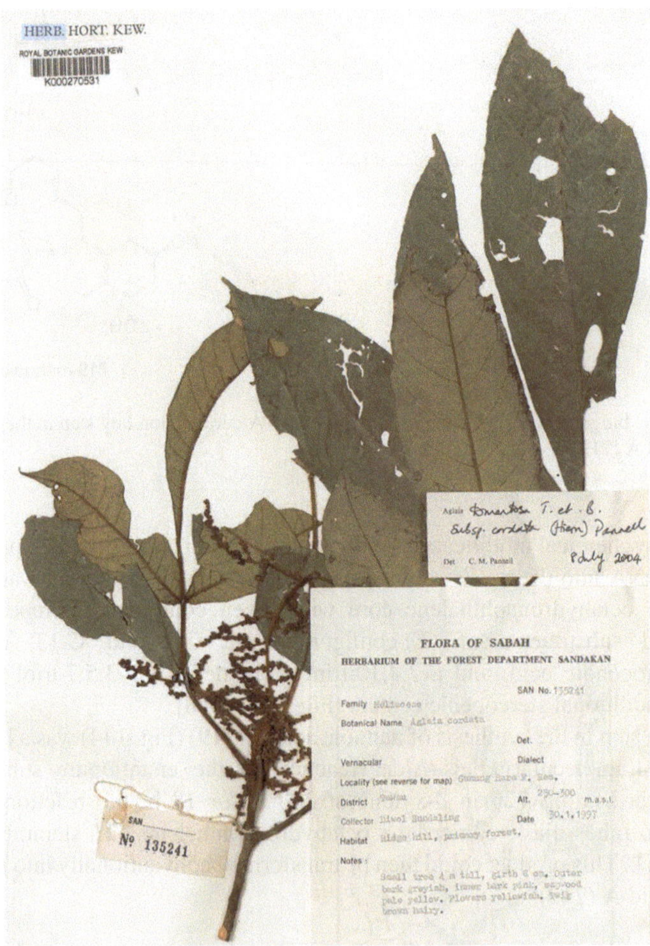


**Plate 91** *Aglaia cordata*. Photograph courtesy RBC, Kew, Creative Commons BY 4.0

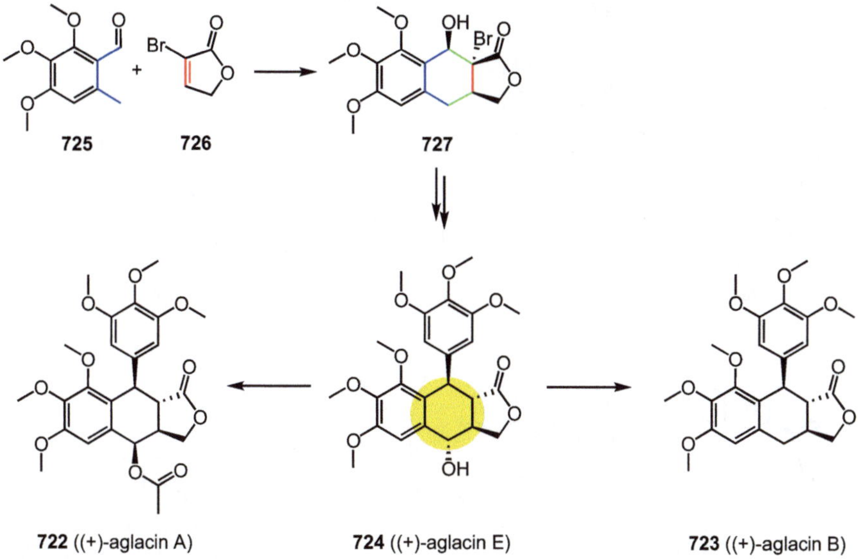


**Fig. 145** An intermolecular asymmetric photoenolization/Diels–Alder reaction key step in the synthesis of the aglacins A, B, and E (**722**–**724**)

α-bromolactone **726** in toluene solution in the presence of 0.5 equivalents of the TADDOL ligand and one equivalent of titanium tetraisopropoxide resulting in 63% yield of the cyclized system **727** (91% *ee*). Both enantiomers were prepared from enantiomers of the TADDOL-titanium ligand system to provide configurational structural determinations for the natural products. From this template **727**, the three natural products, aglacins A, B, and E (**722**–**724**), could then be prepared in 12–13 steps, using conventional methodology. It might be interesting to compare this synthesis of 2021 with the synthesis of the related lignan podophyllotoxin (**102**) of the 1980s, as discussed in Sect. 4.9.

## *8.18 Rhamnofolane, Tigliane, and Daphnane Diterpenoids*

Rhamnofolane, tigliane, and daphnane diterpenoids are three structural families with closely related biosynthetic origins and are classified according to their carbon frameworks. To date, more than 400 compounds of these types have been identified from plants of the Euphorbiaceae and Thymelaeaceae families [433, 434]. Many of these molecules display a wide range of biological activities, such as antiviral, antitumor, analgesic, immunomodulatory, neurotrophic, and tumor-promotion activities, making them pharmacological leads of interest [435]. The carbon skeletons of these diterpenoids have the same 5/7/6-*trans*-fused ring system in common, but the three-carbon substituents attached at the C-13- and C-14-positions of the C-ring

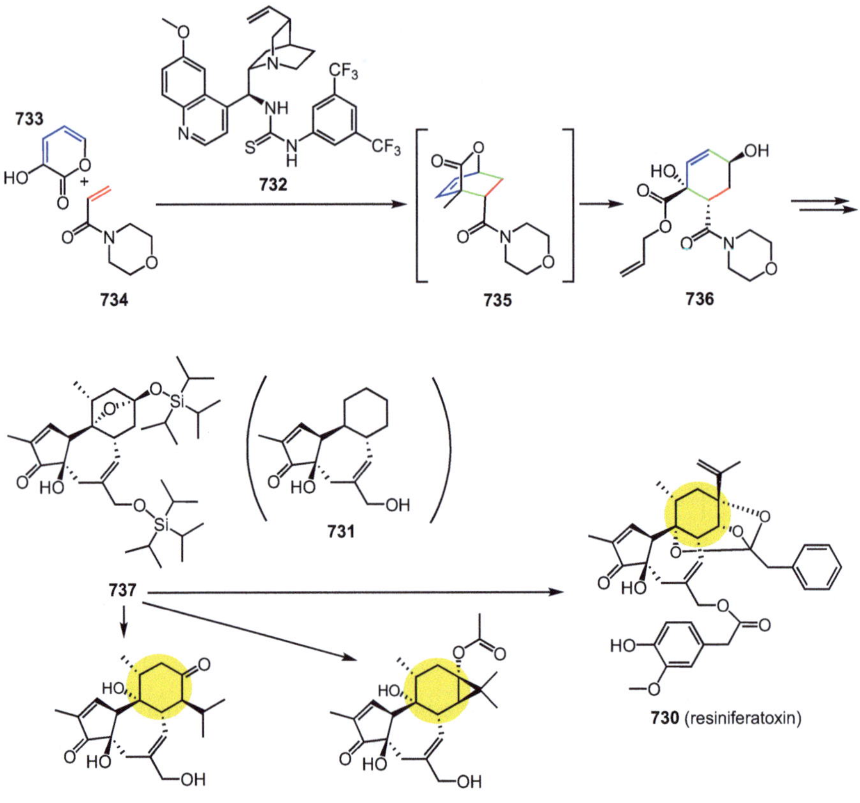


**Fig. 146** The asymmetric Diels–Alder reaction key step to produce the core rings of crotophorbolone (**728**), prostratin (**729**), and resiniferatoxin (**730**)

differ among them. Their structures are further diversified by the presence of various oxygenated functional groups. Accordingly, the synthesis of diterpenoids in these series have posed a formidable challenge for the synthesis community. From the rhamnofolane, tigliane, and daphnane diterpenoids, crotophorbolone (**728**), prostratin (**729**), and resiniferatoxin (**730**) were chosen as representative examples for their generation by synthesis (Fig. 146) [436]. All three have a common formal substructure, **731**, which was the first synthesis target in a suited derivatized form.

The preparation of **731** started with the construction of the six-membered C-ring. The quinine-derived thiourea catalyst **732** [437] promoted an asymmetric Diels–Alder reaction between achiral diene **733** and dienophile **734** in isopropanol, giving rise to the *endo*-adduct oxabicyclo[2.2.2]octene **735** as the major isomer along with minor amounts of the *exo*-adduct. This inseparable mixture was treated in a one-pot fashion with the lithium salt of allyl alcohol. Owing to the difference in the steric environment around the lactone carbonyl group, **735** was converted to its allyl ester

**736** and the *exo* adduct remained unchanged; purification of **736** was facile and resulted in 70% yield, 60% *ee*. Hence, the Diels−Alder reaction installed correctly the vicinal C-8,C-9-stereocenters in one step. Elaboration via conventional steps allowed the formation of the common intermediate **737**, which finally could be transformed into the three targets crotophorbolone (**728**), prostratin (**729**), and resiniferatoxin (**730**).

## *8.19 Euonymine and Euonyminol Octaacetate*

Euonymine (**738**) and euoniminol octaacetate (**739**) are examples of the about 1000 members of a sesquiterpenoid sub-group, called dihydro-$\beta$-agarofurans, which are characteristic compounds found in the Celastraceae family [438]. They are characterized by a *trans*-decalin and tetrahydrofuran ring. This system is further decorated by various functionalities. Euonymine (**738**) a highly oxidized molecule, was isolated from the flowering plant *Euonymus sieboldiana* Blume of the Celastraceae family [439], and euoniminol octaacetate (**739**) from *Maytenus chiapensis* Lundell from the same family (Plate 92) (Fig. 147) [440].

For both targets a common precursor was synthesized by means of an intermolecular triethylamine-accelerated Diels–Alder reaction [441]. Thus, triethylamine facilitated this intermolecular reaction between the diene 3-hydroxy-2-pyrone (**740**) and the dienophile **741** in chloroform solution at room temperature. The electron-rich C-8 of **740** and the electron-poor C-9 of **741** were connected regioselectively, and the C-1-oxy-protecting group of **740** prevented the approach from the same face, resulting in the formation of cycloadducts *endo*-**742** and *exo*-**742** (1:2.4), with stereoselective construction of the hindered C-10 quaternary carbon.

**Plate 92** Left: *Euonymus sieboldiana*. Photograph courtesy A. Minivet, Creative Commons CC0 1.0 Universal Public Domain Dedication. Right: *Maytenus chiapensis*. Photograph courtesy N. R. Marcial, CC BY

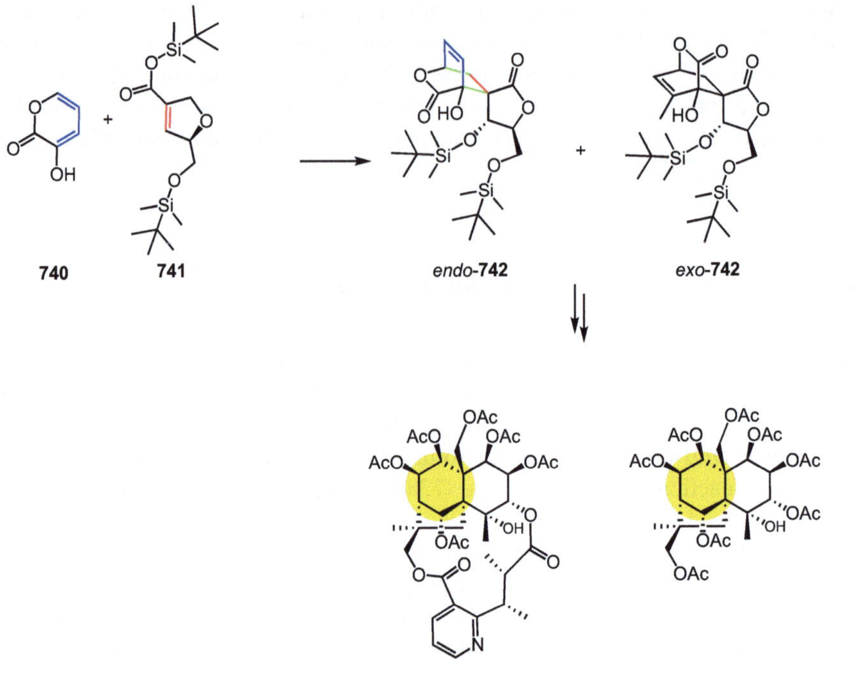


**Fig. 147** The triethylamine-accelerated Diels–Alder reaction key step in the synthesis of euonymine (**738**) and euoniminol octaacetate (**739**)

Further elaboration of the adducts allowed the formation of the targets euonymine (**738**) and euoniminol octaacetate (**739**).

## *8.20 Cephanolides A and B: Da Capo!*

In the beginning of 2021, a synthesis of cephanolides A–D (**694–697**) (Fig. 148) was encountered as an example that was based on an intramolecular inverse electron demand Diels–Alder cycloaddition (Sect. 8.11). By the end of this same year, another synthesis of cephanolides A (**694**) and B (**695**) was published, which had an intermolecular catalytic inverse electron demand Diels–Alder as the key step (Fig. 148) [442]. In comparison with the first example, it displays well the possibility of using proper variations of the methodology to construct a molecule with a desired skeleton.

An asymmetric inverse electron demand Diels–Alder reaction between the electron deficient 2-pyrone **743** and a substituted indene **744** was catalyzed by adding 5 mol% copper bistrifluoroacetate and 6 mol% of the chiral ligand **745** in dichloromethane at 25°C, in the presence of a 4 Å molecular sieve. This produced a

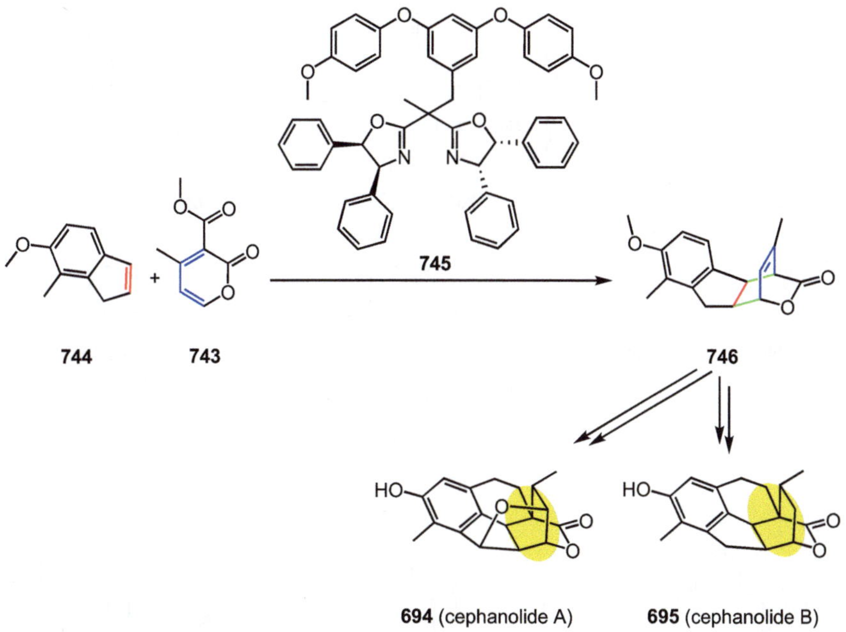

Fig. 148 The asymmetric inverse electron demand Diels–Alder reaction key step in the synthesis of the cephanolides A (**694**) and B (**695**)

99% yield of **746** with 98% *ee* after crystallization. Compound **746** could then be subsequently elaborated into the targets, cephanolides A (**694**) and B (**695**).

## 8.21 *Erectones A and B, and Hyperelodione D*

Polycyclic polyprenylated acylphloroglucinols are a very large family of plant-derived meroterpenoids with a broad spectrum of biological activities [443]. Although most of these contain a bicyclo[3.3.1]nonane ring system (e.g., hyperforin) [444], a growing number of noncanonical members with diverse and often unique polycyclic structures have been isolated [445]. Hence, for example, in 2021 the investigation of a complex di-geranylated tetracycle, hyperelodione D (**747**) (Fig. 149) from *Hypericum elodeoides* Choisy, of the Hypericaceae family (Plate 93), was reported [446]. The proposed structure of **747** has the same scaffold as two further meroterpenoids previously found in *Hypericum erectum* Thunb. (Plate 93), erectones A (**748**) and B (**749**), which possess either prenyl or geranyl side chains [447]. As suggested in the original isolation work [446, 447], the complex tetracyclic structures of **747**–**749** could be biosynthesized by a Diels–Alder based cascade reaction between (*E*)-β-ocimene and a suitable dihydroxyquinone dienophile.

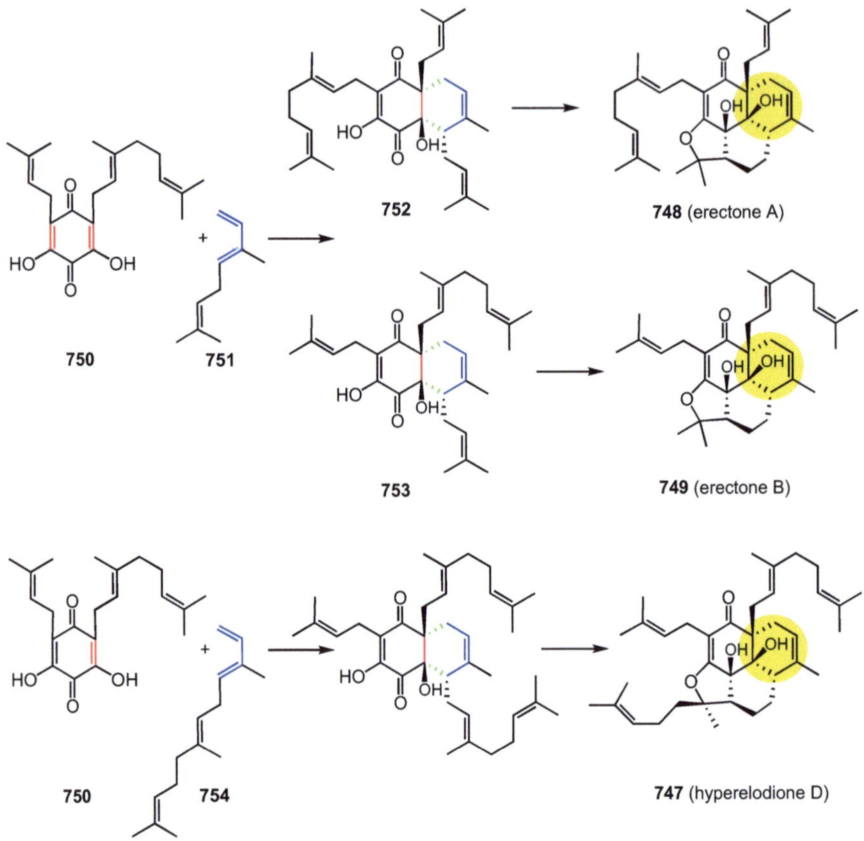


**Fig. 149** The catalyzed intermolecular Diels–Alder reaction key step in the synthesis of erectones A (**748**) and B (**749**), and hyperelodione D (**747**)

**Plate 93** *Hypericum erectum*. Photograph courtesy Daderot, Creative Commons CC0 1.0 Universal Public Domain Dedication (left). *Hypericum elodeoides*. Photograph, Anonymous, Creative Commons CC0 1.0 Universal Public Domain Dedication (right)

For a bioinspired synthesis of erectones A (**748**) and B (**749**), the substrate erectquinone (**750**) was synthesized by conventional means [448]. It was treated with (*E*)-β-ocimene (**751**) at 85°C in benzene solution, in the presence of 10 mol% europium(III)-tris(1,1,1,2,2,3,3-heptafluoro-7,7-dimethyl-4,6-octandionate) as a mild Lewis acid catalyst, which activated the dienophile and thus enhanced its reactivity to favor the *endo* adducts (Fig. 149). These thereby underwent an intermolecular Diels–Alder reaction to intermediates **752** and **753**, followed in the cascade by Prins cyclization and cycloetherification, to yield a mixture of the targets, erectones A (**748**) and B (**749**), which could be separated readily.

In the same way, a bioinspired intermolecular Diels–Alder reaction between erectquinone (**750**) and (*E,E*)-farnesene (**754**), followed by the just-described cascade, provided the six stereocenters and three rings of hyperelodione D (**747**), for which also an accordingly revised structure followed. The additional compound obtained by analogy to the reaction of **750** with **751** had no natural counterpart.

## 8.22 *Lamellodysidine A*

Lamellodysidine A (**755**), exhibiting cytotoxic and antimicrobial activities, was isolated from an extract of the marine sponge *Lamellodysidea herbacea* Keller from the Dysideidae family (Plate 94) [449]. It is a structurally intriguing sesquiterpene featuring a 5/5/6/6-fused tetracyclic skeleton, thereby presenting a challenging target to the synthesis community.

The exemplary key step in the synthesis of lamellodysidine A (**755**) was an intramolecular Diels–Alder reaction embedded in a series of other reactions starting from the chiral pool ((−)-(*R*)-carvone)-originated butenolide **756** (Fig. 150) [450]. Treating this compound in acetic anhydride and *p*-toluene sulfonic acid in toluene under reflux for 16 h produced the desired tetracyclic compound **760** in 70%

**Plate 94** *Lamellodysidea herbacea*. Photograph courtesy A. A. Rahman, Creative Commons Attribution-Share Alike 4.0 International

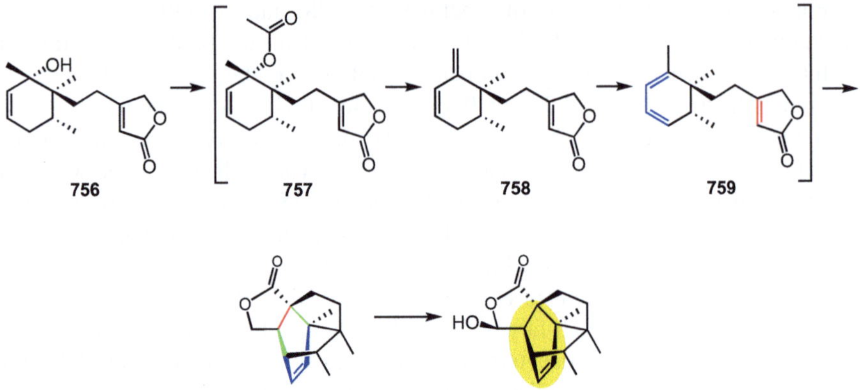


756 757 758 759

760 755 (lamellodysidine A)

**Fig. 150** The intramolecular Diels–Alder reaction key step in the stereoselective synthesis of lamellodysidine A (**755**)

yield. This reaction cascade included the acetylation of **756** to produce **757**, and the elimination of acetic acid to form **758**, which was modified by double bond migration in the equilibrium to the endocyclic diene **759**. The latter underwent the intramolecular Diels–Alder reaction to form **760**. A further step could then converted **760** into the target compound, lamellodysidine A (**755**) finalizing the stereoselective total synthesis.

## *8.23 Kopsane Alkaloids*

The *Kopsia* (for a typical picture of a species of these plants, members of the Apocynanceae family, see Plate 61) alkaloid kopsinine (**558**) was encountered in Sect. 7.23. From the wealth of such alkaloids elucidated to date, four other examples differing in their structures became the synthesis targets of the group of Y. Jia, with the construction of their skeleton occurring by means of a Diels–Alder step [451] (Fig. 151).

*N*-Carbomethoxy-10,22-dioxokopsane (**761**) was isolated from the stembark of a West African tree, *Pleiocarpa mutica* Benth., family Apocynaceae [452], and it was re-isolated from the stems of the Malaysian *Kopsia dasyrachis* Ridl. as a minor base [453]. Epikosanol-10-lactam (**762**) was found in the stem bark of *Aspidosperma duckei* Huber ex Ducke and *A. macrocarpon* Mart. (Brazil). In addition, kopsanone, kopsanol, and epikopsanol were isolated [454]. 10,22-Dioxokopsane (**763**) was purified from the stem bark of *Pleiocarpa mutica* Benth. (Apocynaceae) [452] and from the leaves and stems of *Kopsia dasyrachis* Ridl. [453]. *N*-Methylkopsanone (**764**) was found as a constituent of the bark and stems of *K. dasyrachis* (Apocynaceae) collected in Malaysia [453], and of the kernels of *K. arborea* Blume collected in Hainan, China [455].

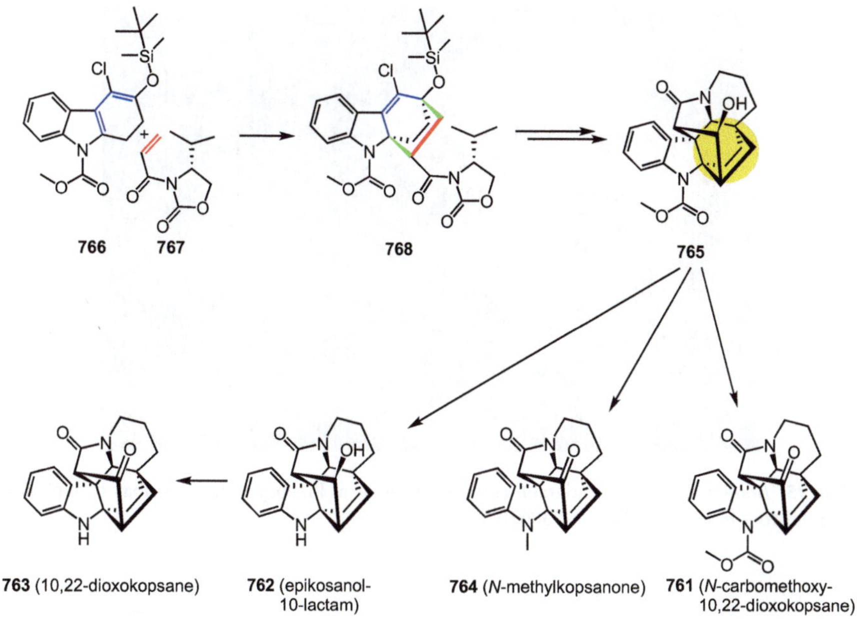


**Fig. 151** The intermolecular Diels–Alder reaction stereoselective key step in the synthesis of (+)-*N*-carbomethoxy-10,22-dioxokopsane (**761**), (−)-epikopsanol-10-lactam (**762**), (+)-10,22-dioxokopsane (**763**), and (+)-*N*-methylkopsanone (**764**)

The synthesis of these four indole alkaloids was envisaged to originate from the common racemic and even enantioenriched bicyclo[2.2.2]octane intermediate **765** (Fig. 151). This compound was prepared by an intermolecular Diels–Alder reaction from the dihydrocarbazolone **766** and the chiral acrolein component **767** as the dienophile by heating in *m*-xylene at 170°C in the presence of 2,6-di-*tert*-butyl-4-methyl-phenol as antioxidant, to yield the bridged derivative **768**. This product was then elaborated into the strategic intermediate **765** for the final stereoselective synthesis steps of the four natural products, (+)-*N*-carbomethoxy-10,22-dioxokopsane (**761**), (−)-epikopsanol-10-lactam (**762**), (+)-10,22-dioxokopsane (**763**), and (+)-*N*-methylkopsanone (**764**).

## *8.24 Cyclocalopin A Core*

From the bitter beech bolete mushroom *Boletus calopus* (Pers.) Vizzini (syn. *Caloboletus calopus* Vizzini) (family Boletaceae) (Plate 95), one of its bitter principles, cyclocalopin A (**769**) (Fig. 152), was isolated in 2002 by the group of W. Steglich [456]. The compound is highly oxygenated and characterized by a tricyclic spiro structure.

Its synthesis was targeted by J. Kim in 2022 using an *exo*-selective Diels–Alder reaction between the bis-γ,δ-lactone **770**, synthesized via a cyclocarbonylation reaction, and the diene **771** by heating in toluene at 110°C for 15 h (Fig. 152)

**Plate 95** *Boletus calopus*. Photograph courtesy Andreas at Pixabay (https://pixabay.com/photos/beautiful-foot-boletus-mushroom-2586702/)

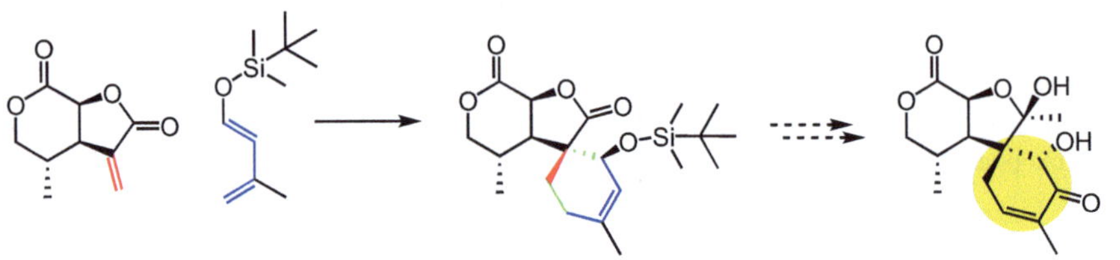

770 771 772 769 (cyclocalopin A)

**Fig. 152** The *exo*-selective Diels–Alder reaction key step in the synthesis of the core **772** of cyclocalopin A (**769**)

[457]. This resulted in the *exo*-compound **772** besides some minor amounts of the *endo*-adduct. The tricyclic core structure **772** then opened the way to the natural cyclocalopins.

## *8.25 (+)-Minovincine*

The monoterpenoid indole alkaloid (+)-minovincine (**773**) was first isolated and characterized from *Tabernaemontana riedelii* (Müll. Arg.) Myers (syn. *T. coriacea* Müll. Arg.) and *T. rigida* Link ex Roem.& Schult. (syn. *T. muricata* Link ex Roem. & Schult.) (Plate 96), all from the Apocynaceae family [458]. Its *cis*-hydroindole scaffold was approached in a concise synthesis route in 2022 by exploring a chiral-*N*,*N′*-dioxide/magnesium bis-trifluoroacetate complex-catalyzed enantioselective inverse electron demand Diels–Alder reaction between 3-carbalkoxy-2-pyrone (**774**) and the cyclic enamine **775** in chloroform solution for 3 h (Fig. 153)

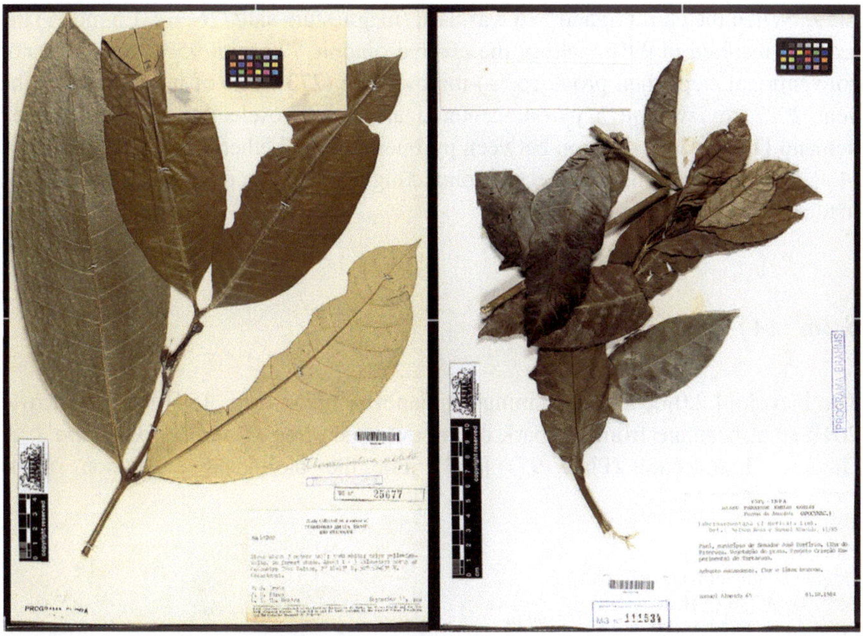

**Plate 96** Left: *Tabernaemontana riedelii*. Herbarium specimen MPEG (MG025677), Wikimedia Commons, CC BY 4.0. Right: *Tabernaemontana muricata*. Herbarium specimen MPEG (MG111543), Wikimedia Commons, CC BY 4.0

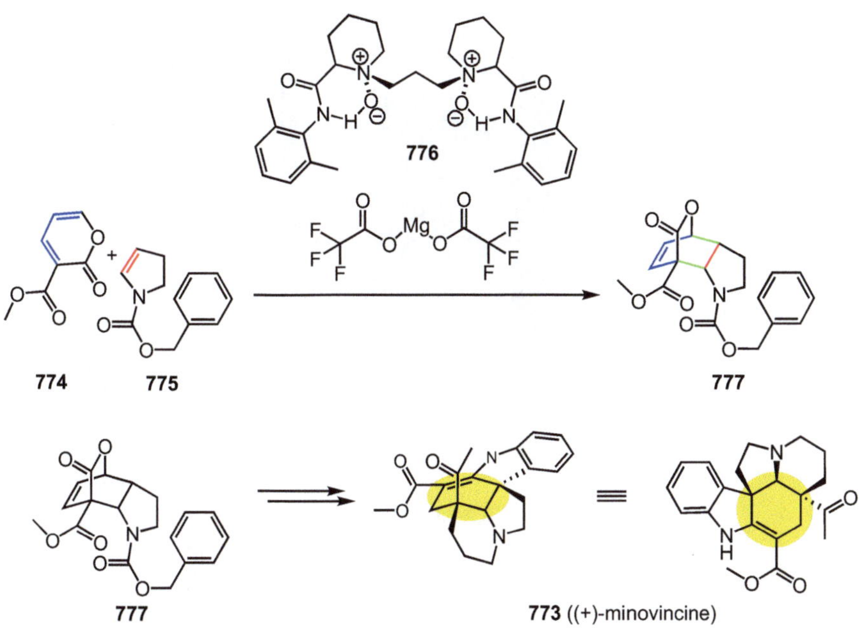


**Fig. 153** The enantioselective inverse electron demand Diels–Alder reaction key step to construct the *cis*-hydroindole scaffold **777** of (+)-minovincine (**773**)

[459]. When the chiral ligand **776** was used (magnesium salt/**776** = 1/1.5 mol%) the reaction resulted in 99% yield of the *cis*-hydroindole **777** with 95% *ee*. A series of conventional steps then provided (+)-minovincine (**773**). It is of interest that in the year 2022 reviews on a pyrone-centered and a stereoselective inverse electron demand Diels–Alder reaction between pyrones and enol ethers also were published [460, 461]. In addition, the aspect of conducting Diels–Alder reactions in sustainable media was also discussed [462].

## *8.26 (+)-Kingianin A*

The bicyclo[4.2.0]octane-containing kingianin A (**778**) (Fig. 154) was isolated in 2010 as a racemate from the bark of the evergreen tree *Endiandra kingiana* J. S. Gamble (Lauraceae) (Plate 97) [463]. For the biological evaluation of its

**Plate 97** *Endiandra kingiana*. Isolectotype BM000799336 (NHM, London) — CC BY 4.0

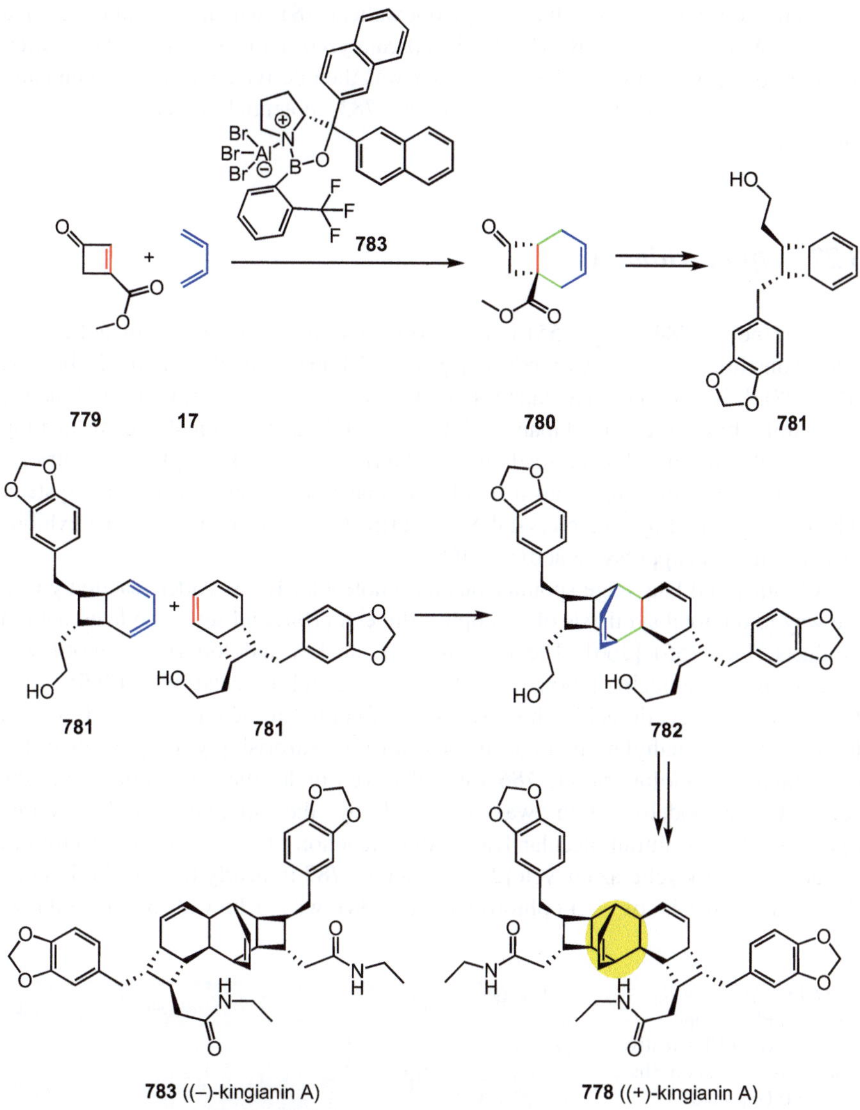


**Fig. 154** The enantioselectively catalyzed Diels–Alder reaction as first key step, followed by a Ledwith-Weitz salt catalyzed radical cation Diels–Alder reaction as the second key step to finally provide (+)-kinginanin A (**778**)

pharmaceutical properties, the synthesis of its enantiomers was considered desirable and thus the group of P. Lu devised a corresponding synthesis [664].

The synthesis of (+)-kingianin A (**778**) commenced with an enantioselectively catalyzed Diels–Alder reaction of cyclobutenone **779** with 1,3-butadiene (**17**) in dichloromethane and **783** as the catalyst, which gave the cyclobuteno derivative **780** in 93% yield with 93% *ee* (Fig. 154). This compound was then elaborated conventionally into the cyclobuteno-cyclohexadiene **781**, which was dimerized in a Ledwith-Weitz salt ($(p\text{-}BrC_6H_4)_3N.SbCl_6$)-catalyzed radical cation Diels–Alder reaction to yield the product **782**. The latter was then derivatized into (+)-kingianin A (**778**), which displayed as its antipode, **783**, potential anticancer activities [463, 464].

## *8.27 Ophiorrhine A*

Ophiorrhine A (**784**) (Fig. 155) is a monoterpene indole alkaloid isolated among other constituents from *Ophiorrhiza japonica* Blume from the family Rubiaceae (Plate 98). It is also known as Japanese snake root, which is a perennial shrub native to central China, Vietnam, Japan, and Taiwan, and has traditional uses in treating various ailments in Chinese medicine. Ophiorrhine A (**784**) displays an intricate polycyclic-fused structure with a highly unusual bridged-spirocyclic ring system, which was secured by single-crystal X-ray diffraction. This natural product exhibits in vitro immunosuppressive activity [465].

A bioinspired late-stage spontaneous intramolecular Diels–Alder reaction paved the way to the total synthesis of (−)-ophiorrhine A (**784**) and several related natural products (Fig. 155) [466]. The formation of the key indolopyridone motif was sought by cyclodehydration of 2-(2-acyl-3-indolyl)-acetamide (**785**) with triethylamine in acetic acid. Thus, **785** was submitted to heating at 125°C with a large excess of triethylamine in acetic acid for 6 h. Surprisingly, only traces of the corresponding indolopyridone **786** were observed under these conditions, but the successful cyclodehydration was followed by the spontaneous bioinspired diastereoselective intramolecular Diels–Alder reaction of transient **786** to produce the desired spirocyclic azabicyclo[2.2.2]octanone **787** in nearly 60% yield. Finally, the total synthesis of (−)-ophiorrhine A (**784**) was achieved by quantitative

**Plate 98** *Ophiorrhiza japonica*. Photograph courtesy KENPEI, Creative Commons Attribution-Share Alike 3.0 Unported

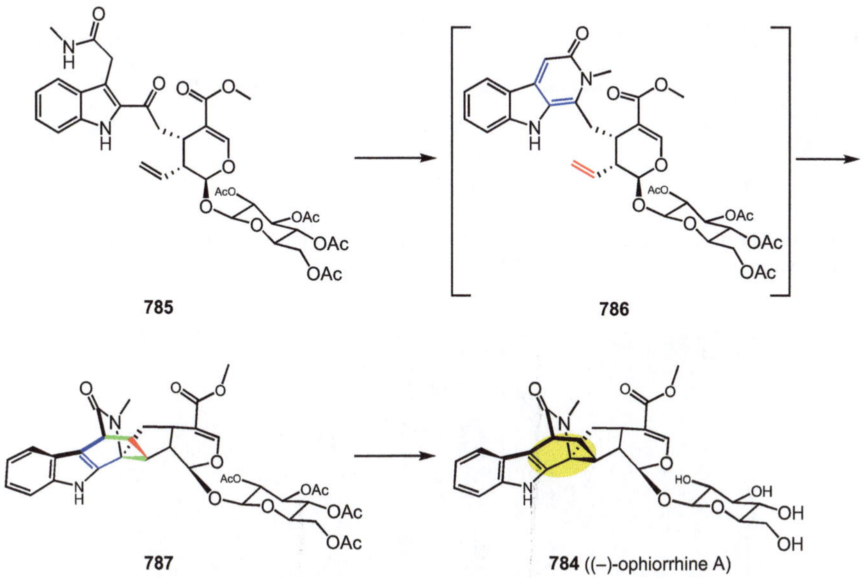


**Fig. 155** The spontaneous bioinspired diastereoselective intramolecular Diels–Alder reaction in the synthesis of (−)-ophiorrhine A (**784**)

methanolysis of the four acetate residues of the glucose moiety. It should be mentioned that the topic of bioinspired Diels–Alder reactions was reviewed in 2022 [467].

## *8.28 Lycopladine A*

The pyrindane alkaloid lycopladine A (**788**) (Fig. 156) which among other activities demonstrated neurotrophic effects, was isolated from the club moss *Lycopodium complanatum* L. (syn. *Diphasiastrum complanatum* (L.) Holub), of the Lycopodiaceae family (Plate 99) [468].

Although several syntheses of lycopladine A (**788**) were already known, as exemplified by that described in Ref. [469], the use of a diastereoselective intramolecular Stille coupling intramolecular Diels–Alder cascade introduced a novel aspect (Fig. 156) [470]. Thus, reacting a dibromolactone with a chiral pyridine stannane under Stille conditions produced the intermediate **789**, which in tetrahydrofuran solution under reflux for 18 h produced directly the tricyclic system **790**. The latter then could be elaborated conventionally into the target, (+)-lycopladine A (**788**).

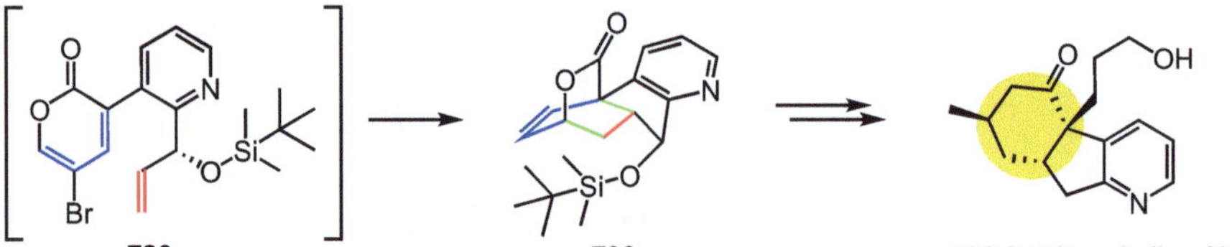


789 790 788 ((+)-lycopladine A)

**Fig. 156** The diastereoselective intramolecular Stille coupling intramolecular Diels–Alder cascade in the synthesis of (+)-lycopladine A (**788**)

**Plate 99** *Lycopodium complanatum* (syn. *Diphasiastrum complanatum*). Photograph courtesy B. Haynold, Creative Commons Attribution-Share Alike 3.0 Unported

## *8.29 Pyritide A2*

Pyritide A2 (**791**) is a member of the pyridine-containing macrocyclic peptides, which was discovered through exploitation of genomic information (genome mining) of the mesophilic bacterium *Micromonospora rosaria* (Wagman et al.) Horan & Brodsky (family Micromonosporaceae) [471]. An interest in its synthesis stems from its engineerable scaffold for pharmaceutically useful derivatives, but no validated medicinal activity is yet available.

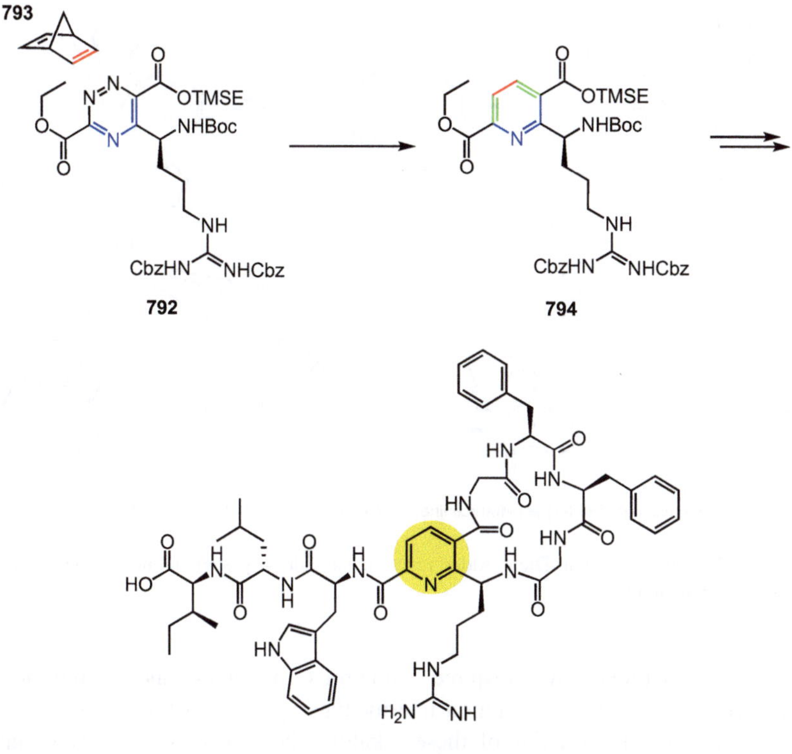


**Fig. 157** The intermolecular aza-Diels–Alder reaction key step in the synthesis of pyritide A2 (**791**). Abbreviations: Cbz = benzyloxycarbonyl, Boc = *tert*-butyloxycarbonyl, TMSE = 2-(trimethylsilyl)ethyl

Again, a bioinspired synthesis of its pyridine ring through an aza-Diels–Alder reaction was targeted (Fig. 157) [472]. The key was to derive the heterocyclic core from the intermediary formed 1,2,4-triazine **792** through an intermolecular aza-Diels–Alder reaction with 2,5-norbornadiene (**793**) (as an ethyne equivalent) to pyridine, as reviewed in Ref. [473], using ethanol as the solvent at 60°C for 21 h to yield nearly 90% of **794**. Further transformations of the product **794** yielded finally pyritide A2 (**791**).

## 8.30 *Aspidosperma Alkaloids*

The *Aspidosperma* alkaloids are a class of monoterpenoid indole alkaloids, which are of significant interest as synthesis targets. Their intriguing densely fused

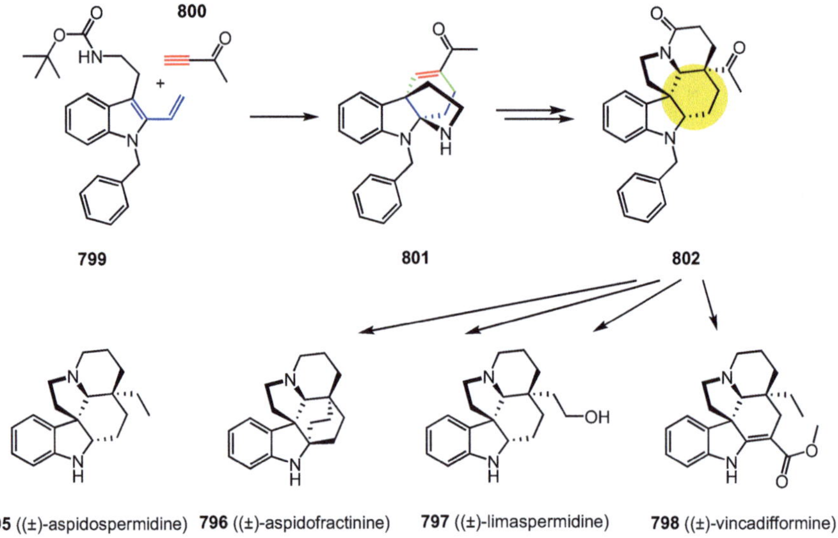


**Fig. 158** The intermolecular Diels–Alder reaction cascade key step in the synthesis of four *Aspidosperma* alkaloids

polycyclic structures have inspired numerous strategies and methodologies [474]. From constructing a common intermediate by means of a Suzuki-Miyaura/ Diels–Alder cascade, a series of these alkaloids became accessible by common methodology: (±)-aspidospermidine (**795**), (±)- aspidofractinine (**796**), (±)-limaspermidine (**797**), and (±)-vincadifformine (**798**) (Fig. 158) [475].

In a one-pot cascade, the vinyl intermediate **799** was first generated by a Suzuki-Miyaura cross-coupling reaction between a 2-bromo-indole derivative and a vinyl borane. In the next step of the cascade, this intermediate **799** was reacted in an intermolecular Diels–Alder reaction with but-3-yn-2-one (**800**) to yield **801**, which was then elaborated conventionally into the common intermediate **802**. The latter was then converted by classical steps into the four targets, (±)-aspidospermidine (**795**), (±)-aspidofractinine (**796**), (±)-limaspermidine (**797**), and (±)-vincadifformine (**798**).

## 8.31 *Matrine Alkaloids*

Members of the *Sophora* genus of plants (Fabaceae), in particular, *Sophora flavescens* Aiton (Plate 100), are used in traditional medicine [476]. A majority of the known biologically active constituents of *Sophora* species belong to the matrine family of alkaloids. These tetracyclic natural products comprise a quinolizidine bicycle linked to a piperidone ring through fusion of another piperidine ring. Of these, matrine (**803**) (Fig. 159) has potential anticancer activity through proliferation

**Plate 100** *Sophora flavescens*. Photograph courtesy Peganum, Creative Commons Attribution-Share Alike 2.0 Generic

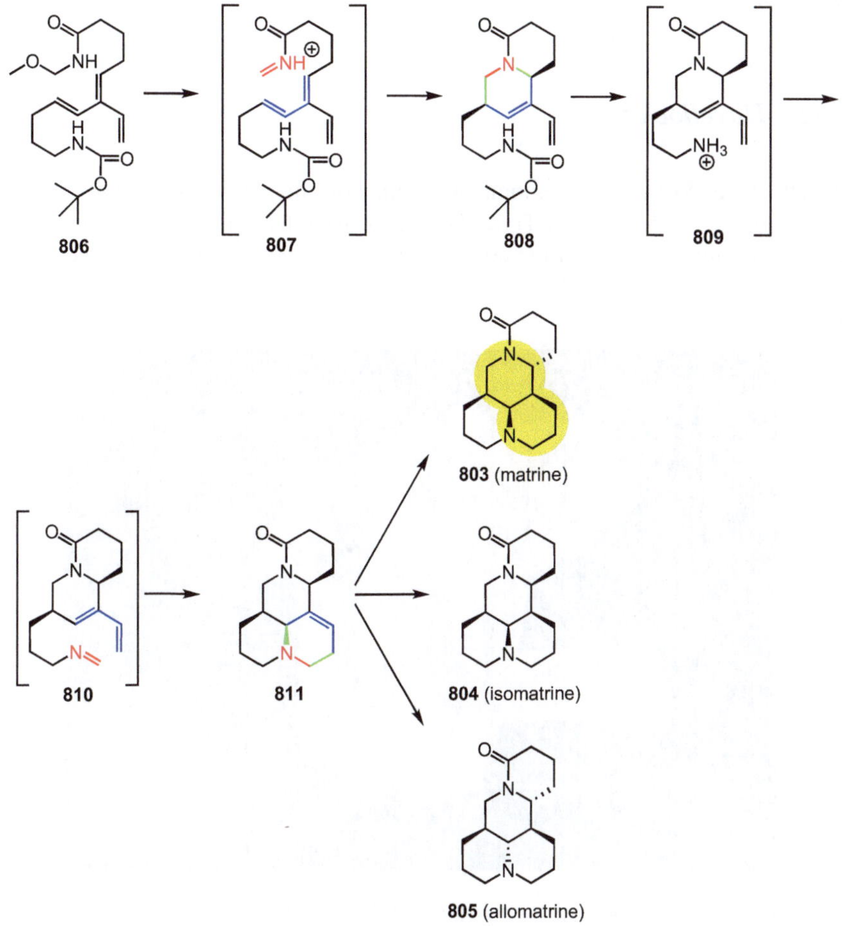


**Fig. 159** The two hetero-intramolecular Diels–Alder reactions are key steps in the synthesis of matrine (**803**), isomatrine (**804**), and allomatrine (**805**)

inhibition and apoptosis induction of various cancer cell lines [477]. It has also antiviral-related activity [478]. Information on the biological activities of isomatrine (**804**) and allomatrine (**805**), which represent diastereomers, and their derivatives is scarce, because of their low natural abundance.

For the synthesis of **803–805**, two hetero-intramolecular Diels–Alder reactions provided the key steps to construct the tetracyclic framework (Fig. 159) [479]. The acyclic dendralene precursor **806** was first reacted with trifluoroacetic acid to provide the putative *N*-acyliminium species **807**. This intermediate cyclized rapidly at ambient temperature into the *cis*-bicycle **808**. Removal of the *tert*-butyloxycarbonyl group from the side chain by exposure to trifluoroacetic acid gave the ammonium salt **809**, which was heated in a sealed tube with degassed formaldehyde to deliver via the iminium dienophile the Diels–Alder cyclized tetracycle **811** as a single diastereomer. Catalytic hydrogenation of **811** then gave a mixture of the natural products matrine (**803**), isomatrine (**804**), and allomatrine (**805**) that could be easily separated.

## *8.32 Havellockate*

Havellockate (**812**), a $C_{20}$-cembranoid diterpenoid first isolated in 1998 from the soft coral, *Sinularia granosa* Tixier-Durivault from the family Sinulariidae (Plate 101), is a prime example of the complexity and synthesis challenge posed by the

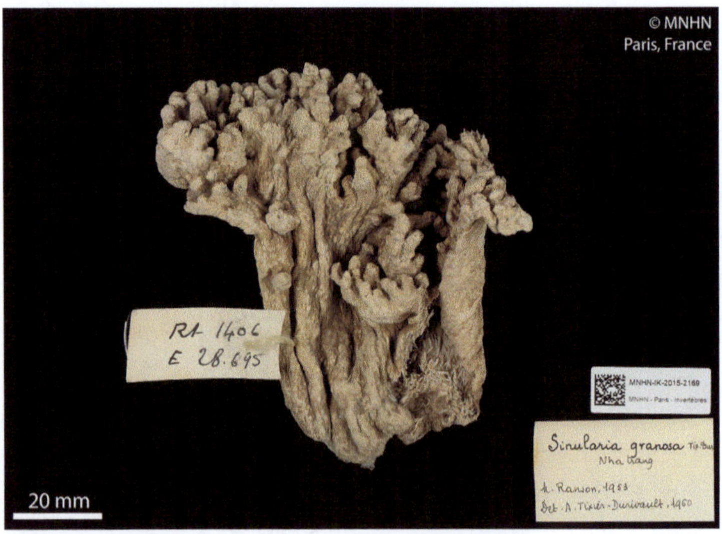


**Plate 101** *Sinularia granosa*, preserved specimen. Photograph courtesy M. Hennion/Muséum national d'Histoire naturelle (MNHN-IK-2000-96), CC BY 4.0

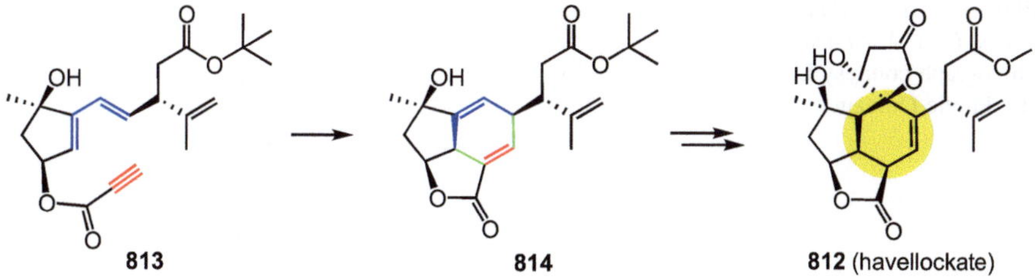


**Fig. 160** The intramolecular Diels–Alder step in the asymmetric synthesis of havellockate (**812**)

polycyclic furanobutenolide-derived cembranoid family [480]. It is characterized by a highly oxygenated *cis*-fused tricyclic core common to other related natural products (see Sect. 7.36 for the related verrubenzospirolactone (**618**)). However, it is adorned with a spiro-fused β-hydroxybutanolide ring, a unique feature not found in other members of the polycyclic furanobutenolide-derived cembranoid family. In addition to these topological features, **812** possesses eight stereogenic centers, of which seven are contiguous around the highly substituted and densely functionalized core.

This formidable synthesis challenge was taken up by the group of B. M. Stoltz in 2022, by designing an intramolecular Diels–Alder step as the key reaction to construct the tricyclic core of havellockate (**812**) (Fig. 160) [481]. Heating the sensitive cyclopentene intermediate **813** at 110°C in a xylene mixture gave the tricyclic compound **814** in 70% yield with >20:1 *dr*. A series of further conventional steps provided finally the natural product havellockate (**812**).

## 8.33 *Atrachinenins A and B*

*Atractylodes lancea* (Thunb.) DC. (syn. *A. chinensis* (Bunge) DC.) from the Asteraceae family (Plate 102), also known as Cang Zhu or black atractylodes, is a flowering plant with a long history of medicinal use in traditional Chinese medicine. Its dried rhizomes are used to invigorate the spleen, promote qi, and have diuretic, antiperspirant, and soothing properties. It is a common ingredient in prescriptions for gastrointestinal issues, and is also used to address dampness, wind-cold, and night blindness [482]. Atrachinenins are a family of structurally intricate meroterpenoid natural products isolated in racemic form from the rhizomes of *Atractylodes chinensis* [483]. The most complex member of the family, atrachinenin B (**816**), features a cage-like pentacyclic ring system with seven contiguous stereocenters and a peroxyhemiacetal bridge. Atrachinenin A (**815**) is devoid of this bridge.

**Plate 102** *Atractylodes lancea* (syn. *A. chinensis*). Photograph courtesy L. C. Ho, Public Domain

The synthesis of **815** and **816**, as conducted by J. H. George and his group, commenced with the product from a geranylation of 4-methoxy-2-methylphenol and succeeding oxidation into the quinone **817** (Fig. 161) [484]. Using a high concentration of lithium perchlorate (2.5 *M* in ethanol), the Grieco conditions for rate-accelerated Diels–Alder reactions [485], this enabled an efficient intermolecular reaction between **817** and (*E*)-β-ocimene (**818**) to give a mixture of all four possible *endo* adducts, with an excess of **819**. This intermediate could be isolated by silver nitrate chromatography and then elaborated conventionally into the natural products atrachinenin B (**816**) and finally atrachinenin A (**815**).

## *8.34 (+)-Lucidumone*

In 2019, Y. X. Cheng and his group discovered the new meroterpenoid lucidumone (**820**) (Fig. 162) after its extraction from the fruiting bodies of the mushroom

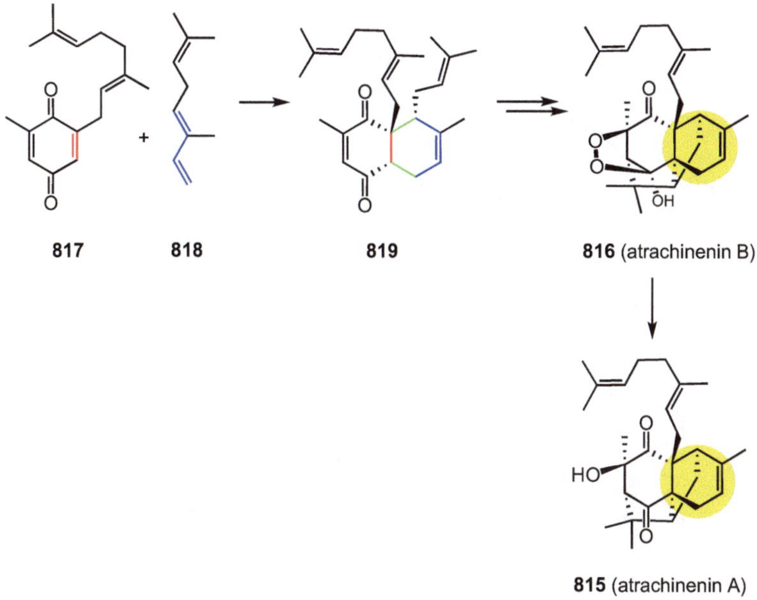


**Fig. 161** The lithium perchlorate rate-accelerated intermolecular Diels–Alder reaction key step in the synthesis of atrachinenin A (**815**) and atrachinenin B (**816**)

**Plate 103** *Ganoderma lucidum*. Photograph courtesy K Molnár, Pixabay

*Ganoderma lucidum* Karst from the Ganodermataceae family (Plate 103) [486]. This natural product displayed an interesting COX-2 inhibitory activity with a relatively good selectivity compared to COX-1, providing it with a promising therapeutic potential for the treatment of inflammation [487]. It displays a complex caged core including a pentacyclic skeleton with 6/5/6/6/5-membered rings, with both a fused

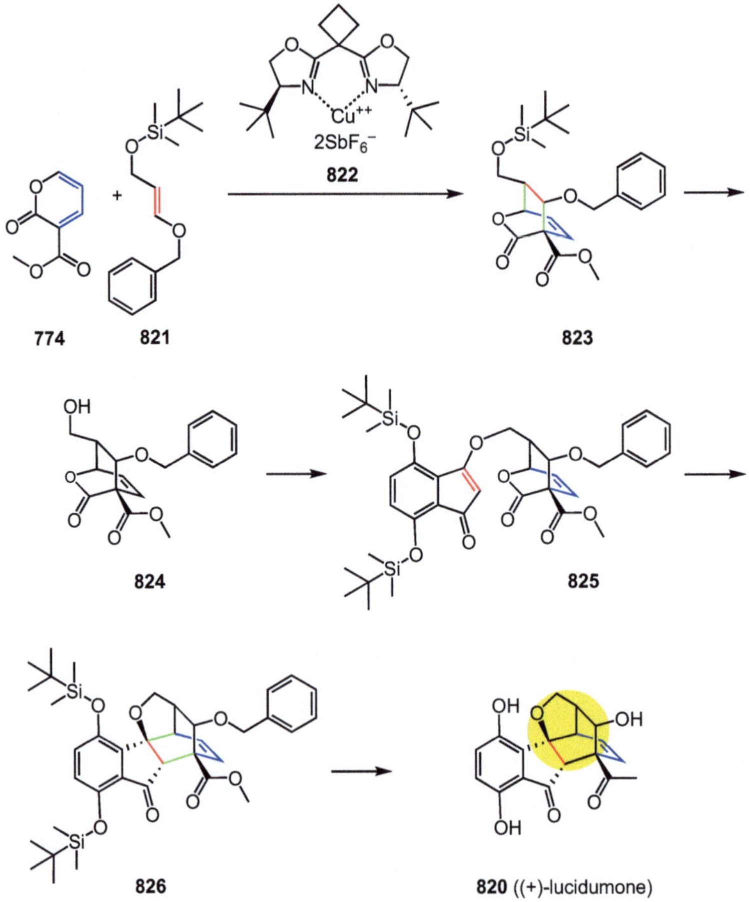


**Fig. 162** The stereoselective intermolecular inverse electron demand Diels–Alder reaction and the retro [4 + 2]/intramolecular Diels–Alder cascade as key steps in the synthesis of (+)-lucidumone (**820**)

and bridged arrangement. Its bicyclo[2.2.2]octane scaffold displays six contiguous stereogenic centers.

An enantioselective synthesis of (+)-lucidumone (**820**) developed by the A. de la Torre group in 2023 involved an intermolecular inverse electron demand Diels–Alder reaction between a 2-pyrone and an acyclic enolether, as described in the preceding year (Fig. 162) [461]. An accordingly masked cyclohexadiene and careful selection of the chiral copper(II) catalyst **822** paved the way to 99% yield with 98% *ee* for the product **823** of this key step, when the 2-pyrone **774** was reacted stereoselectively with the enol ether **821** in dichloromethane/toluene (1/2) in the presence of a 4 Å molecular sieve at −40°C [488].

This bicyclic lactone **823**, after removing one of its protecting groups to give **824**, was reacted with a substituted indandione to yield the second Diels–Alder substrate **825**, which upon heating in chlorobenzene at 140°C allowed the intramolecular retro [4 + 2]/intramolecular Diels–Alder cascade to provide the pentacyclic ketone **826**. A few further conventional steps then led to the targeted (+)-lucidumone (**820**).

## 8.35 *Lissodendoric Acid A*

Lissodendoric acid A (**827**) (Fig. 163) is a structurally highly complex member of the manzamine family, which was encountered already in Sect. 6.18 when describing a synthesis of manzamine A (**245**). It was characterized in 2017 from the marine sponge *Lissodendoryx florida* Koltun, from the family Coelosphaeridae (Plate 104), collected in the Sea of Okhotsk and was shown to reduce reactive oxygen species in a Parkinson's disease model [489].

As seen in Sect. 6.18 through one example, manzamine alkaloids are popular targets among the synthesis community [490]. A delicately balanced synthesis of lissodendoric acid A (**827**) described by N. K. Garg and his group in 2023 was based on the reaction of a stereoselectively constructed precursor of a fleeting strained

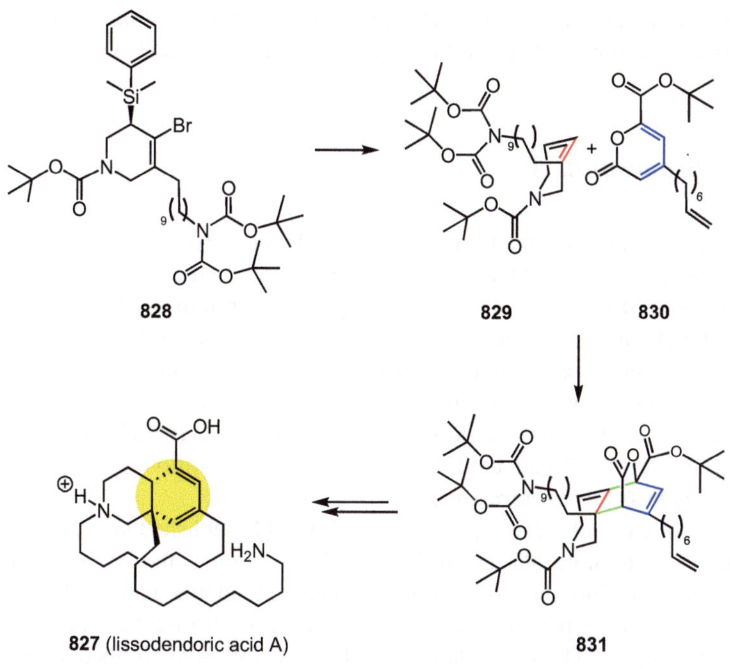


828 829 830 827 (lissodendoric acid A) 831

**Fig. 163** The Diels–Alder reaction between a fleeting strained cyclic allene and a 2-pyrone as key step in the synthesis of lissodendoric acid A (**827**)

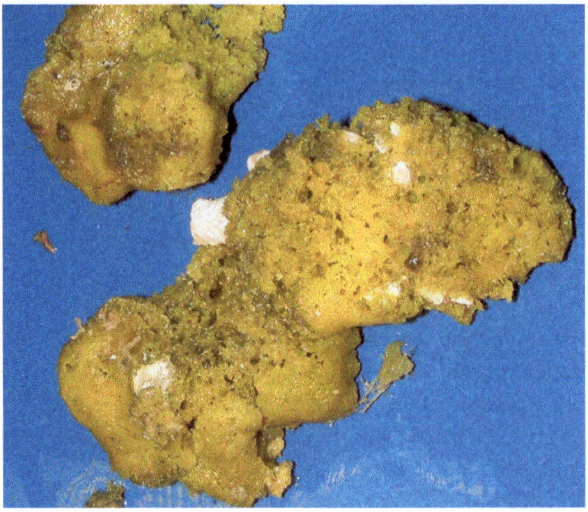

**Plate 104** *Lissodendoryx florida* (PIBOC O47–283). Photo by V. B. Krasokin

cyclic allene intermediate with a pyrone (with pyrones obviously having become favourites as Diels–Alder components) (Fig. 163) [491]. Thus, an enantioenriched allene precursor **828** was simply treated with caesium fluoride, and the short-lived allene **829** that resulted then reacted at $-20$°C in acetonitrile with the 2-pyrone **830** to the desired cycloadduct **831** (*dr* > 20/1, 79% *ee*). In this reaction mixture, some tetrabutylammonium fluoride was added to improve solubility. Further conventional transformations of **831** produced the targeted lissodendoric acid A (**827**).

## *8.36 Bisabosqual A*

In 2001, researchers isolated bisabosqual A (**832**) (Fig. 164) from a culture broth of the filamentous fungus *Stachybotrys* sp. RF-7260, which was obtained from decaying tree leaves. As an unusual type of meroterpenoid, with a unique 2,3,9,9a-tetrahydro-1*H*-benzofuro[4,3,2-*cde*]chromene ring system containing contiguous stereogenic centers and a *cis*-fused skeleton, bisabosqual A (**832**) exhibited inhibitory activity towards microsomal squalene synthase from *Saccharomyces cerevisiae*, as well as on the growth of *Candida albicans* and HepG2 cells. In addition, it showed broad antifungal activity against various yeasts, especially against *Cryptococcus neoformans* [492].

A bioinspired synthesis of (±)-bisabosqual A (**832**) published in 2023 by Y. Tang et al., had an intramolecular hetero-Diels Alder reaction as the key step (Fig. 164) [493]. Upon treating the conventionally prepared intermediate **833** with ferric chloride at room temperature in dichloromethane, a retro 6π-electrocyclization to the intermediate **834** took place, which was followed immediately by an

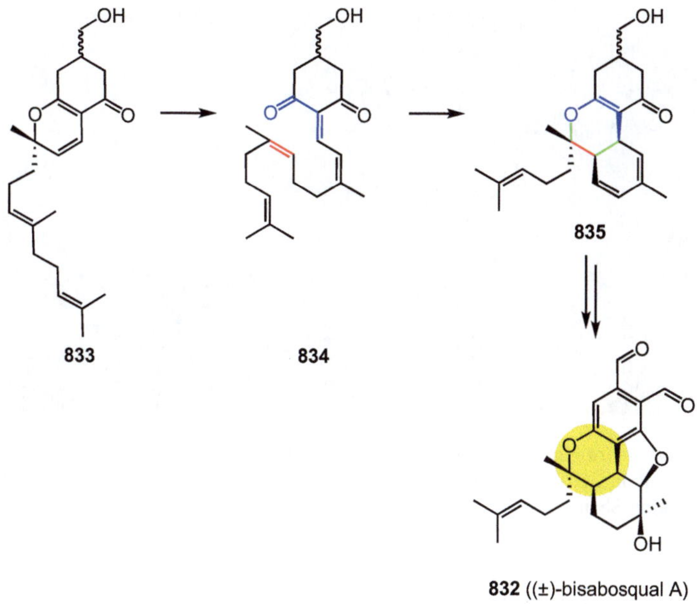


**Fig. 164** The retro 6π-electrocyclization/intramolecular hetero-Diels Alder reaction cascade as key steps in the synthesis of (±)-bisabosqual A (**832**)

intramolecular Diels–Alder cycloaddition to yield the *cis*-fused tricyclic system **835**. Further conventional transformations then led to the targeted (±)-bisabosqual A (**832**).

## *8.37 Sarbracholide and Shizukaol B*

The lindenane sesquiterpenoid dimers sarbracholide (**836**) and shizukaol B (**837**) were isolated from *Sarcandra glabra* subsp. *brachystachys* (Blume) Verdc., from the family Chloranthaceae (Plate 105) and from the evergreen shrub *Chloranthus japonicus* Sieb., also from the Chloranthaceae family. They were found to be active against *Plasmodium falciparum* and to have anti-HIV-1 activity [494, 495].

A biomimetic Diels–Alder reaction between a diene in one of the intermediates and the dienophile in the other part was key to the total syntheses of sarbracholide (**836**) and shizukaol B (**837**) (Fig. 165) [496]. When the stereoselectively built dienophile **838** was treated in neat form in the dark at 30°C with the conventionally prepared diene **839**, a yield of 57% sarbracholide (**836**) resulted. In the same way, reaction of of dienophile **840** with the same diene **839** gave smoothly a 44% yield of shizukaol B (**837**).

**Plate 105** *Sarcandra glabra*. Photograph Natl. Inst. of Ecology, Korea Open Government License Type I: Attribution (left). *Chloranthus japonicus*. Photograph courtesy Alpsdake, Creative Commons Attribution-Share Alike 4.0 International (right)

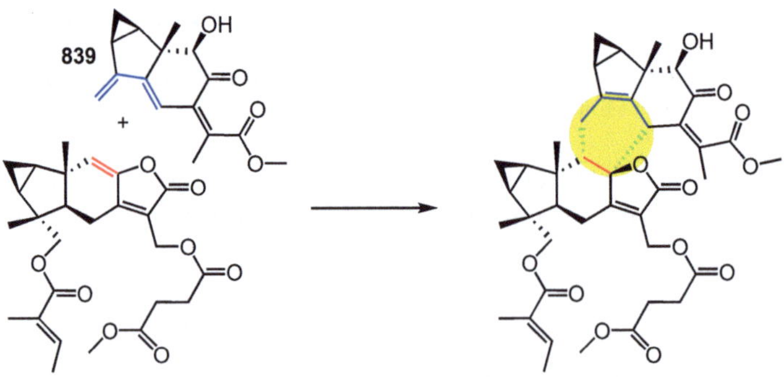


**838** **836** (sarbracholide)

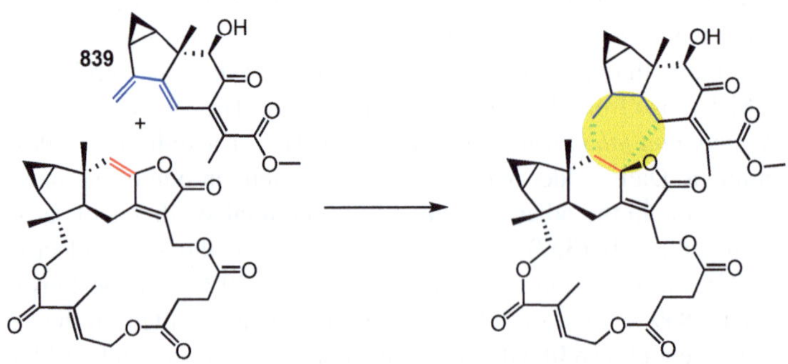


**840** **837** (shizukaol B)

**Fig. 165** The biomimetic Diels–Alder reaction key steps in the asymmetric synthesis of sarbracholide (**836**) and shizukaol B (**837**)

## 8.38 Senepodine F

In 2004, J. Kobayashi and co-workers isolated the *Lycopodium* alkaloid, (−)-senepodine F (**841**), from the clubmoss *Lycopodium chinense* Christ (syn. *Huperzia miyoshiana* Burke) of the family Lycopodiaceae (Plate 106) [497]. As one of the presently known approximately 500 *Lycopodium* alkaloids with varied biological properties [498], an asymmetric synthesis of (−)-senepodine F (**841**) posed a strong challenge.

An organocatalytic enantioselective Diels–Alder reaction proved to be the key procedure to reach this aim, as described by H. Ishikawa and his team (Fig. 166) [499]. The critical ring A/B component of the molecule was synthesized by such an enantioselective Diels–Alder reaction between the 5-nitro-2,3-dihydropyridone derivative **842**, prepared from commercially available *N-tert*-butyloxycarbonyl-β-alanine, and two equivalents of the unsaturated aldehyde **843**. In the presence of 5 mol% diphenylprolinol silyl ether catalyst **844** dissolved in

**Plate 106** *Lycopodium chinense* (syn. *Huperzia miyoshiana*). Photograph courtesy B. Legler, Rocky Mountain Herbarium, University of Wyoming, USA

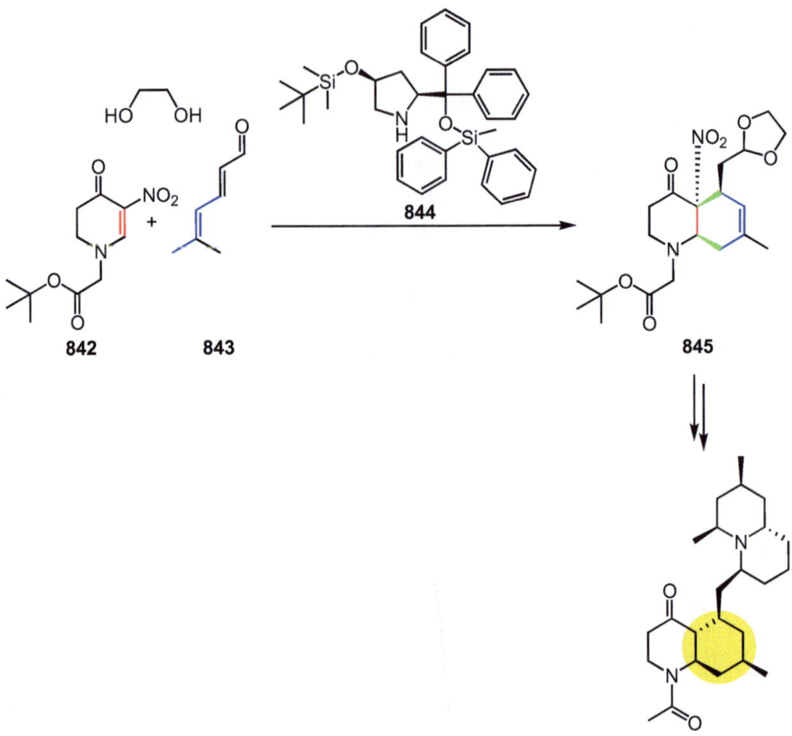


**841** ((−)-senepodine F)

**Fig. 166** The organocatalytic enantioselective Diels–Alder reaction key step in the synthesis of (−)-senepodine F (**841**)

toluene, and on standing at room temperature for 16 h, the reaction provided after concomitant acetalization with ethylene glycol/*p*-toluenesulfonic acid an 88% yield of **845** with $dr = 4.4{:}1$ and 96% *ee*. This methodology had been developed in a prior study [500]. The product **845** was then transformed into the target (−)-senepodine F (**841**), reassigning also the originally derived stereochemistry of this compound at the same time.

## 8.39 *Cinchona Alkaloids*

The *Cinchona* alkaloids, comprising quinine (**846**), quinidine (**847**), cinchonine (**848**), and cinchonidine (**849**) are representative members of the quinoline alkaloids noted for their antimalarial properties and as components in chiral biocatalysis [501–503]. They occur at various abundance levels in the bark of *Cinchona* species like *C. officinalis* L., family Rubiaceae (Plate 107), and *C. ledgeriana* Bern. Moens, *C. succirubra* Pav. ex Klotszch, and *C. calisaya* Wedd.

**Plate 107** *Cinchona officinalis* tree (left) photograph courtesy Luchovaras2501, Creative Commons Attribution-Share Alike 4.0 International, and bark (right) photograph courtesy H. Zell, Creative Commons Attribution-Share Alike 3.0 Unported

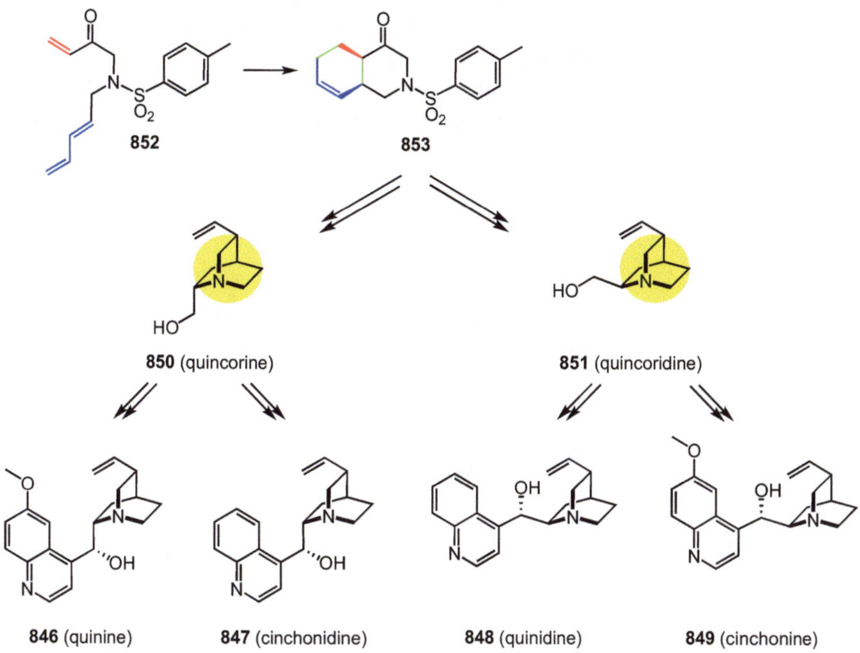


**Fig. 167** The intramolecular Diels–Alder cycloaddition key step in the synthesis of *Cinchona* alkaloids **846–849**

The syntheses of the four natural products **846–849** were based on the preparation of the two diastereomeric quinuclidines quincorine (**850**) and quincoridine (**851**), and an intramolecular Diels–Alder cyclization solved the construction of the critical stereocenters (Fig. 167) [504]. Upon a Dess-Martin oxidation, the enone **852** thus formed, and underwent spontaneously a rapid Diels–Alder cycloaddition to yield the bicycle **853** in 72% isolated yield. The *cis* stereochemistry of **853** was predicted by an *endo* transition state and secured through single-crystal X-ray analysis. Further elaboration of **853** led to the two diastereomers quincorine (**850**) and quincoridine (**851**). These diastereomers were then linked to their quinoline nuclei via an ingenious photoredox-catalyzed deoxygenative arylation procedure to yield the four natural products **846–849**.

## 8.40 *Vermisporin*

Vermisporin (**854**) was isolated from the fungus strain *Ophiobolus vermisporus* L-8 of the family Phaeosphaeriaceae. It shows excellent activities against Gram-positive and anaerobic bacteria, including *Bacillus anthracis*, *Bacteroides* sp., *Propionibacterium acnes*, *Clostridium* sp., *Treponema* sp., *Actinomyces naeslundii*, and *Peptostreptococcus* sp. The structure of **854** is unique, having a *cis*-decalin ring, a tetramic acid ring, perhaps originating from L-valine, and a sterically hindered (*S*, *S*)-configured epoxide group (Fig. 168) [505].

The construction of the functionalized *cis*-decalin ring of vermisporin (**854**) was achieved by a Lewis acid-catalyzed diastereoselective intramolecular Diels–Alder

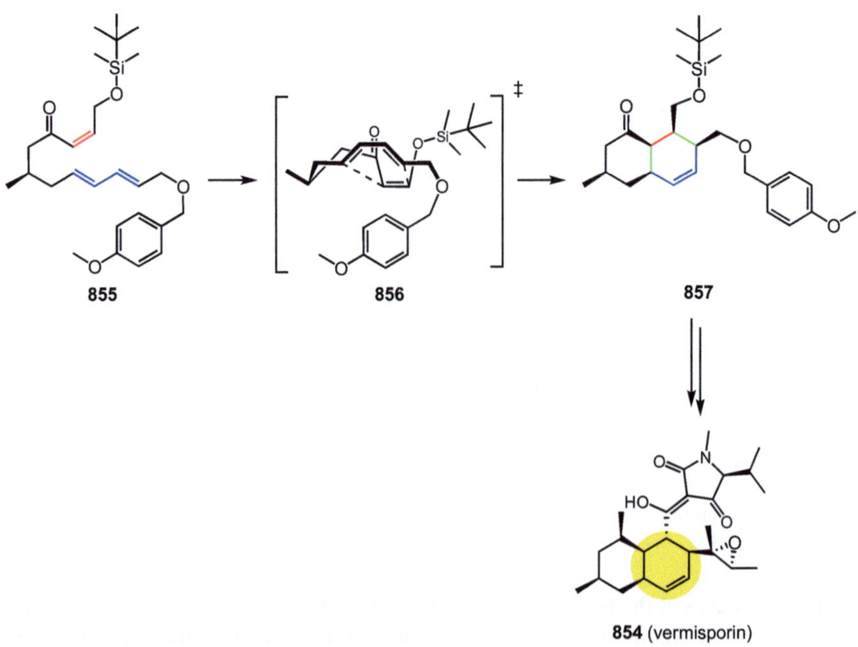


**Fig. 168** The Lewis acid-catalyzed diastereoselective intramolecular Diels–Alder reaction key step in the synthesis of vermisporin (**854**)

reaction, which proceeded via a rare *endo*-boat transition state (Fig. 168) [506]. Thus, the open-chain trienone **855** gave with 0.9 *M* dimethylaluminum chloride as the Lewis acid in heptane at 0°C for 30 min, the desired *cis*-decalin **857** in 52% yield via the transition state **856**. Further conventional stereoselective functionalizations finally led to the target, vermisporin (**854**).

## *8.41 Neocaesalpins A, AA, and K*

Cassane-type diterpenoids have been isolated from medicinal plants of the genus *Caesalpinia*, which comprises about 100 species and is found commonly in tropical and subtropical regions [507]. To date, more than 450 cassane-type diterpenoids have been reported, and their biological properties include anti-inflammatory, anti-malarial, antitumor, antiviral, antimicrobial, and antioxidant activities [508]. The structural complexity and promising biological activity of cassane-type diterpenoids has been of interest to the synthesis community. Of special interest is neocaesalpin A (**858**) isolated in 1996 from the grey nicker liana, *Caesalpinia bonduc* L. (syn. *Guilandina bonducella* L.) from the family Fabaceae (Plate 108), which was the first cassane-type diterpenoid reported with a butenolide ring [509]. Following the first total synthesis of **858**, the synthesis route of E. M. Carreira and his coworkers allowed the syntheses also of the related neocaesalpins AA (**859**) and K [510].

At the heart of this synthesis lies an intermolecular Diels–Alder reaction that rapidly assembles the target framework from commercial materials (Fig. 169). The triene **860** derived from safranal and the furanoquinone **861** reacted at −78°C in hexafluoroisopronal in the presence of catalytic boron trifluoride etherate to give **862** with 85% yield. Having this full carbocyclic cassane skeleton in hand allowed the

**Plate 108** *Caesalpinia bonduc* (syn. *Guilandina bonduc*). Photograph courtesy Vinayara, Creative Commons Attribution-Share Alike 3.0 Unported

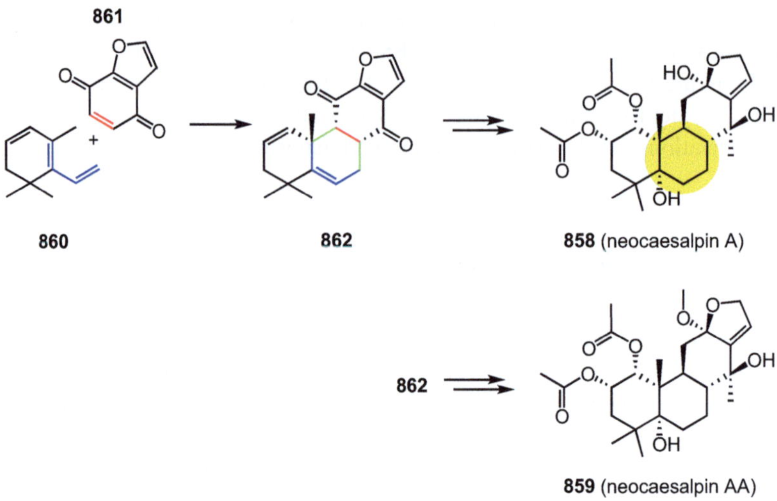


**Fig. 169** The intermolecular Diels–Alder reaction key step in the syntheses of neocaesalpin A (**858**) and AA (**859**)

syntheses of neocaesalpin A (**858**) along with neocaesalpins AA (**859**) and K. However, in the last-mentioned case, its structure had to be reassigned and was found to be identical with that of neocaesalpin AA (**859**).

## *8.42 Sculponin U*

The *ent*-kaurane type diterpenoid sculponin U (**863**) was isolated from the perennial herb *Isodon sculponeatus* (Vaniot) Kudô, of the Lamiaceae family (Plate 109). It has an intriguing tetracyclic structure and exhibits notable antitumor, antiinflammatory, and antibacterial properties [511, 512].

The synthesis of sculponin U (**863**) was based firmly in the initial steps on an intermolecular Diels–Alder reaction between the silyl enolate **861a** and acrolein, providing the cyclohexene derivative **862a** with *dr* = 15:1 (Fig. 170) [513]. This compound was then further elaborated using also a photoinduced radical cascade reaction to yield finally enantiopure sculponin U (**863**).

**Plate 109** *Isodon sculpoenatus*. Photograph courtesy Board of Trustees, RBG Kew, CC BY-NC-SA 3.0

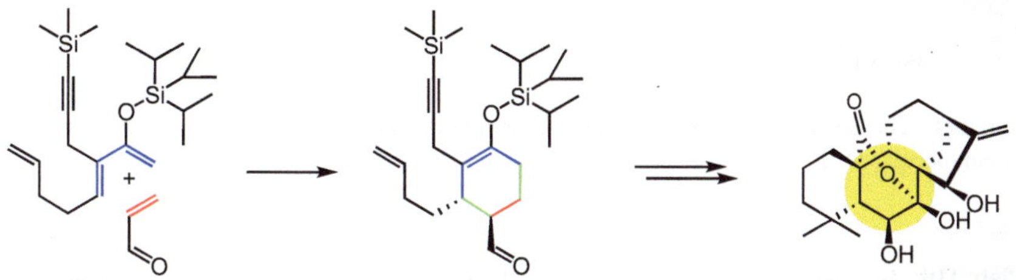


**Fig. 170** The intermolecular Diels–Alder reaction key step in the synthesis of enantiopure sculponin U (**863**)

## *8.43 Vilmoraconitine*

The $C_{19}$-diterpenoid alkaloid vilmoraconitine (**864**) (Fig. 171) displaying a unique skeleton was isolated from the perennial herb *Aconitum vilmorinianum* Kom., family Ranunculaceae (Plate 110), used as a traditional Chinese medicine to treat rheumatism and pains [514]. Another member of the *Aconitum* genus, talatisamine (**655**), belonging to the type I diterpenoids, was encountered previously in Sect. 8.2 in having an intramolecular Diels–Alder step to construct its hexacyclic ring system. Alkaloid **864** is a member of the rearranged type II diterpenoids, which is characterized by an unprecedented heptacyclic backbone including a congested cyclopropane ring and quaternary chiral centers (Fig. 171).

The synthesis of vilmoraconitine (**864**) was most certainly a challenging adventure, which involved two Diels–Alder key steps (Fig. 171) [515]. Upon treatment of the phenol **865** with phenyliodoso diacetate in methanol a tandem oxidative dearomatization/intramolecular Diels–Alder cyclization resulted with 62% yield in the cycloadduct **866**. Further elaboration of this intermediate gave the enol silane

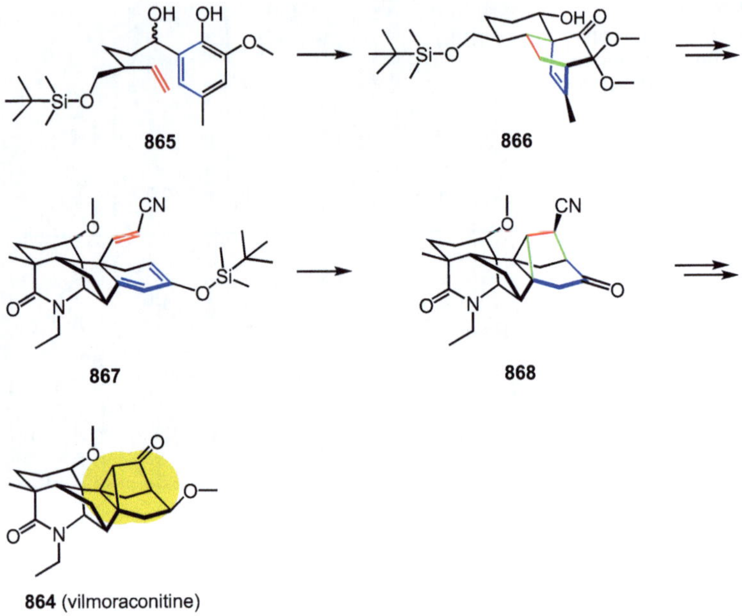


**Fig. 171** The tandem oxidative dearomatization/intramolecular Diels–Alder cyclization and an intramolecular Diels–Alder reaction as key steps in the synthesis of vilmoraconitine (**864**)

**Plate 110** *Aconitum vilmorinianum*, collected in China. California Academy of Sciences, licensed under: http://creativecommons.org/publicdomain/zero/1.0/

**867**. When **867** was heated in xylene at 150°C in the presence of catalytic amounts of *p*-toluenesulfonic acid, the intramolecular Diels–Alder reaction led to 75% yield of the strained cyclopropane **868**, which could then be transformed conventionally to the target compound vilmoraconitine (**864**) in three steps.

## *8.44 (+)-Pedrolide*

The noncytotoxic diterpenoid (+)-pedrolide (**869**) with P-glycoprotein inhibitory properties was isolated in 2021 by the R. J. Ferreira group from *Euphorbia pedroi* Molero & Rovira of the Euphorbiaeae family (Plate 111) [516]. It was proposed to be derived from a tigliane precursor and features an unprecedented 5/5/6/6/3-fused pentacyclic carbon skeleton with an embedded bicyclo[2.2.1]heptane.

The bicyclo[2.2.1]heptane core was constructed via an intramolecular cyclopentadiene-Diels–Alder cycloaddition by M. Fadel and E. M. Carreira in 2023 (Fig. 172) [517]. To this end, a norbornadiene served as an effective surrogate for cyclopentadiene, which is unmasked under mild conditions involving a complex four-step Diels–Alder reaction cascade.

To achieve this goal, the stereoselectively synthesized substrate **870** was reacted in a highly optimized manner with 0.85 equivalents of 3,6-di-2-pyridyl-1,2,4,5-tetrazine in methanol at 0°C for six days to provide 85% of **871**, which was now equipped with the aza-substituted bicycloheptene unit. The route from **870** to **874** is overall a cascade of: (1) an intermolecular Diels–Alder reaction between **870** and the tetrazine to obtain **871** (2) a retro Diels–Alder reaction with loss of nitrogen to yield **872**; (3) a second retro Diels–Alder reaction to afford the cyclopentadiene

**Plate 111** *Euphorbia pedroi*. Photograph courtesy D. T. F. Endresen, Creative Commons Attribution 2.0 Generic

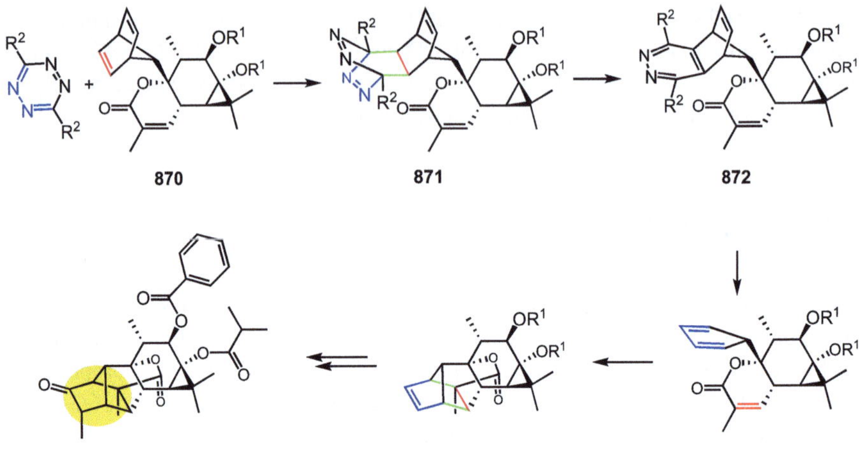


**Fig. 172** The cascade of an intermolecular Diels–Alder reaction, two successive retro Diels–Alder reactions and a further Diels–Alder reaction as the key steps in the enantioselective synthesis of (+)-pedrolide (**869**); $R^1$ = triethylsilyl, $R^2$ = 2-pyridyl

substituent in **873**, which (4) reacts in a further Diels–Alder reaction to the bicycloheptene derivative **874**. Elaboration of **874** finally provided the natural product (+)-pedrolide (**869**).

## *8.45 (+)-Lancilactone C*

Upon searching for potential anti-HIV agents, lancilactone C (**875**) was isolated along with lancilactones A, B, and kadsulactone A from the stems and roots of *Kadsura lancilimba* How. (family Schisandraceae), which have been used traditionally in Chinese folk medicine. Lancilactone C (**875**) inhibited HIV replication in H9 lymphocytes, while it did not show any cytotoxicity [518]. The initially suggested structure of **879** with a unique cycloheptatriene ring offered a good opportunity for synthesizing it via a domino Diels–Alder/elimination/electrocyclization cascade from the cyclopropene derivative **876** plus diene **877** to **878** and finally to the proposed structure **879**, as illustrated in Fig. 173 [519]. However, it turned out that the synthesized compound **679** was not identical with the natural product **875**.

Accordingly, a structure with a benzene ring instead of the cycloheptatriene ring was targeted and the product of this stereoselective synthesis involving an intramolecular Diels–Alder reaction key step proved to be identical with the natural product lancilactone C (**875**). Thus, alkyne **880** was cyclized by heating in toluene under reflux to the cyclohexadiene **881**, which could then be elaborated using conventional synthesis steps into the target **675**. Accordingly, total stereoselective synthesis is still a valuable tool to assign correctly a structure to an isolated natural product.

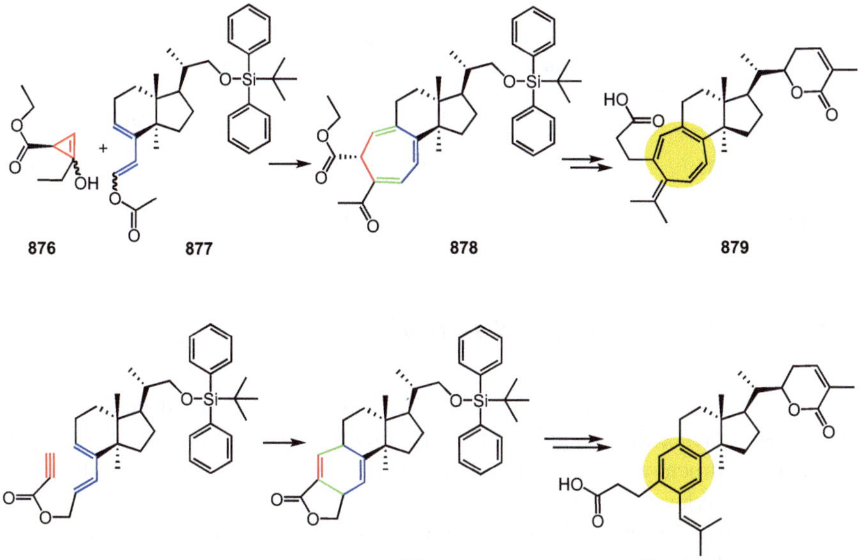


**Fig. 173** The domino intermolecular Diels–Alder/elimination/electrocyclization cascade to synthesize the proposed structure **879** and the intramolecular Diels–Alder reaction leading to the revised structure of (+)-lancilactone C (**875**)

## *8.46 (−)-Vinigrol*

In Sect. 6.38 vinigrol (**334**), isolated from the fungus strain F-5408 of *Virgaria nigra* (Link) Nees of the family Xylariaceae (Plate 43), was already described and discussed as early as 2009 with respect to its non-stereoselective synthesis using inter- and intramolecular Diels–Alder reaction key steps [221]. In 2024, the challenge to synthesize enantioselectively (−)-vinigrol (**882**) was met by using an intramolecular transannular inverse electron demand Diels–Alder strategy (Fig. 174) [520].

On heating the (−)-(*S*)-limonene-derived (chiral pool) pyrone precursor **883** under microwave irradiation at 200°C, this afforded the desired cyclized product **884** of the transannular inverse electron demand Diels–Alder reaction in 52% yield, which worked well even on a gram scale. Further conventional elaboration then led to the targeted (−)-vinigrol (**882**). The ample amounts that became available by this synthesis also allowed the demonstration that it ameliorated collagen-induced rheumatoid arthritis in a mouse model and to investigate its potential anticancer propensity. In 2024, a useful review of inverse electron demand Diels–Alder reactions was published [521].

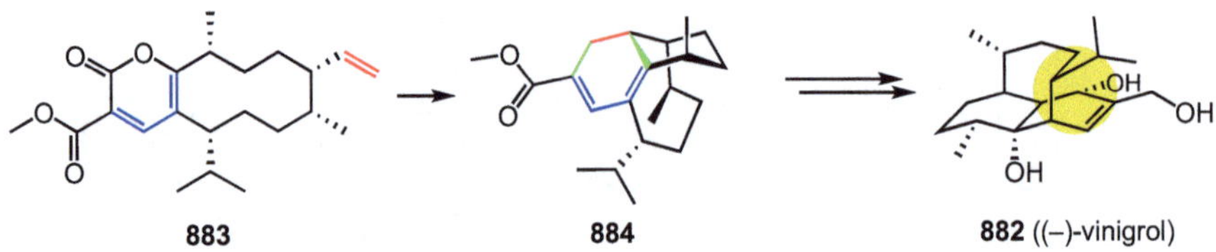


**Fig. 174** The transannular inverse electron demand Diels–Alder reaction key step in the enantioselective synthesis of (−)-vinigrol (**882**)

## *8.47 3-epi-Formicin A and 1-epi-Formicin B*

The *Streptomyces* sp. strain SFA33, associated with wood ants, produces a family of formicins, which exhibit significant antiproliferative activity against cancer cells [522]. Their structure is characterized by a hydrindane-type skeleton with thioesters bearing *N*-acetylcysteamine.

Two of these formicins, 3-*epi*-formicin A (**885**) and 1-*epi*-formicin B (**886**) were prepared by D. S. Reddy and his group, having a delicately balanced Lewis acid-mediated enantioselective intermolecular Diels–Alder reaction in common (Fig. 175) [523]. Accordingly, the dienophile (*E*)-4-oxobut-2-enoate (**887**) and the (*R*)-configured diene **888**, which was derived from commercially available (*R*)-epichlorohydrin, was reacted in the presence of 2.5 equivalents of the Lewis acid catalyst boron trifluoride etherate in dichloromethane starting at −78°C and warming up to 20°C within 4 h, yielding with high diastereoselectivity (95:5, *endo*/*exo*) 75% of the hydrindan **889**. This compound was then elaborated into the targeted 3-*epi*-formicin A (**885**).

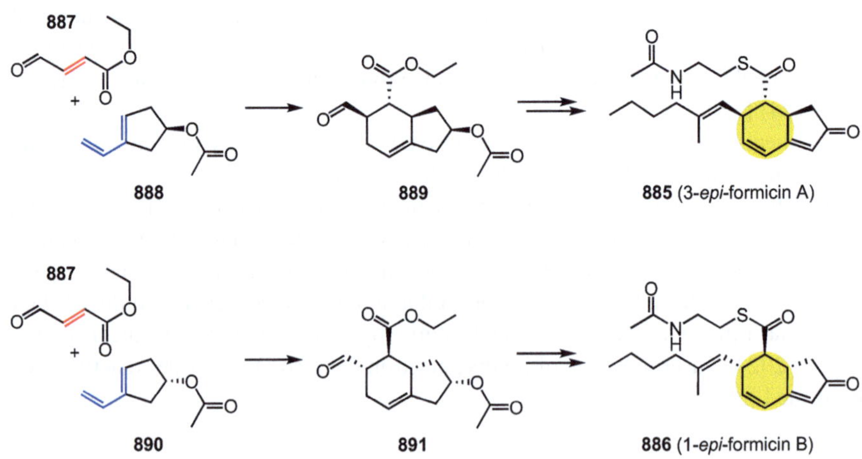


**Fig. 175** The Lewis acid-mediated enantioselective intermolecular Diels–Alder reaction key step in the syntheses of 3-*epi*-formicin A (**885**) and 1-*epi*-formicin B (**886**)

In the same manner, reaction of (*E*)-4-oxobut-2-enoate (**887**) and the (*S*)-configured diene **890** led to the hydrindan **891**, which was transformed by conventional steps into 1-*epi*-formicin B (**886**). The absolute configurations of the hydrindanes were confirmed unambiguously by single-crystal X-ray analysis of a suitable derivative.

## *8.48 Keramaphidin B and Ingenamine*

In 1994, J. Kobayashi and his group isolated the novel pentacyclic alkaloid, keramaphidin B (**892**) (Fig. 176), from an Okinawan marine sponge *Amphimedon* sp. [524]. The isolated **892** proved to be a scalemic mixture ((+)-**892**:(−)-**892** <20:1), and its relative configuration was confirmed by X-ray crystallographic analysis of

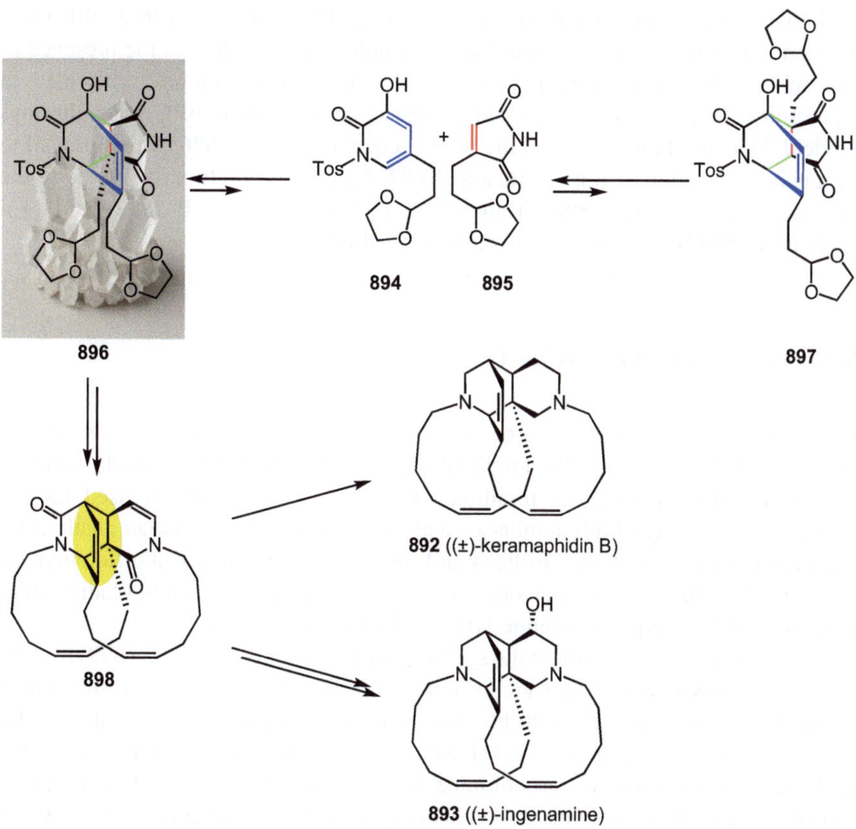


**Fig. 176** The base-catalyzed Diels–Alder reaction with completely controlled regioselectivity by dynamic crystallization as key step in the syntheses of (±)-keramaphidin B (**892**) and (±)-ingenamine (**893**); Tos = toluenesulfonate

the racemic crystals. Soon afterwards, F. Kong and R. J. Andersen isolated keramaphidin B ((+)-**892**) from the marine sponge *Xestospongia ingens* of the Petrosiidae family (a member of this genus was encountered in Sect. 7.33, Plate 68) [525]. It showed cytotoxicity against P388 murine leukemia and KB human epidermoid carcinoma cells. They also reported the isolation of ingenamine ((+)-**893**) from the same sponge. The group of P. Crews reported the isolation of xestocyclamine A [526], for which the structure was revised as the enantiomer of ingenamine ((+)-**893**) in a study by the A. Fürstner group [527]. From the viewpoint of biosynthesis, the manzamine alkaloids (which were already encountered in Sect. 8.35), including **892** and **893** are believed to be produced through the same pentacyclic intermediate generated by the transannular Diels–Alder reaction of dihydropyridine [528]. Their intriguing structure, biosynthesis, and biological activities have inspired the synthesis community.

For the total synthesis of (±)-keramaphidin B (**892**) and (±)-ingenamine (**893**), a base-catalyzed (4-dimethylaminopyridine) Diels–Alder reaction in which the regioselectivity is completely controlled by dynamic crystallization from the equilibrium reaction system, was devised by T. Sato and his team (Fig. 176) [529]. Thus, the pyridone **894** was reacted with the maleimide derivative **895** in the presence of 10 mol% 4-dimethylaminopyridine and 50 mol% lithium chloride in dichloromethane/benzene (1/1) at one molar concentration at 40°C for eight days to yield 77% of white crystals of the pure cyclization product **896**, which proved to be much less soluble than its regioisomer **897**. Further elaboration of **896** led to the bis-macrocyclic product **898**, from which (±)-keramaphidin B (**892**) and (±)-ingenamine (**893**) were derived by conventional reactions.

## 8.49 (−)-Flueggeacosine C

From the deciduous shrub *Flueggea suffruticosa* (Pall.) Baill. (syn. *Securinega ramiflora* (Pall.) Baill.)) of the Phyllanthaceae family (Plate 112), which is one of the 50 fundamental herbs in traditional Chinese medicine, and is used for the treatment of lumbago, limb numbness, and indigestion, the *Securinega* alkaloid (−)-flueggeacosine C (**899**) was isolated and shown to exhibit neuronal differentiation activity [530]. Its structure includes a remote stereogenic benzoquinolizidine fragment and a ring-opened securinine-type alkaloid moiety (Fig. 177).

As the critical key reaction in the synthesis of (−)-flueggeacosine C (**899**) by J. L. Wood and coworkers, a cascade reaction created initially an allene-substituted pyridone, which then underwent the first case of a pyridone/allene intramolecular Diels–Alder reaction (Fig. 177) [531]. Thus, reacting the pyridone **900** with allenic acid and *N,N′*-dicyclohexylcarbodiimide in dichloromethane resulted in the intermediate allenic ester **901**, which upon heating by microwave irradiation to 45°C for 15 h, was cyclized to the butenolide **902** in 42% yield. Further elaboration of this compound using conventional steps gave finally (−)-flueggeacsine C (**899**).

**Plate 112** *Flueggea suffruticosa* (syn. *Securinega ramiflora*). Photograph courtesy Dalgial, Creative Commons Attribution-Share Alike 3.0 Unported

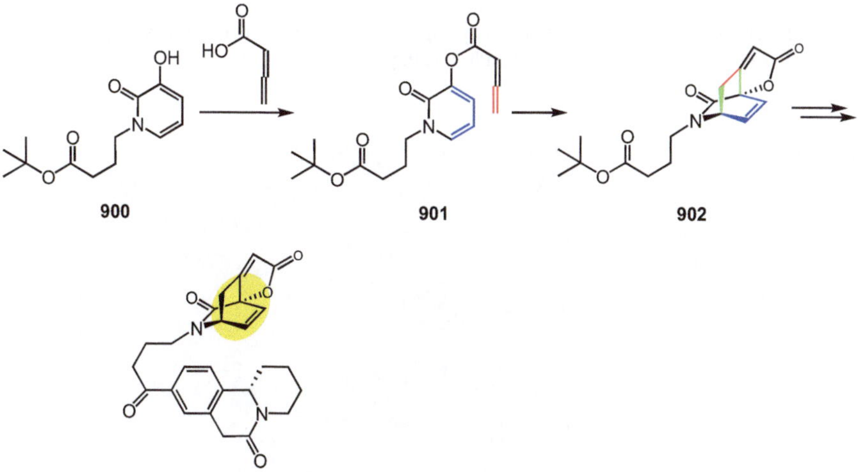


**Fig. 177** The pyridone/allene intramolecular Diels–Alder reaction key step in the synthesis of (−)-flueggeacsine C (**899**)

## 8.50 (+)-Chuanxiongolide L1

Chuanxiongnolide L1 (**903**) (Fig. 178) was isolated and characterized in 2016 from the rhizomes of *Ligusticum chuanxiong* Hort. of the Apiaceae family (Plate 113), which are known as Chuanxiong in traditional Chinese herbal medicine and used to

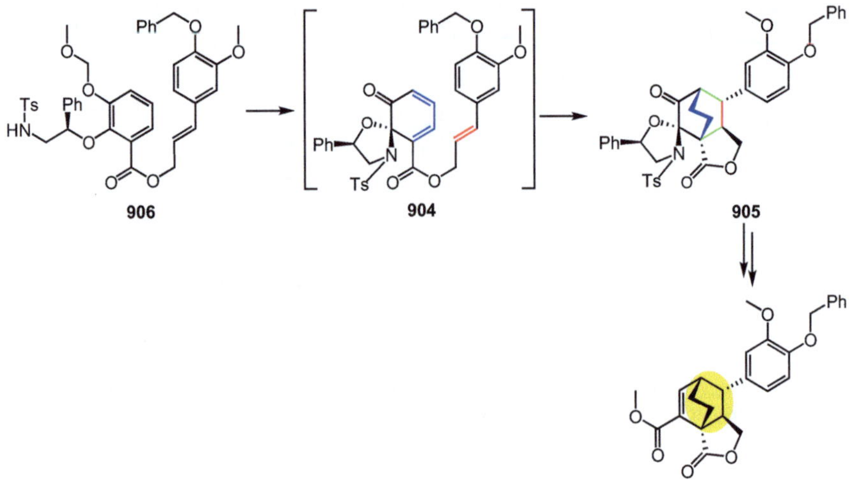


**Fig. 178** The intramolecular asymmetrically triggered Diels–Alder reaction key step in the synthesis of (+)-chuanxiongnolide L1 (**903**); Ph = phenyl, Tos = toluenesulfonate

**Plate 113** *Ligusticum chuanxiong*, cut rhizomes. Photograph courtesy F. Sagisavas, Creative Commons CC0 1.0 Universal Public Domain Dedication

relieve pain and blood stasis. This compound exhibited a potent inhibitory effect against lipopolysaccharide-induced nitric oxide production in RAW 264.7 macrophage cells [532]. A plausible biosynthesis of **903** involves an *endo*-type Diels–Alder addition of coniferyl alcohol and 4,5-dihydro-1,3-isobenzofuran-dione to form a bicyclo[2.2.2]-octane skeleton.

Following this biomimetic idea, the enantioselective synthesis of (+)-chuanxiongnolide L1 (**903**), as published by H. Zheng et al., used as key step an intramolecular asymmetrically triggered Diels–Alder reaction of intermediate **904**, which proceeded at 150°C in the presence of the antioxidant 2,6-di-*tert*-butyl-4-methylphenol to yield **905**, which could be separated readily from its diastereomer that was also produced in minor amounts (Fig. 178) [533]. Intermediate **904** was generated in situ by an asymmetric oxidative dearomatization of **906** with (diacetoxyiodo)benzene. Further elaboration of **905** by conventional means led to (+)-chuanxiongnolide L1 (**903**).

## 8.51 Fortimicin B

Aminoglycosides are highly valued antibiotics, particularly in the treatment of Gram-negative bacterial infections [534]. Fortimicins, among them fortimicin B (**907**) (Fig. 179), isolated from the bacterium *Micromonospora* species MK70

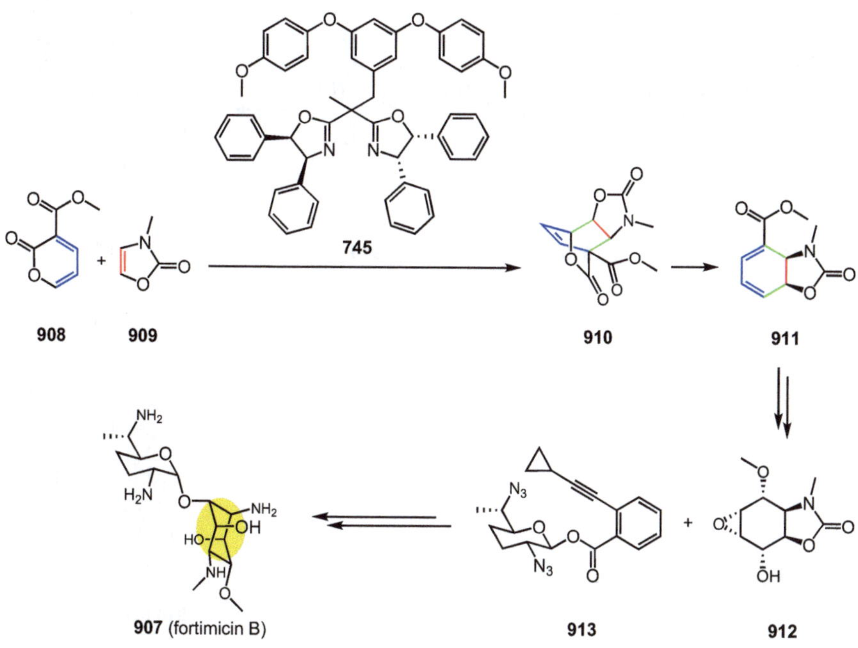


**Fig. 179** The asymmetric catalytic inverse electron demand Diels–Alder reaction key step in the synthesis of fortimicin B (**907**)

(family Micromonosporaceae) represent a special class of these aminoglycosides with a unique pseudodisaccharide skeleton [535].

The key step in the synthesis of fortimicin B (**907**) was an asymmetric catalytic inverse electron demand Diels–Alder reaction between the 2-pyrone **908** as the diene and *N*-methyl-2-oxazolone **909** as dienophile (Fig. 179) [536]. They were dissolved in chloroform and reacted in the presence of a 4 Å molecular sieve at 0°C to produce adduct **910** in 98% yield and 97% *ee*. The ligand **745**, already used some years ago (Sect. 8.20) in the asymmetric synthesis of the cephanolides (Fig. 148), was complexed with copper(II) tetrafluoroborate and utilized as the asymmetric catalyst in this reaction. This was followed by a retro Diels–Alder reaction in chlorobenzene at 152°C to give **911**. Intermediate **911** was then transformed into the fortamin synthon **912** in a classical way. This synthon was next reacted with the also conventionally synthesized 6-*epi*-purpurosamine B fragment **913** to produce fortimicin B (**907**) after several more steps. It should be mentioned that inverse electron demand Diels–Alder reactions were reviewed recently [521].

## 8.52 *Lugdunomycin*

The angucyclinone, lugdunomycin (**914**) (Fig. 180), which exists as a racemate, was isolated from the bacterium *Streptomyces* sp. QL37 (family Streptomycetaceae) in

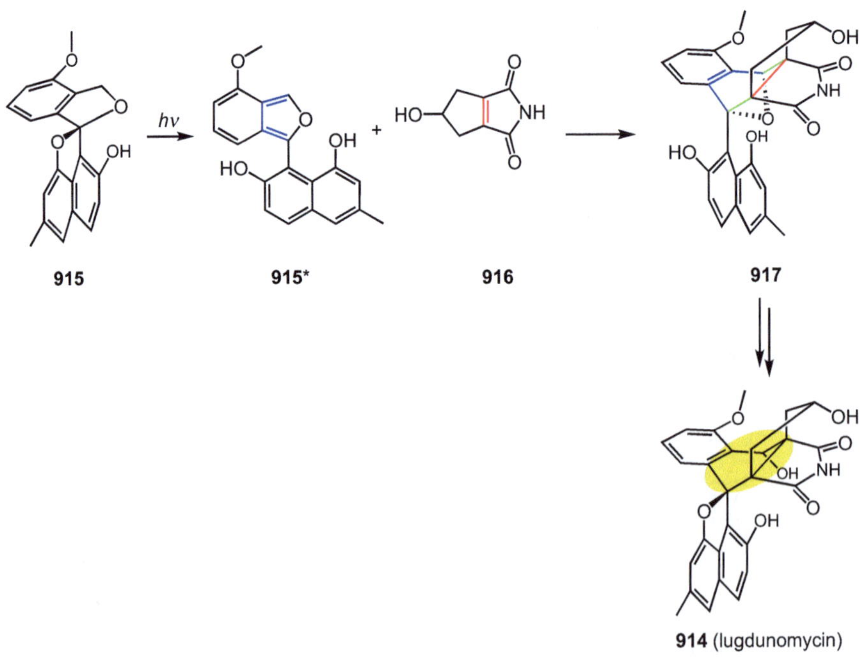


**Fig. 180** The photoinduced intermolecular isobenzofuran Diels–Alder reaction key step in the synthesis of lugdunomycin (**914**)

2019 [537], and found to exhibit selective antibacterial activity against the Gram-positive bacterium *Bacillus subtilis* strain 168 [538]. Its structure contains a unique compact [6-6-5-6-6-5-5]-heptacyclic framework with a benzaza[4.3.3]propellane motif. The carbon framework of lugdunomycin (**914**) and four stereocenters in one ring, of which three are quaternary chiral carbon atoms, has made it a highly challenging target for the chemical synthesis community.

This challenge was met recently by J. Huang and L. Zhu with a photoinduced intermolecular isobenzofuran Diels–Alder reaction key step between elmonin (**915**) and *iso*-maleimycin (**916**) (Fig. 180) [539]. Thus, **915** and six equivalents of **916** were dissolved in dimethylsulfoxide and irradiated with UV light at 365 nm at 500 W (generating the intermediary isobenzofuran **915***) at 30°C for 12 h, to yield 60% **917** as a single diastereomer. Further elaboration of intermediate **917** using conventional methodology resulted in the desired lugdunomycin (**914**).

Note: Looking at the most recent applications of the Diels–Alder reaction as key step in the synthesis of organic natural products, it is evident that there is a steady stream of novel syntheses. Thus, for example, the tetraterpenoid taiwaniaadducts I, J, and L were obtained with an entantioselective late stage intermolecular Diels–Alder reaction as the key reaction [540], and the first total synthesis of the non-classical mulberry Diels–Alder adduct morusalin A was accomplished via a stereoselective intramolecular Diels–Alder reaction as the key step [541].

## 9 Diels–Alderases

The Diels–Alder reaction, a [4 + 2] cycloaddition between a conjugated diene and a dienophile, is as illustrated in the preceding Sections, a cornerstone of natural product synthesis, but has long been postulated to occur also in Nature. Many complex natural products contain polycyclic scaffolds suggestive of such Diels–Alder reactions, and over 150 natural compounds are now thought to arise from Diels–Alder-type cycloadducts in their biosynthesis pathways. For decades, cyclizations in natural product biosynthesis were assumed to occur spontaneously, i.e., non-enzymatically, or via unknown enzymatic assistance. This left open kind of a Diels–Alderase paradox, since despite the prevalence of Diels–Alder reactions in synthesis, enzymes that catalyze true Diels–Alder cycloadditions remained elusive. Recent advances in genome mining, structural biology, and mechanistic analysis have changed this view. Over the last ten years, a growing number of naturally occurring Diels–Alderases have been identified and characterized, firmly establishing that Nature does indeed harness dedicated enzymes to catalyze Diels–Alder reactions in the construction of diverse secondary metabolites [542–545]. These discoveries open up a new frontier in biosynthesis, biocatalysis, and bioinspired synthesis, which was addressed on several occasions in the preceding sections, offering insights into enzyme catalysis of pericyclic reactions and expanding the known repertoire of enzyme fold functions. Thus, it should be meaningful to briefly discuss the specific aspect of enzymatic Diels–Alder synthesis in this section, but without attempting to do this in a comprehensive manner.

The use of pericyclic reactions in Nature extends to various classes, like electrocyclizations, sigmatropic shifts, ene reactions, and notably, cycloadditions. The intramolecular [4 + 2] cycloaddition, often yielding a decalin or bicyclic ring system, is especially important in polyketide, terpenoid, and alkaloid biosynthesis. Classic examples include the formation of octalin or decalin systems in many fungal polyketides and spiro-fused ring systems in certain antibiotics, which were hypothesized to arise via Diels–Alder reactions. Early evidence for enzymatic Diels–Alder reactions came indirectly, as e.g., isotope tracing and stereochemical outcomes in natural product formation, but direct identification of a true Diels–Alderase was long lacking. By the early 2000s, only a handful of enzymes were proposed to facilitate Diels–Alder steps, such as lovastatin synthase and solanapyrone synthase, and even fewer had been purified and studied in vitro. This began to change in the 2010s, culminating in the formal identification of several dedicated Diels–Alderases [546]. These enzymes validated that Diels–Alder reactions in biosynthesis are often enzyme-dependent rather than purely spontaneous, conferring specific stereocontrol and efficiency advantages. Below, an overview of prominent naturally occurring Diels–Alderases discovered in the past decade is summarized, along with the remarkable natural products they help assemble, and information known about how these enzymes achieve catalysis is presented.

Nature appears to utilize Diels–Alderases in multiple contexts, from bacterial antibiotics to fungal toxins and even plant metabolites. These enzymes have diverse protein folds and can be multifunctional, catalyzing Diels–Alder reactions in addition to other reactions, or standalone cyclases. The following points present representative Diels–Alderases examples characterized since about 2010, along with their biosynthesis products and key features:

- Lovastatin nonaketide synthase (LovB) is a megaenzyme polyketide synthase in fungi that cyclizes intramolecularly its polyketide chain via a Diels–Alder step to form the decalin core of lovastatin (**918**) (Fig. 181) first isolated from a strain of *Aspergillus terreus* and used as a cholesterol-lowering drug. In vitro reconstitution showed that LovB produces the specific *endo*-adduct dihydromonacolin L (**919**) with the correct stereochemistry, whereas without the enzyme, a mixture of *endo*/*exo* adducts is obtained. This demonstrated enzymatic control of the cycloaddition outcome in a pharmacologically important natural product [547, 548].
- Solanapyrone synthase (Sol5) is an iterative fungal polyketide synthase that was proposed to catalyze a Diels–Alder reaction forming the bicyclic solanapyrone A (**142**) (Fig. 37). Genetic studies in *Alternaria* fungi suggest Sol5's polyketidesynthase elongation is followed by an enzyme-tethered cyclization to yield the cyclohexene core. Accordingly, Sol5 exemplifies how polyketide

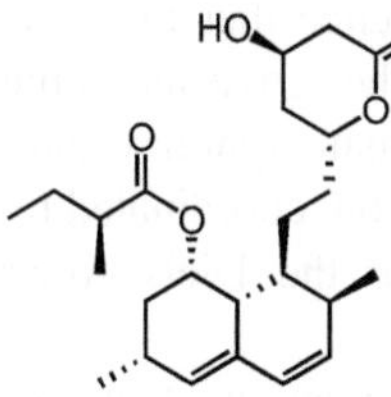

**918** (lovastatin)

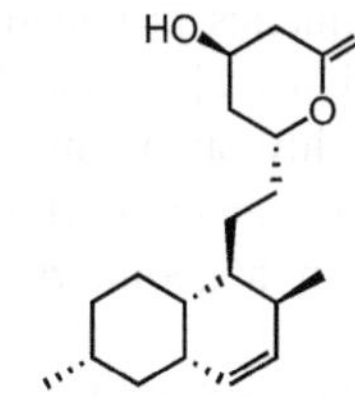

**919** (dihydromonacolin L)

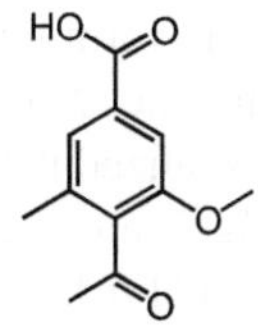

**920** (macrophomic acid)

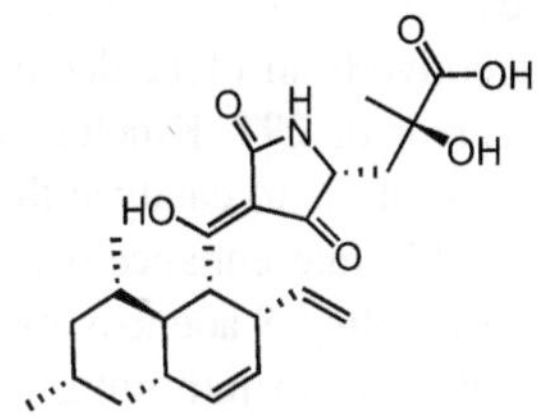

**921** (Sch210972)

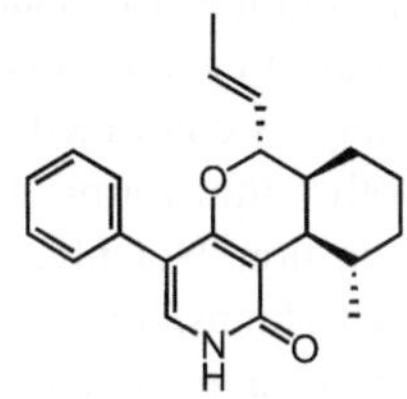

**922** (leporin C)

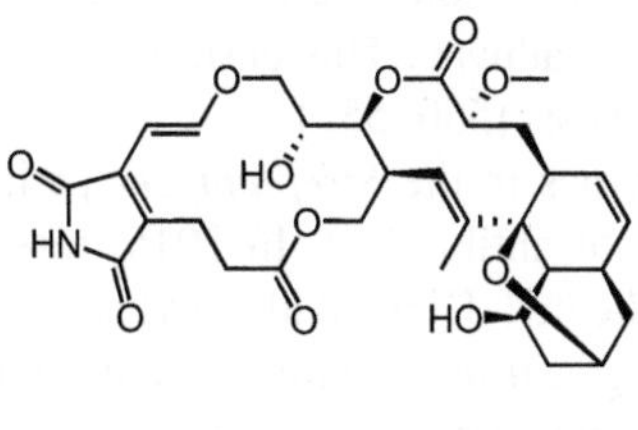

**923** (streptoseomycin)

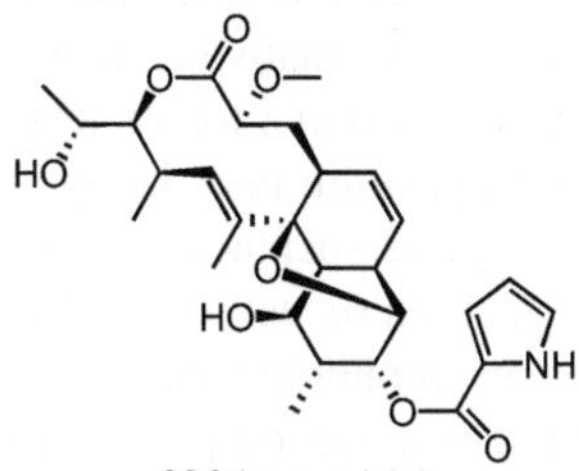

**924** (nargenicin)

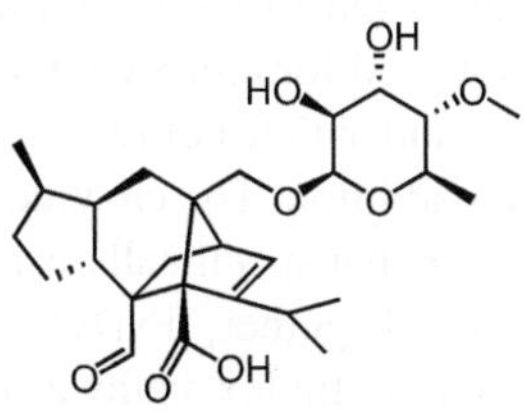

**925** (sordarin)

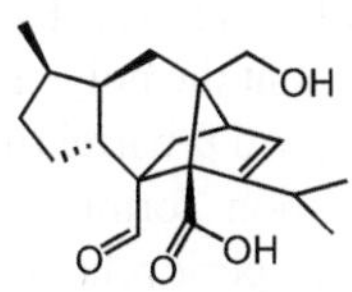

**926** (sordaricin)

**Fig. 181** Selected products **918**–**926** biosynthesized by Diels–Alderases

assembly lines can incorporate pericyclic steps directly into product formation [549].

- Macrophomate synthase is an enzyme from *Macrophoma* fungi that catalyzes an intermolecular Diels–Alder reaction between 2-pyrone and oxaloacetate to form a benzoate (macrophomic acid (**920**) (Fig. 181)). Macrophomate synthase was the first enzyme ever shown to perform a Diels–Alder reaction: it uses the stepwise sequence of oxaloacetate decarboxylation, C–C bond formation via Diels–Alder, and another decarboxylation, to yield the bicyclic product. Inhibition by stable bicyclic analogs and detailed kinetics provided strong evidence that the C–C bond-forming step is a concerted [4 + 2] cycloaddition inside the active site of the enzyme. X-ray crystal structure investigations of macrophomate synthase later visualized the two substrates bound and poised for cycloaddition, coordinated by $Mg^{2+}$ and active-site residues in a geometry favorable for the Diels–Alder step [176].

- SpnF is a stand-alone Diels–Alderase from *Saccharopolyspora spinosa* that produces spinosyn A (**297**) (Fig. 71), a complex polyketide insecticide. Discovered in 2011, SpnF was the first monofunctional enzyme shown to accelerate exclusively a [4 + 2] cycloaddition in secondary metabolism. It converts an 11,12-deoxyspinosyn intermediate into a cyclohexene-containing precursor of **297**. Kinetic assays and mutational analyses confirmed that SpnF's sole activity is to catalyze the cyclization and no other transformations, with a measurable rate enhancement over the uncatalyzed reaction. Notably, SpnF has a fold resembling *S*-adenosylmethionine-dependent methyltransferases, yet it binds no cofactor, an intriguing case of an enzyme evolving a pericyclic function within a common fold. Computational studies indicate that the active site of SpnF preferentially stabilizes a concerted Diels–Alder transition state, steering the reaction along the [4 + 2] pathway and suppressing an alternative competing [6 + 4] pathway. This ensures formation of the correct cycloadduct for spinosyn biosynthesis [550, 551].
- PyrE3 and PyrI4 are two flavoprotein enzymes in the *Streptomyces* pathway to pyrroindomycins (pentacyclic spiro-tetramate antibiotics). In 2015, researchers reported that these act in tandem as dedicated Diels–Alder cyclases: PyrE3 catalyzes an intramolecular [4 + 2] forming a *trans*-decalin system, and PyrI4 then catalyzes a second [4 + 2] forming a spiro-tetramate ring. This enzymatic [4 + 2] cascade constructs the densely fused core of pyrroindomycins in a stereoselective manner. The enzyme PyrE3 was found to be a FAD-dependent enzyme, but with a twist: the FAD cofactor is not used for redox chemistry at all. Structural studies showed FAD is bound in a manner that buttresses the active site, helping orient the polyene substrate for cycloaddition. The enzyme PyrI4, on the other hand, is a β-barrel (lipocalin-like) enzyme that specifically catalyzes the second cyclization forming the spiro-conjugate. Together, PyrE3 and PyrI4 highlighted that pericyclases can be embedded in complex pathways and that Nature can evolve tandem Diels–Alderases to build complex polycycles [552].
- CghA is a lipocalin-like enzyme from the fungus *Chaetomium globosum*, responsible for the decalin ring in the anti-HIV agent Sch210972 (**921**) (Fig. 181) (a tetramate macrolactam). Deletion of *cghA* in the fungus led to a mixture of two stereoisomeric cycloadducts (both *endo* and *exo* octalin diastereomers) instead of the single naturally occurring isomer. Only when CghA was present did the pathway produce exclusively the correct octalin core stereochemistry. This genetic evidence proved CghA acts as a Diels–Alderase that enforces diastereoselectivity in the intramolecular [4 + 2] cyclization step. A 2021 study of CghA, referred to as an octalin-forming Diels–Alderase, revealed that the enzyme inverts the usual *endo*/*exo* selectivity via its active-site environment. In other words, CghA's binding pocket favors the formation of the less

thermodynamically favored isomer, illustrating how an enzyme can override the normal *endo* rule of Diels–Alder reactions to yield the biologically needed product [553].

- LepI is a remarkable multifunctional enzyme from *Aspergillus* fungi that produces leporin C (**922**) (Fig. 181), highlighting the versatility of natural Diels–Alder chemistry. Discovered in 2017, LepI is an *S*-adenosylmethionine-dependent enzyme that performs three reactions in sequence: first, it dehydrates a 2-pyridone precursor to form a reactive conjugated quinone methide intermediate; second, it catalyzes an inverse electron demand hetero-Diels–Alder cycloaddition between that quinone methide (as dienophile) and an alkene, forming a new heterocycle, and third, it also catalyzes an intramolecular [4 + 2] cycloaddition in the same intermediate, followed by a retro-Claisen rearrangement, ultimately yielding the tricyclic leporin C framework. Remarkably, LepI can funnel the reaction along two different pericyclic pathways from a single ambimodal transition state. This ambimodal behavior means the enzyme's active site is arranged such that the intermediate can collapse into either a [4 + 2] adduct or an alternative product, yet with LepI each path gives a specific stereochemical outcome. X-ray crystallography and DFT calculations showed that LepI employs hydrogen-bond catalysis to both facilitate dehydration and to orient/polarize the quinone methide for cycloaddition. Thus, LepI is both a synthase and a pericyclase, illustrating the strategy of Nature of evolving multifunctional enzymes that combine pericyclic chemistry with conventional catalytic steps [554, 555].
- StmD/NgnD The enzyme StmD is a Diels–Alderase from *Streptomyces* that assembles the streptoseomycin (**923**) antibiotic core, while NgnD is a homolog in nargenicin (**924**) (Fig. 181) biosynthesis. These enzymes drew attention for catalyzing a rare [6 + 4] cycloaddition along with the usual [4 + 2]. In vitro, StmD transforms its polyene substrate into two products: a major exo-[6 + 4] adduct and a minor *exo*-[4 + 2] adduct, which can interconvert via a Cope rearrangement. Intriguingly, both cycloadducts lie on the pathway and can eventually lead to the final antibiotic, illustrating an ambimodal pericyclase that initially produces a mixture of pericyclic products. Only a trace of the *endo*-[6 + 4] (thermodynamically most stable) is formed, indicating StmD enforces significant stereoselectivity favoring the exo paths. The existence of an enzymatic [6 + 4] cycloaddition was unprecedented: StmD's dual cyclase activity solved the crystal structure of NgnD, identifying key active-site residues that polarize the transition state to enable the forbidden [6 + 4] mode. Site-directed mutagenesis of those residues could modulate the product ratio, supporting their role in transition state stabilization. The enzymes StmD and NgnD thus expand the definition of Diels–Alderases to include ambimodal pericyclases that can catalyze alternate cycloaddition manifolds [543, 556, 557].
- PdxI/EpiI are two SAM-dependent enzymes from fungi that diverge to perform either a hetero-Diels–Alder or an Alder-ene reaction on a common intermediate. In the fungal genera *Albophoma* Tak. and *Epicoccum* Link, a quinone methide intermediate can undergo either a [4 + 2] cycloaddition or a concerted ene

reaction. Bioinformatic mining predicted two homologous enzymes—later confirmed as PdxI and EpiI. The first enzymatic Alder-ene reaction is catalyzed by one of these enzymes [543]. These findings underscore that pericyclases in Nature can evolve to catalyze different pericyclic reactions on the same substrate, simply by altering the active-site contours and electrostatic environment. The PdxI/EpiI enzymes belong to the same *S*-adenosyl methionine-methyltransferase-like family as LepI and SpnF, suggesting an evolutionary theme of *S*-adenosyl methionine enzyme lineage repurposed for pericyclic catalysis.

- SdnG is A recently discovered Diels–Alderase from fungi that catalyzes a textbook [4 + 2] reaction in terpene biosynthesis. The enzyme SdnG was identified in 2022 as the enzyme that generates the norbornene core of sordaricin (**926**), a terpene precursor to the antifungal sordarin (**925**) (Fig. 181) [558]. Until this discovery, no enzyme was known to catalyze the classic cyclopentadiene + olefin → norbornene Diels–Alder reaction in Nature. The enzyme SdnG, encoded in the *Sordaria* fungus gene cluster, performs an intramolecular [4 + 2] that links a cyclopentadiene moiety and an alkene within the oxidatively cyclized terpene scaffold. Full pathway reconstitution showed that upstream oxidative enzymes create a highly functionalized dienophile that SdnG then cyclizes. Notably, SdnG accelerates the cyclization to such an extent that it outcompetes unwanted side reactions: without SdnG, the activated dienophile tended to undergo off-pathway polymerization or hydration, but in the presence of the enzyme the Diels–Alder product is cleanly formed. This demonstrates a key biological role for Diels–Alderases: by catalyzing the cycloaddition quickly and specifically, they can prevent shunt reactions and stabilize delicate intermediates. The enzyme SdnG has been crystallized (a dimeric α/β enzyme) and shown to be glycosylated, consistent with it being a secreted enzyme in the fungal periplasm.
- Plant Diels–Alderases (Moraceae): In 2024, the first plant Diels–Alderases were reported from the mulberry tree (*Morus alba*) [559]. Mulberry and other Moraceae produce numerous complex phenolic natural products (e.g., Diels–Alder type adducts of prenylated stilbenes) for which their formation was hypothesized to involve enzymatic cycloadditions. *Morus alba* Diels–Alderase enzymes were isolated that catalyze an intermolecular [4 + 2] cycloaddition between an electron-rich dihydrostilbene and an isoprenylated benzofuran, forming natural Diels–Alder adducts (such as mulberrofuran compounds, see Sect. 7.4). These intermolecular plant Diels–Alderases are glycosylated enzymes (consistent with secreted proteins in plants) and belong to the FAD-dependent oxidoreductase superfamily thus raising the question of how an oxidase evolves into a Diels–Alderase. Evolutionary analysis suggests the plant Diels–Alderases diverged from ancestral flavin-dependent oxidative cyclases, losing the oxidative function and repurposing the FAD cofactor for purely structural support. Mutagenesis identified a few key amino acid substitutions that were sufficient to switch an oxidase into a cyclase, by altering the substrate binding mode to favor Diels–Alder coupling without redox chemistry. This finding provided mechanistic insight into enzyme evolution: Nature can recruit existing enzyme scaffolds (even ones with cofactors like FAD) and, through minimal changes, endow

them with novel pericyclic catalytic activity. The *Morus* Diels–Alderases thus not only expand the phylogenetic range of Diels–Alderases into plants, but also shed light on the molecular evolution of cycloaddition catalysts in Nature.

Overall, the last decennium has transformed Diels–Alderases from a speculative idea into an expanding catalog of real enzymes with diverse structures and functions. These studies not only solve longstanding puzzles of natural product biosynthesis but also enrich our fundamental understanding of enzyme catalysis. As we continue to unravel how Nature orchestrates pericyclic reactions, we gain new tools for synthetic biology and green chemistry envisioning a future where engineered Diels–Alderases could be employed to make complex molecules in the lab as elegantly as Nature does in the cell [560]. The discoveries to date are undoubtedly just the tip of the iceberg, and ongoing research promises many more exciting revelations at the interface of organic chemistry and enzymology.

# 10 Conclusions

This volume provides an exploration of the Diels–Alder reaction, tracing its evolution from a classical synthesis tool to a cornerstone of modern organic chemistry. It highlights the versatility of this reaction in constructing complex molecular architectures, its pivotal role in natural product synthesis, and its expanding applications in biomimetic and catalytic methodologies. A few points may summarize this journey through the centuries.

Historical significance: the Diels–Alder reaction, discovered in the early twentieth century, revolutionized organic synthesis by enabling the efficient formation of six-membered rings. Its stereoselectivity and regioselectivity made it indispensable for synthesizing natural products and pharmaceuticals.

Methodological advances: over the decades, innovations such as asymmetric catalysis, organocatalysis, and inverse electron demand variants have expanded the scope of the reaction. These advancements have enabled the synthesis of highly functionalized, stereochemically dense frameworks with precision.

Biomimetic aspects: the reaction has been instrumental in emulating biosynthesis pathways, bridging the gap between chemical synthesis and natural product biogenesis. Examples like chalcomoracin, sorocenol B, and paeoveitol illustrate how biomimetic strategies can replicate the logic of Nature.

Enzymatic catalysis: the discovery of Diels–Alderases has unveiled the ability of Nature to catalyze pericyclic reactions with remarkable efficiency and stereocontrol. These enzymes, found in fungi, bacteria, and plants, offer insights into biosynthesis and open avenues for biocatalysis.

Synthesis challenges: the reaction has been pivotal in tackling synthesis challenges posed by intricate natural products, such as vinigrol, scabrolide, and the manzamine alkaloids. Its ability to construct polycyclic frameworks in a single step underscores its strategic importance.

Future directions: the Diels–Alder reaction continues to evolve, with emerging applications in sustainable chemistry, photochemical activation, and enzyme engineering. Its integration with modern computational tools and cross-coupling strategies promises further breakthroughs.

The Diels–Alder reaction exemplifies the synergy between mechanistic creativity and strategic minimalism in organic synthesis. Its enduring relevance lies in its adaptability, enabling synthesis chemists to construct molecular complexity with elegance and efficiency. As research advances, the role of this reaction as a design principle in retrosynthetic planning and its potential in biocatalysis will undoubtedly expand, solidifying its place as a cornerstone of chemical innovation in the years to come.

**Acknowledgments** Writing needs patience—thank you, Rotraud my love, for yours and the feeling that there is life beyond writing, and thank you, Alexander, for providing me with a valuable library assistant, named ChatGPT 03 Research. I am also extremely grateful to my colleagues Douglas Kinghorn and Simon Gibbons for all the years of editorial adventures and their encouragement to write this Diels–Alder story, and their effort to polish it into shape.

## References

1. Diels O, Alder K (1928) Synthesen in der hydroaromatischen Reihe; 1. Mitteilung: Anlagerungen von "Di-en"-kohlenwasserstoffen. Justus Liebigs Ann Chem 460:98
2. Diels O (1939) Bedeutung der Dien-Synthese für Bildung, Aufbau und Erforschung der Naturstoffe. Prog Chem Org Nat Prod 3:1
3. Cohen A, Warren FL (1937) The synthesis of compounds related to the sterols, bile acids, and oestrus-producing hormones. Part XI. A "diene-synthesis" of phenanthrene and hydrophenanthrene derivatives. J Chem Soc 1937:1315
4. Woodward RB, Sondheimer F, Taub D, Heusler K, McLamore WMJ (1952) The total synthesis of steroids. J Am Chem Soc 74:4223
5. Woodward RB, Bader FE, Bickel H, Frey AJ, Kierstead RW (1956) The total synthesis of reserpine. J Am Chem Soc 78:2023. 1958; Tetrahedron 2:1
6. Woodward RB (1942) The mechanism of the Diels–Alder reaction. J Am Chem Soc 64:3058
7. Corey EJ, Weinshenker NM, Schaaf TK, Huber W (1969) Stereo-controlled synthesis of prostaglandins F-2a and E-2 (*dl*). J Am Chem Soc 91:5675
8. Woodward RB, Hoffmann R (1969) The conservation of orbital symmetry. Angew Chem 8: 781
9. Danishefsky S, Kitahara T (1974) Useful diene for the Diels–Alder reaction. J Am Chem Soc 96:7807
10. Funk RL, Vollhardt KPC (1980) Transition-metal-catalyzed alkyne cyclizations. A cobalt-mediated synthesis of *dl*-estrone. J Am Chem Soc 102:5253
11. Danishefsky S, Hirama M, Fritsch N, Clardy J (1979) Synthesis of disodium prephenate and disodium epiprephenate. Stereochemistry of prephenic acid and an observation on the base-catalyzed rearrangement of prephenic acid to *p*-hydroxyphenyllactic acid. J Am Chem Soc 101:7013
12. Mo Y, Ning L, Luo Z, Yang L, Tang S, Dong S, Zhou Q-L, Feng X (2024) Asymmetric photoenolization/Diels–Alder reaction of 2-methylbenzaldehydes and 2-alkylbenzophenones with chromones. ACS Catal 14:6687
13. Norton JA (1942) The Diels–Alder diene synthesis. Chem Rev 31:319

14. Kloetzel MC (1948) The Diels–Alder reaction with maleic anhydride. Org React 4:1
15. Brieger G, Bennett JN (1980) The intramolecular Diels–Alder reaction. Chem Rev 80:63
16. Sauer J, Sustmann R (1980) Mechanistic aspects of Diels–Alder reactions: a critical survey. Angew Chem Int Ed 19:779
17. Kagan HB, Riant O (1992) Catalytic asymmetric Diels–Alder reactions. Chem Rev 92:1007
18. Winkler JD (1996) Tandem Diels–Alder cycloadditions in organic synthesis. Chem Rev 96: 167
19. Rickborn B (1998) The retro Diels–Alder reaction. Part I. C–C dienophiles. Org React 52:1
20. Rickborn B (1998) The retro Diels–Alder reaction. Part II. Dienophiles with one ore more heteroatoms. Org React 52:223
21. Brocksom TJ, Nakamura J, Ferreira MC, Brocksom U (2001) The Diels–Alder reaction: an update. J Braz Chem Soc 12:597
22. Nicolaou KC, Snyder SA, Montagnon T, Vassilikogiannakis G (2002) The Diels–Alder reaction in total synthesis. Angew Chem Int Ed 41:1668
23. Taşdelen MA (2011) Diels–Alder click reactions: recent applications in polymer and material sciences. Polym Chem 2:2133
24. Jiang X, Wang R (2013) Recent developments in catalytic asymmetric inverse electron demand Diels–Alder reactions. Chem Rev 113:5515
25. Eschenbrenner-Lux, Kumar K, Waldmann H (2014) The asymmetric hetero-Diels–Alder reaction in the syntheses of biologically relevant compounds. Angew Chem Int Ed 53:11146
26. Heravi MM, Ahmadi T, Ghavidel M, Heidari B, Hamidi H (2015) Recent applications of the hetero-Diels–Alder reaction in the total synthesis of natural products. RSC Adv 5:101999
27. Gregoritza M, Brandl FP (2015) The Diels–Alder reaction: a powerful tool for the design of drug delivery systems and biomaterials. Eur J Pharm Biopharm B 97:438
28. Hou ICY, Hu Y, Narita A, Müllen K (2017) Diels–Alder polymerization: a versatile synthetic method toward functional polyphenylenes, ladder polymers and graphene nanoribbons. Polym J 50:3
29. Oliveira BL, Guo Z, Bernardes GJL (2017) Inverse electron demand Diels–Alder reactions in chemical biology. Chem Soc Rev 46:4895
30. Houk KN, Liu F, Yang Z, Seeman JI (2021) Evolution of the Diels–Alder reaction mechanism since the 1930s: Woodward, Houk with Woodward, and the influence of computational chemistry on understanding cycloadditions. Angew Chem Int Ed 60:12660
31. Briou B, Améduri B, Boutevin B (2021) Trends in the Diels–Alder reaction in polymer chemistry. Chem Soc Rev 50:11055
32. Yeingst TJ, Helton AM, Hayes DJ (2024) Applications of Diels–Alder chemistry in biomaterials and drug delivery. Macromol Biosci 24:e2400274
33. Rana A, Mishra A, Awashti SK (2025) Recent advancements in the chemistry of Diels–Alder reaction for total synthsis of natural products: a comprehensive review (2020–2023). RSC Adv 15:4496
34. Albrecht W (1906) Additionsprodukte von Cyklopentadiën und Chinonen. Justus Liebigs Ann Chem 348:31
35. Staudinger H (1912) Die Ketene. Ferdinand Enke, Stuttgart, p 59
36. Alder K, Stein G (1937) Untersuchungen über den Verlauf der Diensynthese. Angew Chem 50:510
37. Bachmann WE, Kloetzel MC (1938) The Diels–Alder reaction between naphthylcyclopentenes and maleic anhydride. J Am Chem Soc 60:2204
38. Gates M, Tschudi G (1952) The synthesis of morphine. J Am Chem Soc 74:1109
39. Gates M (1950) The synthesis of ring systems related to morphine. III. 5,6-Dimethoxy-4-cyanomethyl-1,2-naphthoquinone and its condensation with dienes. J Am Chem Soc 72:228
40. Müller JM, Schlittler E, Bein HJ (1952) Reserpin, der sedative Wirkstoff aus *Rauwolfia serpentina* Benth. Experientia 8:338
41. Woodward RB, Gosteli J, Ernest I, Friary RJ, Nestler G, Raman H, Sitrin R, Suter C, Whitesell JK (1973) A novel synthesis of prostaglandin $F_{2\alpha}$. J Am Chem Soc 95:6853

42. Woodward RB, Hoffmann R (1965) Stereochemistry of electrocyclic reactions. J Am Chem Soc 87:395
43. Plieninger H (1962) Prephenic acid: properties and the present status of its synthesis. Angew Chem Int Ed 1:367–372
44. Breitholle EG, Fallis AG (1976) A total synthesis of (±)-cedrol and (±)-cedrene via an intramolecular Diels–Alder reaction. Can J Chem 54:1991
45. Jung ME, Lowe JA (1977) Synthetic approaches to adriamycin involving Diels–Alder reactions of photochemically generated bisketenes. Total synthesis of islandicin and digitopurpone. J Org Chem 42:2371
46. Gupta R, Chakrabarty SK (2013) Gibberellic acid in plant. Plant Signal Behav 8:e25504
47. Corey EJ, Danheiser RL, Chandrasekaran S, Siret P, Keck GE, Gras J-L (1978) Stereospecific total synthesis of gibberellic acid. A key tricyclic intermediate. J Am Chem Soc 100:8031
48. Corey EJ, Danheiser RL, Chandrasekaran S, Keck GE, Gopalan B, Larsen SD, Siret P, Gras J-L (1978) Stereospecific total synthesis of gibberellic acid. J Am Chem Soc 100:8034
49. Kametani T, Nemoto H, Ishikawa H, Shiroyama K, Fukumoto K (1976) A formal regio- and stereoselective total synthesis of estrone. A convenient synthesis of D-homoestrone. J Am Chem Soc 98:3378
50. Büchi G, Hauser A, Limacher J (1977) The synthesis of khusimone. J Org Chem 42:3323
51. Stevens R (1967) The chemistry of hop contents. Chem Rev 67:19
52. Wilson SR, Mao DT (1978) An intramolecular Diels–Alder route to eudesmane sesquiterpenoids. J Am Chem Soc 100:6289
53. Porter LA (1967) Picrotoxinin and related substances. Chem Rev 67:441
54. Roush WR (1978) Total synthesis of (±)-dendrobine. J Am Chem Soc 100:3599
55. Weinreb SM, Khatri NA, Shringarpure J (1979) Alkaloid synthesis by the intramolecular Diels–Alder reaction. *d*-Coniceine and tylophorine. J Am Chem Soc 101:5073
56. Taber TF, Gunn BP (1979) Control elements in the intramolecular Diels–Alder reaction. Synthesis of (±)-torreyol. J Am Chem Soc 101:3992
57. Lundgren L, Bergström G (1976) Wing scents and scent-released phases in the courtship behavior of *Lycaeides argyrognomon* (Lepidoptera: Lycaenidae). J Chem Ecol 1:399
58. Fruzzetti F, Fidecicchi T, Guevara MMM, Simoncini T (2021) Estetrol: a new choice for contraception. J Clin Med 10:5625
59. Stanczyk FZ, Winer SA, Foidart JM, Archer DF (2024) Comparison of estrogenic components used for hormonal contraception. Contraception 130:110310
60. Dauben WG, Kessel CR, Takemura KH (1980) Simple, efficient total synthesis of cantharidin via a high-pressure Diels–Alder reaction. J Am Chem Soc 102:6893
61. Schiff Jr, PL (2006) Ergot and its alkaloids. Am J Pharm Educ 70:98
62. Kornfeld EC, Fornefeld EJ, Kline GB, Mann MJ, Morrison DE, Jones RG, Woodward RB (1956) The total synthesis of lysergic acid. J Am Chem Soc 78:3087
63. Oppolzer W, Francotte E, Bättig K (1981) Total synthesis of (±)-lysergic acid by an intramolecular imino-Diels-Alder reaction. Helv Chim Acta 64:478
64. Stork G, Clark G, Shiner CS (1981) A New route to 11-oxygenated steroids. J Am Chem Soc 103:4948
65. Stork G, Sherman DH (1982) Efficient de novo construction of the indanpropionic acid precursor of 11-keto steroids. An improved internal Diels–Alder sequence. J Am Chem Soc 104:3758
66. Jung ME, Halweg KM (1984) Stereoselectivity in intramolecular Diels–Alder reactions: a short total synthesis of (±)-estrone. Tetrahedron Lett 25:2121
67. Nemoto H, Nagai M, Fukumoto K, Kametani T (1985) A stereoselective total synthesis of $\Delta^9$-progesterone via an intramolecular cycloaddition reaction. Tetrahedron Lett 26:4613
68. Van Royen LA, Mijngheer R, De Clercq PJ (1985) Intramolecular Diels–Alder reaction with furan-diene. Total synthesis of (±)-11-ketotestosterone and (±)-adrenosterone. Tetrahedron 41:4667

69. Wong CF, LaLonde RT (1970) Sesquiterpene alkaloids of *Nuphar luteum* subsp. *variegatum*. Phytochemistry 9:2417
70. Maurer B, Ohloff G (1976) Zur Kenntnis der stickstoffhaltigen Inhaltsstoffe von *Castoreum*. Helv Chim Acta 59:1169
71. Wiart C (2013) Lead compounds from medicinal plants for the treatment of cancer. Academic Press, San Diego, CA, USA, p 1
72. Hwang YC, Fowler FW (1985) Diels–Alder reaction of 1-azadienes. A total synthesis of deoxynupharidine. J Org Chem 50:2719
73. Gordaliza M, García PA, del Corral JM, Castro MA, Gómez-Zurita MA (2004) Podophyllotoxin: distribution, sources, applications and new cytotoxic derivatives. Toxicon 44:441
74. Jung ME, Lam PY-S, Mansuri MM, Speltz LM (1985) Stereoselective synthesis of an analogue of podophyllotoxin by an intramolecular Diels–Alder reaction. J Org Chem 50:1087
75. Haskell TH, Bartz QR (1959) Actinobolin, a new broad-spectrum antibiotic: isolation and characterization. Antibiot Ann 1958–1959:505
76. Antosz FJ, Nelson DB, Herald DL, Munk ME (1970) The structure and chemistry of actinobolin. J Am Chem Soc 92:4933
77. Kozikowski AP, Konoike T, Nieduzak TR (1986) A total synthesis of actinobolin via the intermolecular Diels–Alder reaction of a threonine-derived diene. J Chem Soc Chem Commun:1350
78. Ihara M, Sudow I, Fukumoto K, Kametani T (1986) Stereoselective total synthesis of testosterone and androsterone via A/B-ring construction of the steroidal ring system by intramolecular Diels–Alder reaction. J Chem Soc Perkin Trans I:117
79. Jones H, Rassmusson GH (1979) Recent advances in the biology and chemistry of vitamin D. Prog Chem Org Nat Prod 39:63
80. Stambulyan H, Minehan TG (2016) A protecting group-free synthesis of (−)-hortonones A–C from the Inhoffen–Lythgoe diol. Org Biomol Chem 14:8728
81. Brandes E, Grieco PA, Garner P (1988) Diastereoselection in an aqueous Diels–Alder reaction: a formal total synthesis of the Inhoffen–Lythgoe diol. J Chem Soc Chem Commun:500
82. Takahashi T, Shimizu K, Doi T, Tsuji J (1988) Macroring contraction methodology. 4. A novel route to steroid A, B, C rings by the transannular Diels–Alder reaction of the 14-membered (*E,E,E*)-macrocyclic triene. J Am Chem Soc 110:2674
83. Misico RI, Nicotra VE, Oberti JC, Barboza G, Gil RR, Burton G (2011) Withanolides and related steroids. Prog Chem Org Nat Prod 94:127
84. Perez-Medrano A, Grieco PA (1991) Synthesis of the highly oxygenated ergostane type steroid (+)-withanolide E. J Am Chem Soc 113:1057
85. Kingston DGI, Jagtap PG, Yuan H, Samala L (2002) The chemistry of taxol and related taxoids. Prog Chem Org Nat Prod 84:53
86. Winkler JD, Kim HS, Kim S (1995) A highly efficient synthesis of taxanes via the tandem Diels–Alder reaction. Tetrahedron Lett 36:687
87. Ohtani I, Moore RE, Runnegar MTC (1992) Cylindrospermopsin: a potent hepatotoxin from the blue-green alga *Cylindrospermopsis raciborskii*. J Am Chem Soc 114:7941
88. Kinnear S (2010) Cylindrospermopsin: a decade of progress on bioaccumulation research. Mar Drugs 8:542
89. Heintzelman GR, Weinreb SM, Parvez M (1996) Imino-Diels–Alder-based construction of a piperidine A-ring unit for the total synthesis of the marine hepatotoxin cyclindrospermopsin. J Org Chem 61:4594
90. Ayer WA, Saeedi-Ghomi MH (1981) 1-Sterpurpurene-3,12,14-triol and 1-sterpurene, metabolites of silver-leaf disease fungus *Stereum purpureum*. Can J Chem 59:2536
91. Strunz GM, Bethell R, Dumas MT, Boyonoski N (1997) On a new synthesis of sterpurene and the bioactivity of some related *Chondrostereum purpureum* sesquiterpene metabolites. Can J Chem 75:742

92. Ichihara A, Tazaki H, Sakamura S (1983) Solanapyrones A, B and C, phytotoxic metabolites from the fungus *Alternaria solani*. Tetrahedron Lett 24:5373
93. Oikawa H, Obayashi T, Katayama K, Suzuki Y, Ichihara A (1998) Total synthesis of (−)-solanapyrone A via enzymatic Diels–Alder reaction of prosolanapyrone. J Org Chem 63:8748
94. Daly JW (1982) Alkaloids of neotropical poison frogs (Dendrobatidae). Prog Chem Org Nat Prod 41:205
95. Plikus MV, Astrowski AA (2014) Deadly hairs, lethal feathers—convergent evolution of poisonous integument in mammals and birds. Exp Dermatol 23:466
96. Kurosu M, Marcin LR, Grinsteiner TJ, Kishi Y (1998) Total synthesis of (±)-batrachotoxin A. J Amer Chem Soc 120:6627
97. Muntwyler R, Keller-Schierlein W (1972) Stoffwechselprodukte von Mikroorganismen. 107. Mitteilung. Die Struktur des Chlorothricins, eines neuartigen Makrolid Antibiotikums. Helv Chim Acta 55:2071
98. Brufani M, Cerrini S, Fedely W, Mazza F, Muntwyler R (1972) Stoffwechselprodukte von Mikroorganismen. 108. Mitteilung. Kristallstrukturanalyse des Chlorothricolid-methylesters. Helv Chim Acta 55:2094
99. Roush WR, Sciotti RJ (1998) Enantioselective total synthesis of (−)-chlorothricolide via the tandem inter- and intramolecular Diels–Alder reaction of a hexaenoate intermediate. J Am Chem Soc 120:7411
100. Abe N, Onoda R, Shirahata K, Kato T, Woods MC, Kitahara Y (1998) The structure of bakkenolide A. Tetrahedron Lett 1968:369
101. Naya K, Hayashi M, Takagi I, Nakamura S, Kobayashi M (1972) The structural elucidation of sesquiterpene lactones from *Petasites japonicus* Maxim. Bull Chem Soc Jpn 45:3673
102. Fischer NH, Olivier EJ, Fischer HD (1979) The biogenesis and chemistry of sesquiterpene lactones. Prog Chem Org Nat Prod 38:47
103. Jamieson GR, Reid EH, Turner BP, Jamieson AT (1976) Bakkenolide-A. Its distribution in *Petasites* species and cytotoxic properties. Phytochemistry 15:1713
104. Nawrot J, Koul O, Isman MB, Harmatha J (1991) Naturally occurring antifeedants: effects on two polyphagous lepidopterans. J Appl Entomol 112:194
105. Back TG, Payne JE (1999) A concise total synthesis of (±)-bakkenolide A by means of an intramolecular Diels–Alder reaction. Org Lett 1:663
106. Pinhey JT, Ritchie E, Taylor WC (1961) The chemical constituents of *Himantandra* (*Galbuimima*) species. IV. The structures of himbacine, himbeline, himandravine, and himgravine. Aust J Chem 14:106
107. Malaska MJ, Fauq AH, Kozikowski AP, Aagaard PJ, McKinney M (1995) Chemical modification of ring C of himbacine: discovery of a pharmacophoric element for $M_2$-selectivity. Bioorg Med Chem Lett 5:61
108. Miller JH, Aagaard PJ, Gibson VA, McKinney MJ (1992) Binding and functional selectivity of himbacine for cloned and neuronal muscarinic receptors. J Pharmacol Exp Ther 263:663
109. Davies RJ, Wang Y, Asberom T, Doller D, Wong J, Leone D, McPhail AT (1999) Total synthesis of (+)-himbacine and (+)-himbeline. J Org Chem 64:1932
110. Takao K-I, Munakata R, Tadano K-I (2005) Recent advances in natural product synthesis by using intramolecular Diels–Alder reactions. Chem Rev 105:4779
111. Juhl M, Tanner D (2009) Recent applications of intramolecular Diels–Alder reactions to natural product synthesis. Chem Soc Rev 38:2983
112. Biard JF, Guyot S, Roussakis C, Verbist JF, Vercauteren J, Weber JF, Boukef K (1994) Lepadiformine, a new marine cytotoxic alkaloid from *Clavelina lepadiformis* Müller. Tetrahedron Lett 35:2691
113. Patil AD, Freyer AJ, Reichwein R, Carte B, Killmer LB, Faucette L, Johnson RK, Faulkner DJ (1997) Fasicularin, a novel tricyclic alkaloid from the ascidian *Nephteis fasicularis* with selective activity against a DNA repair-deficient organism. Tetrahedron Lett 38:363

114. Abe H, Aoyagi S, Kibayashi C (2000) First total synthesis of the marine alkaloids (±)-fasicularin and (±)-lepadiformine based on stereo-controlled intramolecular acylnitroso-Diels-Alder reaction. J Am Chem Soc 122:4583
115. Dabrah TT, Kaneko T, Massefski Jr, W, Whipple EB (1997) CP-225,917 and CP-263,114: novel ras farnesylation inhibitors from an unidentified fungus. 2. Structure elucidation. J Am Chem Soc 119:1594
116. Nicolaou KC, Jung J-K, Yoon WH, He Y, Zhong Y-L, Baran PS (2000) The absolute configuration and asymmetric total synthesis of the CP molecules (CP-263,114 and CP-225,917, phomoidrides B and A). Angew Chem Int Ed 39:1829
117. Schuep W, Blount JF, Williams TH, Stempel A (1978) Production of a novel red pigment, rubrolone, by *Streptomyces echinoruber* sp. nov. II. Chemistry and structure elucidation. J Antibiot 31:1226
118. Yan Y, Yang J, Yu Z, Yu M, Ma Y-T, Wang L, Su C, Luo J, Horsman GP, Huang S-X (2016) Non-enzymatic pyridine ring formation in the biosynthesis of the rubrolone tropolone alkaloids. Nat Commun 7:13083
119. Boger DL, Ichikawa S, Jiang H (2000) Total synthesis of the rubrulone aglycon. J Am Chem Soc 122:12169
120. Nalbandov O, Yamamoto RT, Fraenkel GS (1964) Insecticides from plants, nicandrenone, a new compound with insecticidal properties, isolated from *Nicandra physaloides*. J Agric Food Chem 12:55
121. Stoltz BM, Kano T, Corey EJ (2000) Enantioselective total synthesis of nicandrenones. J Am Chem Soc 122:9044
122. Lin W-H, Ye W-H, Xu R-S (1992) Chemical studies of new *Stemona* alkaloids, IV. Studies on new alkaloids from *Stemona tuberosa*. J Nat Prod 55:571
123. Jacobi PA, Lee K (2000) Total syntheses of (±)- and (−)-stemoamide. J Am Chem Soc 122:4295
124. Burreson BJ, Scheuer PJ, Finer J, Clardy J (1975) 9-Isocyanopupukeanane, a marine invertebrate allomone with a new sesquiterpene skeleton. J Am Chem Soc 97:4763
125. Ho T-L, Kung L-R, Chein R-J (2000) Total synthesis of 9-isocyano-pupukeanane. J Org Chem 65:5774
126. Blanchflower SE, Banks RM, Everett JR, Reading C (1993) Further novel metabolites of the paraherquamide family. J Antibiot 46:1355
127. Stocking EM, Sanz-Cervera JF, Williams RM (2000) Total synthesis of VM55599. Utilization of an intramolecular Diels–Alder cycloaddition of potential biogenetic relevance. J Am Chem Soc 122:1675
128. Velten R, Klostermeyer D, Steffan B, Steglich W, Kuschel A, Anke T (1994) The mniopetals, new inhibitors of reverse transcriptase from a *Mniopetalum* species (Basidomycetes). 2. Structure elucidation. J Antibiot 47:1017
129. Suzuki Y, Nishimaki R, Ishikawa M, Murata T, Takao K-I, Tadano K-I (2000) Total synthesis of (−)-mniopetal E, a novel biologically intriguing drimane sesquiterpenoid. J Org Chem 65:8595
130. Jauch J (2001) Synthetic studies towards mniopetals (II). A short and efficient synthesis of mniopetal E. Synlett:87
131. Hufford CD, Guerrero RD, Doorenbos NJ (1976) Two new diterpenes from *Stemodia maritima* L. J Pharm Sci 95:778
132. Toró A, Nowak P, Deslongchamps P (2000) Transannular Diels–Alder entry into stemodanes: first asymmetric total synthesis of (+)-maritimol. J Am Chem Soc 122:4526
133. Rodríguez AD, Ramírez C (2000) A marine diterpene with a novel tetracyclic framework from the West Indian gorgonian octocoral *Pseudopterogorgia elisabethae*. Org Lett 2:507
134. Nicolaou KC, Vassilikogiannakis G, Mägerlein W, Kranich R (2001) Total synthesis of colombiasin A. Angew Chem Int Ed 40:2482

135. Johnson TW, Corey EJ (2001) Enantiospecific synthesis of the proposed structure of the antitubercular marine diterpenoid pseudopteroxazole: revision of stereochemistry. J Am Chem Soc 123:4475
136. Zhan X, Fang X, Xu M, Lei Y, Wu Z, Hu X (2019) Enantioselective total synthesis of pseudopteroxazole and ileabethoxazole. Angew Chem Int Ed 58:7845
137. Lee TW, Corey EJ (2001) Enantioselective total synthesis of eunicenone A. J Am Chem Soc 123:187
138. Silverton JV, Kabuto C, Buck KT, Cava MP (1977) Structure of imerubrine, a novel condensed tropolone-isoquinoline alkaloid. J Am Chem Soc 99:6708
139. Menachery MD, Cava MP (1980) Grandirubrine, a new tropoloisoquinoline alkaloid. Heterocycles 14:943
140. Lee JC, Cha JK (2001) Total synthesis of tropoloisoquinolines: imerubine, soimerubine, and grandirubine. J Am Chem Soc 123:3243
141. Gunasekera SP, McCarthy PJ, Kelly-Borges M, Lobkovsky E, Clardy JJ (1996) Dysidiolide: a novel protein phosphatase inhibitor from the Caribbean sponge *Dysidea etheria* de Laubenfels. J Am Chem Soc 118:8759
142. Miyaoka H, Kajiwara Y, Hara Y, Yamada Y (2001) Total synthesis of natural dysidolide. J Org Chem 66:1429
143. Pinar M, Rodriguez B, Alemany A (1978) Gummiferolic acid, a new *ent*-atis-16-ene diterpenoid from *Margotia gummifera*. Phytochemistry 17:1637
144. Villalobos N, Martin L, Macias MJ, Mancheno B, Grande M (1994) Gibberelin-like activity of some tetracyclic diterpenoids from *Elaeoselinum* species and their derivatives. Phytochemistry 37:635
145. Cross BE, Norton KJ (1965) New metabolites of *Gibberella fujikuori*. Part VIII. Gibberellin $A_{12}$. J Chem Soc:1570
146. Toyota M, Yokota M, Ihara M (2001) Remarkable control of radical cyclization processes of cyclic enyne: total syntheses of (±)-methyl gummiferolate, (±)-methyl 7$\beta$-hydroxykaurenoate, and (±)-methyl 7-oxokaurenoate and formal synthesis of (±)-gibberellin $A_{12}$ from a common synthetic precursor. J Am Chem Soc 123:1856
147. Pelletier PP, Caventou JB (1819) Note sur un nouvel alkali. Ann Chim Phys 8:323
148. Bonjoch J, Solé D (2000) Synthesis of strychnine. Chem Rev 100:3455
149. Cannon JS, Overman LE (2012) Is there no end to the total syntheses of strychnine? Lessons learned in strategy and tactics in total synthesis. Angew Chem 51:4288
150. Hua Y (2022) Recent advances in total synthesis of strychnine (2017–2022). BIO Web Conf 55:01008
151. Bodwell GJ, Li J (2002) A concise formal total synthesis of (±)-strychnine by using a transannular inverse electron demand Diels–Alder reaction of a [3](1,3)indolo[3](3,6) pyridazinophane. Angew Chem Int Ed 41:3261
152. Golden JE, Aubé J (2002) A combined intramolecular Diels–Alder/intramolecular Schmidt reaction: formal synthesis of (±)-stenine. Angew Chem Int Ed 41:4316
153. Ginn JD, Padwa A (2002) Total synthesis of (±)-stenine using the IMDAF cycloaddition of a 2-methylthio-5-amido-substituted furan. Org Lett 4:1515
154. Magnier E, Langlois Y (1998) Manzamine alkaloids, syntheses and synthetic approaches. Tetrahedron 54:6201
155. Sakai R, Higa T, Jefford CW, Bernardinelli G (1986) Manzamine A, a novel antitumor alkaloid from a sponge. J Am Chem Soc 108:6404
156. El Sayed KA, Kelly M, Kara UAK, Ang KKH, Katsuyama I, Dunbar DC, Khan AA, Hamann MT (2001) New manzamine alkaloids with potent activity against infectious diseases. J Am Chem Soc 123:1804
157. Humphrey JM, Liao Y, Ali A, Rein T, Wong Y-L, Chen H-J, Courtney AK, Martin SF (2002) Enantioselective total syntheses of manzamine A and related alkaloids. J Am Chem Soc 124: 8584

158. Fu X, Hossain MB, Schmitz FJ, van der Helm D (1997) Longithorones, unique prenylated *para-* and *meta-*cyclophane type quinones from tunicate *Aplidium longithorax.* J Org Chem 62:3810
159. Layton ME, Morales CA, Shair MD (2002) Biomimetic synthesis of (−)-longithorone. J Am Chem Soc 124:773
160. Stewart M, Fell PM, Blunt JW, Munro MHG (1997) Avarol and related compounds from the New Zealand marine sponge *Dysidea* sp. Aust J Chem 50:341
161. Taber DF, Nakajima K, Xu M, Rheingold AL (2002) Lactone-directed intramolecular Diels–Alder cyclization: synthesis of *trans*-dihydroconfertifolin. J Org Chem 67:4501
162. Sato B, Muramatsu H, Miyauchi M, Hori Y, Takase S, Hino M, Hashimoto S, Terano H (2000) A new antimitotic substance, FR182877. I. Taxonomy, fermentation, isolation, physico-chemical properties and biological activities. J Antibiot 53:123
163. Kingston DGI (2001) Taxol, a molecule for all seasons. Chem Commun:867
164. Vosburg DA, Vanderwal CD, Sorensen EJ (2002) A synthesis of (+)-FR182877, featuring tandem transannular Diels-Alder reactions inspired by a postulated biogenesis. J Am Chem Soc 124:4552
165. Vanderwal CD, Vosburg DA, Weiler S, Sorensen EJ (2003) An enantioselective synthesis of FR182877 provides a chemical rationalization of its structure and affords multigram quantities of its direct precursor. J Am Chem Soc 125:5393
166. Kim AI, Rychnovsky SD (2003) Unified strategy for the synthesis of (−)-elisapterosin B and (−)-colombiasin A. Angew Chem Int Ed 42:1267
167. Rodríguez AD, Ramírez C, Rodríguez II, Barnes CL (2000) Elisapterosins A and B: rearranged diterpenes possessing an unprecedented cagelike framework. J Org Chem 65:1390
168. Heckrodt TJ, Mulzer J (2003) Total synthesis of elisabethin A: intramolecular Diels–Alder reaction under biomimetic conditions. J Am Chem Soc 125:4680
169. Strobel GA (2002) Microbial gifts from rain forests. Can J Plant Pathol 24:14
170. Lee JC, Strobel GA, Lobkovsky E, Clardy J (1996) Torreyanic acid: a selectively cytotoxic quinone dimer from the endophytic fungus *Pestalotiopsis microspora.* J Org Chem 61:3232
171. Russell JR, Huang J, Anand P, Kucera K, Sandoval AG, Dantzler KW, Hickman DS, Jee J, Kimovec FM, Koppstein D, Marks DH, Mittermiller PA, Núñez SJ, Santiago M, Towne MA, Vishnevetsky M, Williams NE, Núñez Vargas MC, Boulanger L-A, Bascom-Slack C, Strobel SA (2011) Biodegradation of polyester polyurethane by endophytic fungi. Appl Environ Microbiol 77:6076
172. Li C, Johnson RP, Porco Jr, JA (2003) Total synthesis of the quinone epoxide dimer (+)-torreyanic acid: application of a biomimetic oxidation/electrocyclization/Diels-Alder dimerization cascade. J Am Chem Soc 125:5095
173. Li C, Lobkovsky E, Porco Jr, JA (2000) Total synthesis of (±)-torreyanic acid. J Am Chem Soc 122:10484
174. Joshi KC, Singh P, Taneja S, Cox PJ, Howie RA, Thomson RH (1982) New terpenoid aldehydes from *Kigelia pinnata*: crystal structure of pinnatal. Tetrahedron 38:2703
175. Onegi B, Kraft C, Kohler I, Freund M, Jenett-Siems K, Siems K, Beyer G, Melzig MF, Bienzle U, Eich E (2002) Antiplasmodial activity of naphthoquinones and one anthraquinone from *Stereospermum kunthianum.* Phytochemistry 60:39
176. Ose T, Watanabe K, Mie T, Honma M, Watanabe H, Yao M, Oikawa H, Tanaka I (2003) Insight into a natural Diels–Alder reaction from the structure of macrophomate synthase. Nature 422:185
177. Malerich JP, Trauner D (2003) Biomimetic synthesis of (±)-pinnatal and (±)-sterekunthal A. J Am Chem Soc 125:9554
178. Stocking EM, Williams RM (2003) Chemistry and biology of biosynthetic Diels–Alder reactions. Angew Chem Int Ed 42:3078
179. Jackson M, Karwowski JP, Theriault RJ, Rasmussen RR, Hensey DM, Humphrey PE, Swanson SJ, Barlow GJ, Premachandran U, McAlpine JB (1995) Macquarimicins, microbial

metabolites from *Micromonospora*. I. Discovery, taxonomy, fermentation and biological properties. J Antibiot 48:462
180. Hochlowski JE, Mullally MM, Henry R, Whittern DM, McAlpine JB (1995) Maquarimicins, microbial metabolites from *Micromonospora* II. Isolation and structural elucidation. J Antibiot 48:467
181. Tanaka M, Nara F, Yamasato Y, Masuda-Inoue S, Doi-Yoshioka H, Kumakura S, Enokita R, Ogita T (1999) Maquarimicin A inhibits membrane-bound neutral sphingomyelinase from rat brain. J Antibiot 52:670
182. Shindo K, Matsuoka M, Kawai H (1996) Studies on cochleamycins, novel antitumor antibiotics. J Antibiot 49:241
183. Sato B, Muramatsu H, Miyauchi M, Hori Y, Takase S, Hino M, Hashimoto S, Terano H (2000) A new antimitotic substance, FR182877 II. The mechanism of action. J Antibiot 53:123
184. Munakata R, Katakai H, Ueki T, Kurosaka J, Takao K-I, Tadano K-I (2003) Total synthesis of (+)-macquarimicin A. J Am Chem Soc 125:14722
185. Munakata R, Katakai H, Ueki T, Kurosaka J, Takao K-I, Tadano K-I (2004) Total synthesis of macquarimicins using an intramolecular Diels-Alder approach inspired by a biosynthetic pathway. J Am Chem Soc 126:11254
186. Shindo K, Kawai H (1992) Novel antibiotics, cochleamycins A and B. J Antibiot 45:292–295
187. Tatsuta K, Narazaki F, Kashiki N, Yamamoto J-I, Nakano S (2003) The first total synthesis of cochleamycin A and determination of the absolute structure. J Antibiot 56:584
188. Dineen TA, Roush WR (2004) Total synthesis of cochleamycin A. Org Lett 6:2043
189. Sung P-J, Chen M-C (2002) The heterocyclic natural products of gorgonian corals of genus *Briareaum* exclusive of briarane diterpenoids. Heterocycles 57:1705
190. Crimmins MT, Brown BH (2004) An intramolecular Diels–Alder approach to the eunicelins: enantioselective total synthesis of ophirin B. J Am Chem Soc 126:10264
191. Larsen TO, Frydenvang K, Frisvad JC, Christophersen C (1998) UV-guided isolation of alantrypinone, a novel *Penicillium* alkaloid. J Nat Prod 61:1154
192. Chen Z, Fan J, Kende AS (2003) Total synthesis of (±)-alantrypinone by hetero Diels–Alder reaction. J Org Chem 69:79
193. Kirst HA, Michel KH, Martin JW, Creemer LC, Chio EH, Yao RC, Nakatsukasa WA, Boeck LD, Occolowitz JL, Paschal JW, Deeter JB, Jones ND, Thompson GD (1991) A83543A–D, unique fermentation-derived tetracyclic macrolides. Tetrahedron Lett 32:4839
194. Mergott DJ, Frank SA, Roush WR (2004) Total synthesis of (−)-spinosyn A. Proc Natl Acad Sci USA 101:11955
195. Kuramoto M, Hayashi K, Yamaguchi K, Yada M, Tsuji T, Uemura D (1998) Structure-activity relationship of norzoanthamine exhibiting significant inhibition of osteoporosis. Bull Chem Soc Jpn 71:771
196. García-García P, Reyes R, Évora C, Delgado A, Fernández JJ, Daranas AH (2022) Osteoprotective effect of the marine alkaloid norzoanthamine on an osteoporosis model in ovariectomized rat. Biomed Pharmacother 147:112631
197. Miyshita M, Sasaki M, Hattori I, Sakai M, Tanino K (2004) Total synthesis of norzoanthamine. Science 305:495
198. Oikawa H, Yokota T, Ichihara A, Sakamura S (1989) Structure and absolute configuration of solanapyrone D: a new clue to the occurrence of biological Diels–Alder reactions. J Chem Soc Chem Commun:1284
199. Wilson RM, Jen WS, MacMillan DWC (2005) Enantioselective organocatalytic intramolecular Diels–Alder reactions. The asymmetric synthesis of solanapyrone D. J Am Chem Soc 127:11616
200. Igarashi M, Hayashi C, Homma Y, Hattori S, Kinoshita N, Hamada M, Takeuchi T (2000) Tubelactomicin A, a novel 16-membered lactone antibiotic, from *Nocardia* sp. I. J Antibiot 53: 1096
201. Igarashi M, Nakamura H, Naganawa H, Takeuchi T (2000) Tubelactomicin A, a novel 16-membered lactone antibiotic, from *Nocardia* sp. II. J Antibiot 53:1102

202. Motozaki T, Sawamura K, Suzuki A, Yoshida K, Ueki T, Ohara A, Munakata R, Takao K-I, Tadano K-I (2005) Total synthesis of (+) tubelactomicin A. 1. Stereoselective synthesis of the lower-half segment by an intramolecular Diels-Alder approach. Org Lett 7:2261
203. Motozaki T, Sawamura K, Suzuki A, Yoshida K, Ueki T, Ohara A, Munakata R, Takao K-I, Tadano K-I (2005) Total synthesis of (+)-tubelactomicin A. 2. Synthesis of the upper-half segment and completion of the total synthesis. Org Lett 7:2265
204. Snyder SA, Corey EJ (2006) Concise total syntheses of palominol, dolabellatrienone, β-araneosene, and isoedunol via an enantioselective Diels–Alder macrobicyclization. J Am Chem Soc 128:740
205. Paddon-Row M, Anderson CD, Houk KN (2009) Computational evaluation of enantioselective Diels–Alder reactions mediated by Corey's cationic oxazaborolidine catalysts. J Org Chem 74:861
206. Cichewicz RH, Lim K-C, McKerrow JH, Nair MG (2002) Kwanzoquinones A–G and other constituents of *Hemerocallis fulva* 'Kwanzo' roots and their activity against the human pathogenic trematode *Schistosoma mansoni*. Tetrahedron 58:8597
207. Shiao S-H, Shao B-R, He Y-X, Yang Y-Q, Yang HZ, Chang Y-C (1962) Studies on *Hemerocallis thunbergii* Baker II. Effectiveness of *Hemerocallis thumbergii* in oral treatment of experimental schistosomiasis in mice. Acta Pharm Sin 9:218
208. Cichewicz RH, Zhang Y, Seeram NP, Nair MG (2004) Inhibition of human tumor cell proliferation by novel anthraquinones from daylilies. Life Sci 74:1791
209. Tietze LF, Gericke KM, Güntner C (2006) First total synthesis of the bioactive anthraquinone kwanzoquinone C and related natural products by a Diels–Alder approach. Eur J Org Chem 2006:4910
210. Bister B, Bischoff D, Strobele M, Riedlinger J, Reicke A, Wolter F, Bull AT, Zahner H, Fiedler HP, Süssmuth RD (2004) Abyssomicin C — a polycyclic antibiotic from a marine *Verrucosispora* strain as an inhibitor of the *p*-aminobenzoic acid/tetrahydrofolate biosynthesis pathway. Angew Chem Int Ed 43:2574
211. Riedlinger J, Reicke A, Zähner H, Krismer B, Bull AT, Maldonado LA, Ward AC, Goodfellow M, Bister B, Bischoff D, Süssmuth RD, Fiedler HP (2004) Abyssomicins, inhibitors of the *para*-aminobenzoic acid pathway produced by the marine *Verrucosispora* strain AB-18-032. J Antibiot 57:271
212. Nicolaou KC, Harrison ST (2006) Total synthesis of abyssomicin C and *atrop*-abyssomicin C. Angew Chem Int Ed 45:3256
213. Nicolaou KC, Harrison ST (2007) Total synthesis of abyssomicin C, *atrop*-abyssomicin C, and abyssomicin D: implications for natural origins of *atrop*-abyssomicin C. J Am Chem Soc 129:429
214. Zapf CW, Harrison BA, Drahl C, Sorensen EJ (2005) A Diels–Alder macrocyclization enables an efficient asymmetric synthesis of the antibacterial natural product abyssomicin C. Angew Chem Int Ed 44:6533
215. Snider BB, Zou YF (2005) Synthesis of the carbocyclic skeleton of abyssomicins C and D. Org Lett 7:4939
216. Couladouros EA, Bouzas EA, Magos AD (2006) Formal synthesis of abyssomicin C. Tetrahedron 62:5272
217. Balskus EP, Jacobsen EN (2007) Asymmetric catalysis of the transannular Diels–Alder reaction. Science 317:1736
218. Uchida I, Ando T, Fukami N, Yoshida K, Hashimoto M, Tada T, Koda S, Morimoto Y (1987) The structure of vinigrol, a novel diterpenoid with antihypertensive and platelet aggregation-inhibitory activities. J Org Chem 52:5292
219. Ando T, Tsurumi Y, Ohata N, Uchida I, Yoshida K, Okuhara M (1988) Vinigrol, a novel antihypertensive and platelet aggregation inhibitory agent produced by a fungus, *Virgaria nigra*. I. Taxonomy, fermentation, isolation, physicochemical, and biological properties. J Antibiot 41:25

220. Maimone TJ, Voica A-F, Baran PS (2008) A concise approach to vinigrol. Angew Chem Int Ed 47:3054
221. Maimone TJ, Shi J, Ashida S, Baran PS (2009) Total synthesis of vinigrol. J Am Chem Soc 131:17066
222. De Munari S, Barassi P, Cerri A, Fedrizzi G, Gobbini M, Mabilia M, Melloni P (1998) A new approach to the design of novel inhibitors of $Na^+$,$K^+$-ATPase: 17$\alpha$-substituted *seco*-$\Delta^5$-β-androstane as cassaine analogues. J Med Chem 41:3033
223. Dalma G (1935) New alkaloids in *Erythrophleum guineense*. Ann Chim Appl 25:569
224. Turner RB, Herzog EG, Morin RB, Riebel A (1959) Structure of cassaic acid. Tetrahedron Lett 1:7
225. Phoenix S, Reddy MS, Deslongchamps P (2008) Total synthesis of (+)-cassaine via transannular Diels–Alder reaction. J Am Chem Soc 130:13989
226. Kupchan SM, Court WA, Dailey RG Jr, Gilmore CJ, Bryan RF (1972) Triptolide and tripdiolide, novel antileukemic diterpenoid triepoxides from *Tripterygium wilfordii*. J Am Chem Soc 94:7194
227. Brinker AM, Ma J, Lipsky PE, Raskin I (2007) Medicinal chemistry and pharmacology of genus *Tripterygium* (Celastraceae). Phytochemistry 68:732
228. Miller NA, Willis AC, Sherburn MS (2008) Formal total synthesis of triptolide. Chem Commun:1226
229. Tsukamoto S, Kawabata T, Kato H, Greshock TJ, Hirota H, Ohta T, Williams RM (2009) Isolation of antipodal (−)-versicolamide B and notoamides L–N from a marine-derived *Aspergillus* sp. Org Lett 11:1297
230. Greshock TJ, Grubbs AW, Jiao P, Wicklow DT, Gloer JB, Williams RM (2008) Isolation, structure elucidation, and biomimetic total synthesis of versicolamide B, and the isolation of antipodal (−)-stephacidin A and (+)-notoamide B from *Aspergillus versicolor* NRRL 35600. Angew Chem Int Ed 47:3573
231. Miller KA, Tsukamoto S, Williams RM (2009) Asymmetric total syntheses of (+)- and (−)-versicolamide B and biosynthetic implications. Nat Chem 1:63
232. Tanaka N, Suzuki T, Matsumura T, Hosoya Y, Nakada M (2009) Total synthesis of (−)-FR182877 through tandem IMDA–IMHDA reactions and stereoselective transition-metal-mediated transformations. Angew Chem Int Ed 48:2580
233. Weimar M, Dürner G, Bats JW, Göbel MW (2010) Enantioselective synthesis of (+)-estrone exploiting a hydrogen bond-promoted Diels–Alder reaction. J Org Chem 75:2718
234. Quinkert G, Del Grosso M, Döring A, Döring W, Schenkel RI, Bauch M, Dambacher GT, Bats JW, Zimmermann G, Dürner G (1995) Intramolecular carbenoid insertions into C($sp^3$)–H bonds in conformationally constrained systems. Construction of bridged polycycles from 4-alkyl- and 4-aryl-1,2,4-triazoles and of bicyclo[3.2.1]octanes from cyclic ketones and tosylhydrazones. Helv Chim Acta 78:1345
235. Dalko PI, Moisan L (2004) In the golden age of organocatalysis. Angew Chem Int Ed 43:5138
236. List B (2007) Organocatalysis. Chem Rev 107:5413
237. Dane E, Schmitt J (1939) Synthesen in der hydroaromatischen Reihe. V. Die Synthese eines Isomeren des Östrons. Liebigs Ann Chem 537:246
238. Mackay EG, Sherburn MS (2015) The Diels–Alder reaction in steroid synthesis. Synthesis 47: 1
239. Wei X, Rodríguez II, Rodríguez AD, Barnes CL (2007) Caribenols A and B, sea whip-derived norditerpenes with novel tricarbocyclic skeletons. J Org Chem 72:7386
240. Bloom BR, Murray CJL (1992) Tuberculosis: commentary on a re-emergent killer. Science 257:1055
241. Liu L-Z, Han J-C, Yue G-Z, Li C-C, Yang Z (2010) Asymmetric total synthesis of caribenol A. J Am Chem Soc 132:13608
242. Löbermann F, Mayer P, Trauner D (2010) Biomimetic synthesis of (−)-pycnanthuquinone C through the Diels–Alder reaction of a vinyl quinone. Angew Chem Int Ed 49:6199

243. Laird DW, Poole R, Wikström M, van Altena IA (2007) Pycnanthuquinone C, an unusual 6,6,5-tricyclic geranyltoluquinone from the Western Australian brown alga *Cystophora harveyi*. J Nat Prod 70:671
244. Appleton DR, Chuen CS, Berridge MV, Webb VL, Copp BR (2009) Rossinones A and B, biologically active meroterpenoids from the Antarctic ascidian, *Aplidium* species. J Org Chem 74:9195
245. Zhang Z, Chem J, Yang Z, Tang Y (2010) Rapid biomimetic total synthesis of (±)-rossinone B. Org Lett 12:5554
246. Nomura T, Hano Y (1994) Isoprenoid-substituted phenolic compound of Moraceae plants. Nat Prod Rep 20:5
247. Takasugi M, Nagao S, Masamune T, Shirata A, Takahashi K (1980) Structure of bruceantin, a new antitumor quassinoid, and its derivatives. Chem Lett 9:1573
248. Fukai T, Kaitou K, Terada S (2005) Antimicrobial activity of 2-arylbenzofurans from *Morus* species against methicillin-resistant *Staphylococcus aureus*. Fitoterapia 76:708
249. Zhang Q-J, Tang Y-B, Chen R-Y, Yu D-Q (2007) Three new cytotoxic Diels–Alder-type adducts from *Morus australis*. Chem Biodivers 4:1533
250. Gunanwan C, Rizzacasa MA (2010) Mulberry Diels–Alder adducts: synthesis of chalcomoracin and mulberrofuran C methyl ethers. Org Lett 12:1388
251. Marco-Contelles J, do Carmo Carreiras CM, Rodríguez C, Villarroya M, García AG (2006) Synthesis and pharmacology of galanthamine. Chem Rev 106:116
252. Akram MN, Verpoorte R, Pomahačová B (2021) Methods for the analysis of galanthamine and its extraction from laboratory to industrial scale. S Afr J Bot 136:51
253. Chaplin DA, Fraser N, Tiffiin DL (2000) Processes for the preparation of derivatives of galanthamine (Sanochemia Pharmazeutika AG), US Pat no. 6 043359A
254. Kodama S, Hamashima Y, Nishide K, Node M (2004) Total synthesis of (−)-galanthamine by remote asymmetric induction. Angew Chem Int Ed 43:2659
255. Scott LJ, Goa KL (2000) Galantamine: a review of its use in Alzheimer's disease. Drugs 60: 1095
256. Chang JH, Kang H-U, Jung I-H, Cho C-G (2010) Total synthesis of (±)-galanthamine via a C3-selective Stille coupling and IMDA cycloaddition cascade of 3,5-dibromo-2-pyrone. Org Lett 12:2016
257. Cao S-G, Sim K-Y, Goh SH (2000) (−)-Ledol from *Calophyllum teysmanii*: structure and stereochemistry. Nat Prod Lett 14:447
258. Kir'yalov NP (1949) Toward an elucidation of the structure of ledol. Zh Obshch Khim 19: 2123. (Chem Abstr 44:3969c)
259. Thamapipol S, Kündig EP (2011) Intramolecular Diels–Alder reactions using chiral ruthenium Lewis acids and application in the total synthesis of *ent*-ledol. Org Biomol Chem 9:7564
260. Saito F, Kuramochi K, Nakazaki A, Mizushina Y, Sugawara F, Kobayashi S (2006) Synthesis and absolute configuration of (+)-pseudodeflectusin: structural revision of aspergione B. Eur J Org Chem 2006:4796
261. Sato Y, Kuramoch K, Suzuki T, Nakazaki A, Kobayashi S (2011) The second generation synthesis of (+)-pseudodeflectusin. Tetrahedron Lett 52:626
262. Wright AE, Pomponi SA, McConnell OJ, Kohmoto S, McCarthy PJ (1987) (+)-Curcuphenol and (+)-curcudiol, sesquiterpene phenols from shallow and deep water collection of the marine sponge *Didiscus flavus*. J Nat Prod 50:976
263. Martin MJ, Berrue F, Amade P, Fernandez R, Francesch A, Reyes F, Cuevas C (2005) Halogenated helianane derivatives from the sponge *Spirastrella hartmani*. J Nat Prod 68:1554
264. Green JC, Jiménez-Alonso S, Brown ER, Pettus TRR (2011) Total synthesis and repudiation of the helianane family. Org Lett 13:5500
265. Jeong S-J, Kim N-Y, Kim D-H, Kang T-H, Ahn N-H, Miyamoto T, Higuchi R, Kim Y-C (2000) Hyaluronidase inhibitory active 6*H*-dibenzo[*b,d*]pyran-6-ones from the feces of *Trogopterus xanthipes*. Planta Med 66:76

266. The Dictionary of Chinese Drugs (1985) Shanghai Scientific and Technical Publishers: Shogakukan Tokyo, p 875
267. Pottie IR, Nandaluru PR, Bodwell GJ (2011) An inverse electron-demand Diels–Alder-based total synthesis of urolithin M7. Synlett 15:2245
268. Dong M, Cong B, Yu S-H, Sauriol F, Huo FC-H, Shi Q-W, Gu Y-C, Zamir LO, Kiyota H (2008) Echinopines A and B: sesquiterpenoids possessing an unprecedented skeleton from *Echinops spinosus*. Org Lett 10:701
269. Liu X-Y, Qin Y (2015) Chemical synthesis of the echinopine sesquiterpenoids. Nat Prod Commun 10:2173
270. Peixoto PA, Severin R, Tseng C-C, Chen DY-K (2011) Formal asymmetric synthesis of echinopine A and B. Angew Chem Int Ed 50:3013
271. Chee CF, Lee YK, Buckle MJC, Abd Rahman N (2011) Synthesis of (±)-kuwanone V and (±)-dorsterone methyl ethers via Diels–Alder reaction. Tetrahedron Lett 52:1797
272. Hano Y, Yamanaka J, Nomura T, Momose Y (1995) Sorocenols A and B, two new isoprenylated phenols from the root bark of *Sorocea bonplandii* Baillon. Heterocycles 41:1035
273. Cong H, Porco JA Jr (2012) Total synthesis of (±)-sorocenol B employing nanoparticle catalysis. Org Lett 14:2516
274. Nkunya MHH, Jonker SA, de Gelder R, Wachira SW, Kihampa C (2004) (±)-Schefflone: a trimeric monoterpenoid from the root bark of *Uvaria scheffleri*. Phytochemistry 65:399
275. Liao D, Li H, Lei X (2012) Efficient generation of *ortho*-quinone methide: application to the biomimetic syntheses of (±)-schefflone and tocopherol trimers. Org Lett 14:18
276. Maly M, Benes F, Binova Z, Zlechovcova M, Kastanek P, Hajslova J (2023) Effective isolation of cannabidiol and cannabidiolic acid free of psychotropic phytocannabinoids from hemp extract by fast centrifugal partition chromatography. Anal Bioanal Chem 415:4827
277. Rhee MH, Vogel Z, Barg J, Bayewitch M, Levy R, Hanus L, Breuer A, Mechoulam R (1997) Cannabinol derivatives: binding to cannabinoid receptors and inhibition of adenylylcyclase. J Med Chem 40:322
278. Nandaluru PR, Bodwell GJ (2012) Multicomponent synthesis of 6*H*-dibenzo[*b*,*d*]pyranones and a total synthesis of cannabinol. Org Lett 14:310
279. Mishra R, Tritch HR, Pandey RC (1985) Defucogilvocarcin V, a new antibiotic from *Streptomyces arenae* 2064: isolation, characterization, partial synthesis and biological activity. J Antibiot 38:1280
280. Nandaluru PR, Bodwell GJ (2012) An inverse electron demand Diels–Alder-based total synthesis of defucogilvocarin V and some C-8 analogues. J Org Chem 77:8028
281. Brophy GC, Mohandas J, Slaytor M, Sternhell S, Watson TR, Wilson LA (1969) Novel lignans from a *Cinnamomum* sp. from Bougainville. Tetrahedron Lett 1969:5159
282. Constantin M-A, Conrad J, Merişor E, Koschorreck K, Urlacher VB, Beifuss U (2012) Oxidative dimerization of (*E*)- and (*Z*)-2-propenylsesamol with $O_2$ in the presence and absence of laccases and other catalysts: selective formation of carpanones and benzopyrans under different reaction conditions. J Org Chem 77:4528
283. Cherkasov OA, Tolkachev ON (2002) *Narcissus* and other Amaryllidaceae as sources of galanthamine. In: Hanks G (ed) Narcissus and Daffodil. CRC Press, Boca Raton, FL, p 242
284. Jung Y-G, Lee S-C, Cho H-K, Darvatkar NB, Song J-Y, Cho C-G (2013) Total syntheses of (±)-lycorane and (±)-1-deoxylycorine. Org Lett 15:132
285. Ohta T, Kita T, Kobayashi N, Obara Y, Nakahata N, Ohizumi Y, Takaya Y, Oshima Y (1998) Scabronine A, a novel diterpenoid having potent inductive activity of the nerve growth factor synthesis, isolated from the mushroom *Sarcodon scabrosus*. Tetrahedron Lett 39:6229
286. Shi X-W, Liu L, Gao J-M, Zhang A-L (2011) Cyathane diterpenes from Chinese mushroom *Sarcodon scabrosus* and their neurite outgrowth-promoting activity. Eur J Med Chem 46:3112
287. He X, Garcia KC (2004) Structure of nerve growth factor complexed with the shared neurotrophin receptor P75. Science 304:870
288. Hefti F (1997) Pharmacology of neurotrophic factors. Annu Rev Pharmacol Toxicol 37:239

289. Kirik D, Georgievska B, Bjorklund A (2004) Localized striatal delivery of GDNF as a treatment for Parkinson disease. Nat Neurosci 7:105
290. Pollack SJ, Harper SJ (2002) Small molecule Trk receptor agonists and other neurotrophic factor mimetics. Curr Drug Targ CNS Neurol Disord 1:59
291. Kobayakawa Y, Nakada M (2013) Total syntheses of (−)-scabronines G and A, and (−)-episcabronine A. Angew Chem Int Ed 52:7569
292. Jensen SR, Kirk O, Nielsen BJ (1987) 9-Hydroxy-substituted iridoids from *Gelsemium sempervirens*. Phytochemistry 26:1725
293. Li Y-S, Matsunaga K, Kato R, Ohizumi Y (2001) Potentiation of nerve growth factor-induced elongation of neurites by gelsemiol and 9-hydroxysemperoside aglycone in PC12D cells. J Pharm Pharmacol 53:915
294. Burch P, Binaghi M, Scherer M, Wenthel C, Bossert D, Eberhardt L, Neuburger M, Scheiffle P, Gademann K (2013) Total synthesis of gelsemiol. Chem Eur J 19:2589
295. Messana I, Ferrari F (1988) Structure of a new Diels–Alder type adduct, brosimone A. Tetrahedron 44:6693
296. Messana I, Ferrari F, de Mello JF, do Carmo Mesquita de Araujo M (1989) Constituents of *Brosimopsis oblongifolia*. 4. Structures of two new Diels-Alder type adducts, brosimone B and brosimone D. Heterocycles 29:683
297. Qi C, Cong H, Cahill KJ, Müller P, Johnson RP, Porco Jr, JA (2013) Biomimetic dehydrogenative Diels–Alder cycloadditions: total syntheses of brosimones A and B. Angew Chem Int Ed 52:8345
298. Jin Z (2011) Amaryllidaceae and *Sceletium* alkaloids. Nat Prod Rep 28:1126
299. McLachlan A, Kekre N, McNulty J, Pandey S (2005) Pancratistatin: a natural anti-cancer compound that targets mitochondria specifically in cancer cells to induce apoptosis. Apoptosis 10:619
300. Cho H-K, Lim H-Y, Cho C-G (2013) (*E*)-β-Borylstyrene in the Diels–Alder reaction with 3,5-dibromo-2-pyrone for the syntheses of (±)-*epi*-pancratistatin and (±)-pancratistatin. Org Lett 15:5806
301. Toyota M, Yoshida T, Kan Y, Takaoka S, Asakawa Y (1996) (+)-Cavicularin: a novel optically active cyclic bibenzyl-dihydrophenanthrene derivative from the liverwort *Cavicularia densa* Steph. Tetrahedron Lett 37:4745
302. Zhao P, Beaudry CM (2014) Enantioselective and regioselective pyrone Diels–Alder reactions of vinyl sulfones: total synthesis of (+)-cavicularin. Angew Chem Int Ed 53:10500
303. Crow WD, Michael M (1955) The alkaloids of *Kopsia longiflora* Merrill. I. Isolation of the alkaloids. Aust J Chem 8:129
304. Xie J, Wolfe AL, Boger DL (2013) Total synthesis of kopsinine. Org Lett 15:868
305. Lee K, Boger DL (2015) Total synthesis of (−)-kopsinine. Tetrahedron 71:3741
306. Nguyen NT, Nguyen MTT, Nguyen HX, Dang PH, Dibwe DF, Esumi H, Nguyen SA (2017) Constituents of the rhizomes of *Boesenbergia pandurata* and their antiausterity activities against the PANC-1 human pancreatic cancer line. J Nat Prod 80:141
307. Li X, Han J, Jones AX, Lei X (2016) Chiral boron complex-promoted asymmetric Diels–Alder cycloaddition and its application in natural product synthesis. J Org Chem 81:458
308. Li Y, Niu S, Sun B, Liu S, Liu X, Che Y (2010) Cytosporolides A–C, antimicrobial meroterpenoids with a unique peroxylactone skeleton from *Cytospora* sp. Org Lett 12:3144
309. Takao K-I, Noguchi S, Sakamoto S, Kimura M, Yoshida K, Tadano K-I (2015) Total synthesis of (+)-cytosporolide C via a biomimetic hetero-Diels–Alder reaction. J Am Chem Soc 137: 15971
310. Gan P, Smith MW, Braffman NR, Snyder SA (2016) Pyrone Diels–Alder routes to indolines and hydroindolines: syntheses of gracilamine, mesembrine, and $\Delta^7$-mesembrenone. Angew Chem Int Ed 55:3625
311. Jeffs PW, Hawks RL, Farrier DS (1970) Structures of the mesembranols and the absolute configuration of mesembrine and related alkaloids. J Am Chem Soc 91:3831

312. Patnala S, Kanfer I (2015) Medicinal use of *Sceletium*: characterization of phytochemical components of *Sceletium* plant species using HPLC with UV and electrospray ionization-tandem mass spectroscopy. J Pharm Pharm Sci 18:414
313. Ünver N, Kaya Gİ (2005) An unusual pentacyclic dinitrogenous alkaloid from *Galanthus gracilis*. Turk J Chem 29:547
314. Nagasawa K, Odagi M, Yamamoto Y (2017) Enantioselective total synthesis of (+)-gracilamine and preliminary cytotoxicity. Angew Chem Int Ed 56:7358
315. Gao J, Radwan MM, León F, Dale OR, Husni AS, Wu Y, Lupien S, Wang X, Manly SP, Hill RA, Dugan FM, Cutler HG, Cutler SJ (2013) *Neocosmospora* sp.-derived resorcylic acid lactones with in vitro binding affinity for human opioid and cannabinoid receptors. J Nat Prod 76:2174
316. Ahmad S, Dray A (2004) Novel G protein-coupled receptors as pain targets. Curr Opin Investig Drugs 5:67
317. Zhang Y, Dlugosch M, Jubermann M, Banwell MG, Ward JS (2015) Total syntheses of the resorcylic acid lactone neocosmosin A and its enantiomer. J Org Chem 80:4828
318. Lee JH, Cho C-G (2016) Total synthesis of (−)-neocosmosin A via intramolecular Diels-Alder reaction of 2-pyrone. Org Lett 18:5126
319. Xie ZL, Wang XH (2005) Clinical study of modified Sini powder for 38 cases of dysthymic disorder. Chin J Inf Tradit Chin Med 12:8
320. Liang WJ, Geng CA, Zhang XM, Chen H, Yang CY, Rong GQ, Zhao Y, Xu HB, Wang H, Zhou NJ, Ma YB, Huang XY, Chen J (2014) Paeoveitol, a pair of new norditerpene enantiomers from *Paeonia veitchii*. Org Lett 16:424
321. Zhan Y, Guo Y, Li Z, Xie Z (2016) Biomimetic total synthesis of paeoveitol. Org Lett 18:4578
322. Xiao W-L, Zhu H-J, Shen Y-H, Li R-T, Li S-H, Sun H-D, Zheng Y-T, Wang R-R, Lu Y, Wang C, Zheng Q-T (2005) Lancifodilactone G: a unique nortriterpenoid isolated from *Schisandra lancifolia* and its anti-HIV activity. Org Lett 7:2145
323. Xiao W-L, Zhu H-J, Shen Y-H, Li R-T, Li S-H, Sun H-D, Zheng Y-T, Wang R-R, Lu Y, Wang C, Zheng QT (2006) Additions and corrections: Lancifodilactone G: a unique nortriterpenoid isolated from *Schisandra lancifolia* and its anti-HIV activity. Org Lett 8:801
324. Shi Y-M, Xiao W-L, Pu J-X, Sun H-D (2015) Triterpenoids from the Schisandraceae family: an update. Nat Prod Rep 32:367
325. Liu D-D, Sun TW, Wang KY, Lu Y, Zhang S-L, Li Y-H, Jiang Y-L, Chen J-H, Yang Z (2017) Asymmetric total synthesis of lancifodilactone G acetate. J Am Chem Soc 139:5732
326. Mevers E, Saurí J, Liu Y, Moser A, Ramadhar TR, Varlan M, Williamson RT, Martin GE, Clardy J (2016) Homodimericin A: a complex hexacyclic fungal metabolite. J Am Chem Soc 138:2324
327. Peterson EA, Overman LE (2004) Contiguous stereogenic quaternary carbons: a daunting challenge in natural products synthesis. Proc Natl Acad Sci USA 101:11943
328. Feng J, Lei X, Guo Z, Tang Y (2017) Total synthesis of homodimericin A. Angew Chem Int Ed 56:7895
329. Leonti M, Casu L, Sanna F, Bonsignore L (2009) A comparison of medicinal plant use in Sardinia and Sicily—De Materia Medica revisited? J Ethnopharmacol 121:255
330. Alipour G, Dashti S, Hosseinzadeh H (2014) Review of pharmacological effects of *Myrtus communis* L. and its active constituents. Phytother Res 28:1125
331. Liu CL, Ang S, Huang X-J, Tian H-Y, Deng Y-Y, Zhang D-M, Wang Y, Ye W-C, Wang L (2016) Meroterpenoids with new skeletons from *Myrtus communis* and structure revision of myrtucommulone K. Org Lett 16:4004
332. Li T-Z, Geng C-A, Yin X-J, Yang T-H, Chen X-L, Huang X-Y, Ma Y-B, Zhang X-M, Chen JJ (2017) Catalytic asymmetric total synthesis of (+)- and (−)-paeoveitol via a hetero-Diels–Alder reaction. Org Lett 19:429
333. Zhang YL, Ge HM, Zhao W, Dong H, Xu Q, Li SH, Zhang J, Song YC, Tan RX (2008) Unprecedented immunosuppressive polyketides from *Daldinia eschscholzii*, a mantis-associated fungus. Angew Chem Int Ed 47:5823

334. Zhang YL, Zhang J, Jiang N, Lu YH, Wang L, Zhang GF, Xu Q, Ge HM, Ma J, Song YC, Tan RX (2011) Immunosuppressive polyketides from mantis-associated *Daldinia eschscholzii*. J Am Chem Soc 133:5931
335. Barnes EC, Jumpathong J, Lumyong S, Voigt K, Hertweck C (2016) Daldionin, an unprecedented binaphthyl derivative, and diverse polyketide congeners from a fungal orchid endophyte. Chemistry 22:4551
336. Blunt JW, Copp BR, Keyzers RA, Munro MHG, Prinsep MR (2016) Marine natural products. Nat Prod Rep 33:382
337. Chinta BS, Siraswar A, Baire B (2017) The dehydro Diels–Alder (DDA) reaction based approach to isofuranonaphthalenone, nodulones A–C and xestolactone A. Tetrahedron 73: 4178
338. Wessig P, Müller G (2008) The dehydro-Diels–Alder reaction. Chem Rev 108:2051–2063
339. Li W, Zhang J (2016) Recent progress in dehydro(genative) Diels–Alder reaction. Chem Eur J 22:1558
340. Liu X, Luo H-B, Huang Y-Y, Bao J-M, Tang G-H, Chen Y-Y, Wang J, Yin S (2014) Selaginpulvilins A–D, new phosphodiesterase-4 inhibitors with an unprecedented skeleton from *Selaginella pulvinata*. Org Lett 16:282
341. Huang Y, Liu X, Wu D, Tang G, Lai Z, Zheng X, Yin S, Luo HB (2017) The discovery, complex crystal structure, and recognition mechanism of a novel natural PDE4 inhibitor from *Selaginella pulvinata*. Biochem Pharmacol 130:51
342. Chinta BS, Baire B (2017) Formal total synthesis of selaginpulvilin D. Org Biomol Chem 15: 5908
343. Castillo M, Loyola LA, Morales G, Singh I, Calvo C, Rolland HL, MacLean DB (1976) The alkaloids of *L. magellanicum* and the structure of magellanine. Can J Chem 54:2893
344. McGee P, Bétournay G, Barabé F, Barriault L (2017) A 11-steps total synthesis of magellanine through a gold(I)-catalyzed dehydro Diels–Alder reaction. Angew Chem Int Ed 56:6280
345. Cheng S-Y, Huang K-J, Wang S-K, Wen Z-H, Chen P-W, Duh C-Y (2010) Antiviral and anti-inflammatory metabolites from the soft coral *Sinularia verruca*. J Nat Prod 73:771
346. Yuan W, Cheng S, Fu W, Zhao M, Li X, Cai Y, Dong J, Huang K, Gustafson KR, Yan P (2016) Structurally diverse metabolites from the soft coral *Sinularia verruca* collected in the South China Sea. J Nat Prod 79:1124
347. Lam HC, Pepper HP, Sumby CJ, George JH (2017) Biomimetic total synthesis of (±)-verrubenzospirolactone. Angew Chem Int Ed Engl 56:8532
348. Stauffacher D, Niklaus P, Tscherter H, Weber HP, Hofmann A (1969) Cycloclavin, ein neues Alkaloid aus *Ipomoea hildebrandtii*. Mutterkornalkaloide 71. Tetrahedron 25:5879
349. Furuta T, Koike M, Abe M (1982) Isolation of cacloclavine from the culture broth of *Aspergillus japonicus* SAITO. Agric Biol Chem 46:1921
350. Körber K, Song D, Rheinheimer J, Kaiser JF, Dickhaut J, Narine A, Culbertson DL, Thompson S, Rieder J (2014) Cycloclavine and derivatives thereof for controlling invertebrate pests. WO2014096238 A1
351. McCabe SR, Wipf P (2017) Eight-step enantioselective total synthesis of (−)-cycloclavine. Angew Chem Int Ed 56:324
352. Gavagnin M, Carbone M, Nappo M, Mollo E, Roussis V, Cimino G (2005) First chemical study of anaspidean *Syphonota geographica*: structure of degraded sterols aplykurodinone-1 and -2. Tetrahedron 61:617
353. Ortega MJ, Zubía E, Salvá J (1997) 3-*epi*-Aplykurodinone B, a new degraded sterol from *Aplysia fasciata*. J Nat Prod 60:488
354. Lee J-H, Cho C-G (2018) H-Bonding mediated asymmetric intramolecular Diels–Alder reaction in the formal synthesis of (+)-aplykurodinone-1. Org Lett 20:7312
355. Klayman DL (1985) Qinghaosu (artemisinin): an antimalarial drug from China. Science 228: 1049
356. Ziffer H, Highet RJ, Klayman DL (1997) Artemisinin: an endoperoxidic antimalarial from *Artemisia annua* L. Prog Chem Org Nat Prod 72:121–214

357. Tu Y (2011) The discovery of artemisinin (qinghaosu) and gifts from Chinese medicine. Nat Med 17:1217
358. Ma Z, Woon CY, Liu C-G, Cheng J-T, You M, Sethi G, Wong AL-A, Ho PC-L, Zhang D, Ong P, Wang L, Goh B-C (2021) Repurposing artemisinin and its derivatives as anticancer drugs: a chance or a challenge. Front Pharmacol 12:828856358
359. Krieger J, Smeilus T, Kaiser M, Seo E-J, Efferth T, Giannis A (2018) Total synthesis and biological investigation of (−)-artemisinin: the antimalarial activity of artemisinin is not stereospecific. Angew Chem Int Ed 57:8293
360. Luo J-G, Wang X-B, Xu Y-M, U'Ren JM, Arnold AE, Kong L-Y, Gunatilaka AAL (2014) Delitschiapyrone A, a pyrone–naphthalenone adduct bearing a new pentacyclic ring system from the leaf-associated fungus *Delitschia* sp. FL1581. Org Lett 16:5944
361. Hyde KD, Steinke TS (1996) Two new species of *Delitschia* from submerged wood. Mycoscience 37:99
362. Kurasawa K, Kwon E, Kuwahara S, Enomoto M (2018) Bioinspired total synthesis of delitchiapyrone A. Org Lett 20:4645
363. Li Y, Pattenden G (2011) Perspectives on the structural and biosynthetic interrelationships between oxygenated furanocembranoids and their polycyclic congeners found in corals. Nat Prod Rep 28:1269
364. Anjaneyulu ASR, Venugopal MJRV, Sarada P, Rao GV, Clardy J, Lobkovsky E (1998) Sarcophytin, a novel tetracyclic diterpenoid from the Indian Ocean soft coral *Sarcophyton elegans*. Tetrahedron Lett 39:135
365. Bogdanov A, Hertzer C, Kehraus S, Nietzer S, Rohde S, Schupp PJ, Wagele H, König GM (2016) Defensive diterpene from the Aeolidoidean *Phyllodesmium longicirrum*. J Nat Prod 79:611
366. Nannini LJ, Nemat SJ, Carreira EM (2018) Total synthesis of (+)-sarcophytin. Angew Chem Int Ed 57:823
367. Ainsworth AM, Chicarelli-Robinson MI, Copp BR, Fauth U, Hylands RJ, Holloway JA, Latif M, O'Beirne GB, Porter N, Renno DV, Richards M, Robinson N (1995) Xenovulene A, a novel GABA-benzodiazepine receptor-binding compound produced by *Acremonium strictum*. J Antibiot 48:568
368. Thomas P, Sundaram H, Krishek SJ, Chazot P, Xie X, Bevan P, Brocchini SJ, Latham CJ, Charlton P, Moore M, Lewis SJ, Thornton DM, Stephenson FA, Smart TG (1997) Regulation of neuronal and recombinant GABA(A) receptor ion channels by xenovulene A, a natural product isolated from *Acremonium strictum*. J Pharmacol Exp Ther 282:513
369. Li P-J, Dräger G, Kirschning A (2019) A general biomimetic hetero-Diels–Alder approach to the core skeletons of xenovulene A and the sterhirsutins A and B. Org Lett 21:998
370. Yoshida T, Mori K, He G (1995) Inulavosin, a new thymol dimer with piscicidal activity from *Inula nervosa*. Heterocycles 41:1923
371. Kim TH, Hatano T, Okamoto K, Yoshida T, Kanzaki H, Arita M, Ito H (2017) Antifungal and ichthyotoxic sesquiterpenoids from *Santalum album* heartwood. Molecules 22:1139
372. Gong P-X, Li H-J, Wang M, Cheng Y-F, Wu Y-C (2019) Synthesis of natural product inulavosin via $Ga(OTf)_3$-catalyzed hetero Diels–Alder dimerization of salicyl alcohol derivative. Nat Prod Res 33:2911
373. Zhang Q, Di Y-T, Li C-S, Fang X, Tan C-J, Zhang Z, Zhang Y, He H-P, Li S-L, Hao X-J (2009) Daphenylline, a new alkaloid with an unusual skeleton, from *Daphniphyllum longeracemosum*. Org Lett 11:2357
374. Xu B, Wang B, Xun W, Qiu FG (2019) Total synthesis of (−)-daphenyllene. Angew Chem Int Ed 58:5754
375. von Rague Schleyer P, Wu JI, Cossío FP, Fernández I (2014) Aromaticity in transition structures. Chem Soc Rev 43:4909
376. Telan LA, Firestone RA (1999) Heavy atom effects reveal diradical intermediates. I. An aqueous Diels–Alder reaction. Tetrahedron 55:14269
377. Fukui K (1975) Theory of orientation and stereoselection. Springer Verlag, Berlin

378. Ess DH, Jones GO, Houk KN (2006) Conceptual, qualitative, and quantitative theories of 1,3-dipolar and Diels–Alder cycloadditions used in synthesis. Adv Synth Catal 348:2337
379. Geerlings P, Ayers PW, Toro-Labbé A, Chattaraj PK, de Proft F (2012) Conceptual, qualitative, and quantitative theories of 1,3-dipolar and Diels–Alder cycloadditions used in synthesis. Acc Chem Res 45:683
380. Fernández I, Franking G (2019) The Diels–Alder reaction from the EDA-NOVC perspective: a re-examination of the frontier molecular orbital model. Eur J Org Chem 2019:478
381. Sheu J, Ahmed AF, Shiue R, Dai C, Kuo Y (2002) Scabrolides A–D. Four new norditerpenoids isolated from the soft coral *Sinularia scabra*. J Nat Prod 65:1904
382. Thao NP, Nam NH, Cuong NX, Quang TH, Tung PT, Dat LD, Chae D, Kim S, Koh S, Kiem PV, Minh CV, Kim YH (2013) Anti-inflammatory norditerpenoids from the soft coral *Sinularia maxima*. Bioorg Med Chem Lett 23:228
383. Hafeman NJ, Loskot SA, Reimann CE, Pritchett BP, Virgil SC, Stoltz BM (2020) The total synthesis of (−)-scabrolide A. J Am Chem Soc 142:8585
384. Khaimova MA, Palamareva MD, Mollov NM, Krestev VP (1971) The structures of talatisamine and cammaconine by correlation with isotalatizidine. Tetrahedron 27:819
385. Wang Y, Song M, Hou L, Yu Z, Chen H (2012) The newly identified $K^+$ channel blocker talatisamine attenuates beta-amyloid oligomers-induced neurotoxicity in cultured cortical neurons. Neurosci Lett 518:122
386. Kamakura D, Todoroki H, Urabe D, Hagiwara K, Inoue M (2020) Total synthesis of talatisamine. Angew Chem Int Ed 59:479
387. Moczydlowski EG (2013) The molecular mystique of tetrodotoxin. Toxicon 63:165
388. Tahara Y (1909) Studies on the puffer-fish toxin. J Pharm Soc Jpn 1909:587
389. Mosher HS, Fuhrmann FA, Buchwald HD, Fischer HG (1964) Tarichatoxin — tetrodotoxin: a potent neurotoxin: a non-protein substance isolated from the California newt is the same as the toxin from the puffer fish. Science 144:1100
390. Shen H, Liu D, Wu K, Lei J, Yan N (2019) Structures of human Nav1.7 channel in complex with auxiliary subunits and animal toxins. Science 363:1303
391. Kudo Y, Yotsu-Yamashita M (2019) Isolation and biological activity of 8-epitetrodotoxin and the structure of a possible biosynthetic shunt product of tetrodotoxin, Cep-226A, from the newt *Cynops ensicauda popei*. J Nat Prod 82:1656
392. Murakami K, Toma T, Fukuyama T, Yokoshima S (2020) Total synthesis of tetrodotoxin. Angew Chem Int Ed 59:6253
393. Birch AJ, Wright JJ (1969) The brevianamides: a new class of fungal alkaloid. J Chem Soc D:644
394. Paterson RRM, Simmonds MJS, Kemmelmeier C, Blaney WM (1990) Effects of brevianamide A, its photolysis product brevianamide D, and ochratoxin A from two *Penicillium* strains on the insect pests *Spodoptera frugiperda* and *Heliothis virescens*. Mycol Res 94:538
395. Godfrey RC, Green NJ, Nichol GS, Lawrence AL (2020) Total synthesis of brevianamide A. Nat Chem 12:615
396. Wang YF, Shi QW, Dong M, Kiyota H, Gu Y-C, Cong B (2011) Natural taxanes: developments since 1828. Chem Rev 111:7652
397. Schiff PB, Fant J, Horwitz SB (1979) Promotion of microtubule assembly in vitro by taxol. Nature 277:665
398. Malik S, Cusido RM, Mirjalili MH, Moyano E, Palazon J, Bonfill M (2011) Production of the anticancer drug taxol in *Taxus baccata* suspension cultures: a review. Process Biochem 46:23
399. Huo C-H, Su X-H, Wang Y-F, Zhang X-P, Shi Q-W, Kiyota H (2007) Canataxpropellane, a novel taxane with a unique polycyclic carbon skeleton (tricyclotaxane) from the needles of *Taxus canadensis*. Tetrahedron Lett 48:2721
400. Schneider F, Samarin K, Zanella S, Gaich T (2020) Total synthesis of the complex taxane diterpene canataxpropellane. Science 367:676

401. Gomi S, Imamura K-I, Yaguchi T, Kodama Y, Minowa N, Koyama M (1994) PF1018, a novel insecticidal compound produced by *Humicola* sp. J Antibiot 47:571
402. Zalucki MP, Shabbir A, Silva R, Adamson D, Liu S-S, Furlong MJ (2012) Estimating the economic cost of one of the world's major insect pests, *Plutella xylostella* (Lepidoptera: Plutellidae): just how long is a piece of string? J Econ Entomol 105:1115
403. Quintela-Varela H, Jamieson CS, Shao Q, Houk KN, Trauner D (2020) Bioinspired synthesis of (−)-PF1018. Angew Chem Int Ed 59:5263
404. Okanya PW, Mohr KI, Gerth K, Jansen R, Müller R (2011) Marinoquinolines A–F, pyrroloquinolines from *Ohtaekwangia kribbensis* (Bacteroidetes). J Nat Prod 74:603
405. Osano M, Jhaveri DP, Wipf P (2020) Formation of 6-azaindoles by intramolecular Diels–Alder reaction of oxazoles and total synthesis of marinoquinoline A. Org Lett 22:2215
406. Chan WR, Taylor DR, Willis CR (1967) Crotonin, a furanoid norditerpene from *Croton lucidus* L. Chem Commun:191
407. Chan WR, Taylor DR, Willis CR (1968) Terpenoids from the Euphorbiaceae. Part I. The structure of crotonin, a norditerpene from *C. lucidus* L. J Chem Soc C:2781
408. Puebla P, Correa SX, Guerrero M, Carron R, Feliciano AS (2005) New *cis*-clerodane diterpenoids from *Croton schiedeanus*. Chem Pharm Bull 53:328
409. Si X-G, Zhang ZM, Zheng C-G, Li Z-T, Cai Q (2020) Enantioselective synthesis of *cis*-decalin derivatives by the inverse electron demand Diels–Alder reaction of 2-pyrones. Angew Chem Int Ed 59:18412
410. Briggs CK, Highet PF, Highet RJ, Wildman WC (1956) Alkaloids of the Amaryllidaceae. VII. Alkaloids containing the hemiacetal or lactone group. J Am Chem Soc 78:2899
411. Jeffs PW, Hansen JF, Döpke W, Bienert M (1971) The structure and stereochemistry of clivonine. Tetrahedron 27:5065
412. Wang C-D, Chen Q, Shin S, Cho CG (2020) Total synthesis of (±)-clivonine via Diels–Alder reactions of 3,5-dibromo-2-pyrone. J Org Chem 85:1003
413. Hayakawa Y, Kobayashi T, Izawa M (2013) Indanostatin, a new neuro-protective compound from *Streptomyces* sp. J Antibiot 66:731
414. Thawai C, Bunbamrung N, Pittayakhajonwut P, Chongruchiroj S, Pratuangdeijkul J, He Y-W, Tadtong S, Sareedenchai V, Prombutra P, Qian Y (2020) A novel diterpene agent isolated from *Microbispora hainanensis* strain CSR-4 and its in vitro and in silico inhibition effects on acetylcholine esterase enzyme. Sci Rep 10:11058
415. Dissanayake I, Hart JD, Becroft EC, Sumby CJ, Newton CG (2020) Bisketene equivalents as Diels–Alder dienes. J Am Chem Soc 142:13328
416. Sara AA, Saeed A, Kalesse M (2022) Recent applications of the Diels–Alder reaction in the synthesis of natural products (2017–2020). Synthesis 54:975
417. Fan Y-Y, Xu J-B, Liu H-C, Gan L-S, Ding J, Yue J-M (2017) Cephanolides A–J, cepholan-type diterpenoids from *Cephalotaxus sinensis*. J Nat Prod 80:3159
418. Haider M, Sennari G, Eggert A, Sarpong R (2021) Total synthesis of the *Cepahalotaxus* norditerpenoids (±)-cephanolides A–D. J Am Chem Soc 143:2710
419. Jiang Z-Y, Yu Y-J, Huang C-G, Huang X-Z, Hu Q-F, Yang G-Y, Wang H-B, Zhang X-Y, Li G-P (2015) Icetexane diterpenoids from *Perovskia atriplicifolia*. Planta Med 81:241
420. Parvez A, Choudhary MI, Akhter F, Noorwala M, Mohammad FV, Hasan NM, Zamir T, Ahmad VU (1992) Perovskone: a triterpene with a novel carbon skeleton from *Perovskia abrotanoides*. J Org Chem 57:4339
421. Wu Y-B, Ni Z-Y, Shi Q-W, Dong M, Kiyota H, Gu Y-C, Cong B (2012) Constituents from *Salvia* species and their biological activities. Chem Rev 112:5967
422. Yang B, Wen G, Zhang Q, Hou M, He H, Gao S (2021) Asymmetric total synthesis and biosynthetic implications of perovskones, hydrangenone, and hydrangenone B. J Am Chem Soc 143:6370
423. Smith GN (1968) Strictosidine: a key intermediate in the biogenesis of indole alkaloids. Chem Commun:912

424. Anthony SM, Tona V, Zou Y, Morill LA, Billingsley JM, Lim M, Tang Y, Houk KN, Garg NK (2021) Total synthesis of (−)-strictosidine and interception of aryne natural product derivatives "strictosidine" and "strictosamidyne". J Am Chem Soc 143:7471
425. Wang RB, Ma S-G, Jamieson CS, Gao R-M, Liu Y-B, Li Y, Wang X-J, Li Y-H, Houk KN, Qu J, Yu S-S (2021) Library construction of stereochemically diverse isomers of spirooliganin: their total synthesis and antiviral activity. Chem Sci 12:7003
426. Wan L-S, Nian Y, Ye C-J, Shao L-D, Peng X-R, Shao L-D, Geng C-A, Zuo Z-L, Li X-N, Yang J, Zhou M, Qiu M-H (2016) Three minor diterpenoids with three carbon skeletons from *Euphorbia peplus*. Org Lett 18:2166
427. Yuan P, Gerlinger CKG, Herberger J, Gaich T (2021) Ten-step asymmetric total synthesis of (+)-pepluanol A. J Am Chem Soc 143:11934
428. Williams DE, Dalisay DS, Li F, Amphlett J, Maneerat W, Chavez MAG, Wang YA, Matainaho T, Yu W, Brown PJ, Arrowsmith CH, Vedadi M, Andersen RJ (2013) Nahuoic acid A produced by a *Streptomyces* sp. isolated from a marine sediment is a selective SAM-competitive inhibitor of the histone methyltransferase SETD8. Org Lett 15:414
429. Guillade L, Mora P, Villar P, Alvarez R, de Lera AR (2021) Total synthesis of nahuoic acid A via a putative biogenetic intramolecular Diels–Alder (IMDA) reaction. Chem Sci 12:15169
430. Wang B-G, Ebel R, Nugroho BW, Prijono D, Frank W, Steube KG, Hao X-J, Proksch P (2001) Aglacins A–D, first representatives of a new class of aryltetralin cyclic ether lignans from *Aglaia cordata*. J Nat Prod 64:1521
431. Wang B-G, Ebel R, Wang C-Y, Edrada RA, Wray V, Proksch PJ (2004) Aglacins I–K, three highly methoxylated lignans from *Aglaia cordata*. J Nat Prod 67:682
432. Xu M, Hou M, He H, Gao S (2021) Asymmetric total synthesis of aglacins A, B, and E. Angew Chem 133:16791
433. Vasas A, Hohmann J (2014) *Euphorbia* diterpenes: isolation, structure, biological activity, and synthesis (2008–2012). Chem Rev 114:8579
434. Wang H-B, Wang X-Y, Liu L-P, Qin G-W, Kang T-G (2015) Tigliane diterpenoids from the Euphorbiaceae and Thymelaeaceae families. Chem Rev 115:2975
435. Newman DJ, Cragg GM (2020) Natural products as sources of new drugs over the nearly four decades from 01/1981 to 09/2019. J Nat Prod 83:770
436. Hirose A, Watanabe A, Ogino K, Nagamoto M, Inoue M (2021) Unified total syntheses of rhamnofolane, tigliane, and daphnane diterpenoids. J Am Chem Soc 143:12387
437. Wang Y, Li H, Wang Y-Q, Liu Y, Foxman BM, Deng L (2007) Asymmetric Diels-Alder reactions of 2-pyrones with a bifunctional organic catalyst. J Am Chem Soc 129:6364
438. Gao J-M, Wu W-J, Zhang J-W, Konishi Y (2007) The dihydro-β-agarofuran sesquiterpenoids. Nat Prod Rep 24:1153
439. Sugiura K, Shizuri Y, Wada H, Yamada K, Hirata Y (1971) Isolation of three new alkaloids from *Euonymus sieboldiana* Blume and their structures. Tetrahedron Lett 12:2733
440. Núñez MJ, Cortés-Selva F, Bazzocchi IL, Jiménez IA, González AG, Ravelo AG, Gavin JA (2003) Absolute configuration and complete assignment of $^{13}C$ NMR data for new sesquiterpenes from *Maytenus chiapensis*. J Nat Prod 66:572
441. Wang Y, Nagai T, Watanabe I, Hagiwara K, Inoue M (2021) Total synthesis of euonymine and euonyminol octaacetate. J Am Chem Soc 143:21037
442. Lu Y, Xu M-M, Zhang Z-M, Zhang J, Cai Q (2021) Catalytic asymmetric inverse electron demand Diels–Alder reactions of 2-pyrones with indenes: total syntheses of cephanolides A and B. Angew Chem Int Ed 60:26610
443. Yang X-W, Grossman RB, Xu G (2018) Research progress of polycyclic polyprenylated acylphloroglucinols. Chem Rev 118:3508
444. Xiao C-Y, Mu Q, Gibbons S (2020) The phytochemistry and pharmacology of *Hypericum*. Prog Chem Org Nat Prod 112:85
445. Ciochina R, Grossman RB (2006) Polycyclic polyprenylated acylphloroglucinols. Chem Rev 106:3963

446. Qiu DR, Zhou M, Liu X-Z, Chen J-J, Wang G-H, Lin T, Yu F-R, Ding R, Sun C-L, Tian W-J, Chen HF (2021) Cytotoxic polyprenylated phloroglucinol derivatives from *Hypericum elodoides* Choisy modulating the transactivation of RXRa. Bioorg Chem 107:104578
447. An TY, Hu L-H, Chen Z-L, Sim K-Y (2002) Erectones A and B, two dome-shaped polyprenylated phloroglucinol derivatives, from *Hypericum erectum*. Tetrahedron Lett 43:163
448. Franov LJ, Hart JD, Pullella GA, Sumby CJ, George JH (2022) Bioinspired total synthesis of erectones A and B, and the revised structure of hyperelodione D. Angew Chem Int Ed 61: e202200420
449. Torii M, Kato H, Hitora Y, Angkouw ED, Mangindaan REP, de Voogd NJ, Tsukamoto S (2017) Lamellodysidines A and B, sesquiterpenes isolated from the marine sponge *Lamellodysidea herbacea*. J Nat Prod 80:2536
450. Kamo S, Kurosawa H, Matsuzawa A, Sugita K (2022) Total synthesis of (−)-lamellodysine A via an intramolecular Diels–Alder reaction. Org Lett 24:921
451. Qin B, Lu Z, Jia Y (2022) Divergent total synthesis of four kopsane alkaloids: *N*-carbomethoxy-10,22-dioxokopsane, epikosanol-10-lactam, 10,22-dioxokopsane, and *N*-methylkopsanone. Angew Chem Int Ed 61:e202201712
452. Achenbach H, Biemann K (1965) 10,22-Dioxokopsane, *N*-methyl-10,22-dioxokopsane, and *N*-carbomethoxy-10,22-dioxokopsane—three new alkaloids of *Pleiocarpa mutica* Benth. J Am Chem Soc 87:4944
453. Kam TS, Subramaniam G, Chen W (2004) Alkaloids from *Kopsia dasyrachis*. Phytochemistry 51:159
454. Filho JMF, Gilbert B, Kitagawa M, Paes Leme LA, Durham LJ (1966) Four heptacyclic alkaloids from *Aspidosperma* species. J Chem Soc C:1260
455. Chen C, Ding X, Han L, Zhus W, Huang K, Hao X, Zhang Y (2021) Two new monoterpenoid indole alkaloids from the kernels of *Kopsia arborea*. Nat Prod Res 37:1047
456. Hellwig V, Dasenbrock J, Gräf C, Kahner L, Schumann S, Steglich W (2002) Calopins and cyclocalopins—bitter principles from *Boletus calopus* and related mushrooms. Eur J Org Chem 2002:2895
457. Yu W, Song J, Yu SY, Kim J (2022) Cyclocarbonylation of allenyl glyoxylate strategy to build the tricyclic core of cyclocalopin A. Org Lett 24:2242
458. Cava MP, Tjoja SS, Ahmed QA, Da Rocha AI (1968) Alkaloids of *Tabernaemontana riedelii* and *T. rigida*. J Org Chem 33:1055
459. Zhang F, Ren B-T, Zhou Y, Liu Y, Feng X (2022) Enantioselective construction of *cis*-hydroindole scaffolds via an asymmetric inverse electron-demand Diels–Alder reaction: application to the formal total synthesis of (+)-minovincine. Chem Sci 13:5562
460. Rammohan A, Khasanov AF, Kopchuk DS, Gunasekar D, Zyryanov GV, Chupakhin ON (2022) Assessment on facile Diels–Alder approach of α-pyrone and terpenoquinone for the expedient synthesis of various natural scaffolds. Nat Prod Bioprospect 12:12
461. Huang G, Guillot R, Kouklovsky C, Maryasin B, de la Torre A (2022) Diastereo- and enantioselective inverse electron demand Diels–Alder cycloaddition between 2-pyrones and acylic enol ethers. Angew Chem Int Ed 61:e202208185
462. Soares MI, Cardoso AL, Pinho e Melo TMVD (2022) Diels–Alder cycloaddition reactions in sustainable media. Molecules 27:1304
463. Leverrier A, Tran Huu Dau ME, Retailleau P, Awang K, Guéritte F, Litaudon M (2010) Kingianin A: a new natural pentacyclic compound from *Endiandra kingiana*. Org Lett 12: 3638
464. Zhang J, Yan P, Wang Z, Zhao J, Chen Q, Lu P (2022) Total synthesis of (+)-kingianin and an enantioselective cycoaddition of strained cyclobutenone. Synthesis 54:1977
465. Feng T, Duan K-T, He S-J, Wu B, Zheng Y-S, Ai H-L, Li Z-H, He J, Zuo J-P, Liu J-K (2018) Two cationic indole alkaloids from *Ophiorrhiza japonica* and their xanthine oxidase inhibitory activity. Org Lett 20:7926

466. Cao W, Dou Y, Kouklovsky C, Vincent G (2022) Total synthesis of ophiorrhine A, G and ophiorrhiside E featuring a bioinspired intramolecular Diels–Alder cycloaddition. Angew Chem Int Ed 61:e202209135
467. Liu J, Xi S, Tang Y (2022) Biomimetic Diels–Alder reactions in natural product synthesis: a personal retrospect. Synlett 33:836
468. Ishiuchi K, Kubota T, Morita H, Kobayashi J (2006) Lycopladine A, a new $C_{16}N$ alkaloid from *Lycopodium complanatum*. Tetrahedron Lett 47:3287
469. Hartrampf FWW, Trauner D (2017) Total synthesis of lycopladine A and carinatine A via base-mediated carbocyclization. J Org Chem 82:8206
470. Sim K-H, ul Ansari T, Parl Y-G, Jeong Y, Oh S-H, Min H-W, Jeon D-Y, Kim H, Cho C-H (2022) A regio-and diastereoselective Stille coupling/intramolecular Diels–Alder cascade for the generation of fused pyridines and application in the synthesis of (+)-lycopladine A and (−)-lycoposerramine R. Angew Chem Int Ed 61:e202212016
471. Hudson GA, Hooper AR, DiCaprio AJ, Sarlah D, Mitchell DA (2020) Structure prediction and synthesis of pyridine-based macrocyclic peptide natural products. Org Lett 23:253
472. Hooper AR, Oštrek A, Milian-Lopez A, Sarlah D (2022) Bioinspired total synthesis of pyritide A2 through pyridine ring synthesis. Angew Chem Int Ed 61:e202212299
473. Zhang F-G, Chen Z, Tang X, Ma J-A (2021) Triazines: syntheses and invers electron demand Diels–Alder reactions. Chem Rev 121:14555
474. Saxton JE (1998) Synthesis of the *Aspidosperma* alkaloids. In: Cordell GA (ed) The alkaloids: chemistry and biology, vol 50. Academic Press, San Diego, CA, p 343
475. Cain DL, Anderson NA, Cordes DB, Slawin MZ, Watson AJB (2022) Total synthesis of (±)-aspidospermidine, (±)- aspidofractinine, (±)-limaspermidine, and (±)-vincadifformine via a cascade and common intermediate strategy. J Org Chem 87:15559
476. Aly SH, Elissawy AM, Eldahshan OA, Elshanawany MA, Efferth T, Singab ANB (2019) The pharmacology of the genus *Sophora* (Fabaceae): an updated review. Phytomedicine 64: 153070
477. Zhou Y, Wu Y, Deng L, Chen L, Zhao D, Lv L, Chen X, Man J, Wang Y, Shan H, Lu Y (2014) The alkaloid matrine of the root of *Sophora flavescens* prevents arrhythmogenic effect of ouabain. Phytomedicine 21:931
478. Sun N, Sun P, Lv H, Sun Y, Guo J, Wang Z, Luo T, Wang S, Li H (2016) Matrine displayed antiviral activity in porcine alveolar macrophages co-infected by porcine reproductive and respiratory syndrome virus and porcine circovirus type 2. Sci Rep 6:24401
479. Magann NL, Westley E, Sowden MJ, Gardiner MG, Sherburn MS (2022) Total synthesis of matrine alkaloids. J Am Chem Soc 144:19695
480. Anjaneyulu ASR, Venugopal MJRV, Sarada P, Clardy J, Lobkovsky E (1998) Havellockate, a novel *seco* and spiro lactone diterpenoid from the Indian Ocean soft coral *Sinularia granosa*. Tetrahedron Lett 39:139
481. Hafeman NJ, Chan M, Fulton TJ, Alexy EJ, Loslot SA, Virgil SC, Stoltz BM (2022) Asymmetric total synthesis of havelolockate. J Am Chem Soc 144:20232
482. Wang W, Jiang Y, Song B, Tang X, Wu H, Jin Z, Chen L (2023) Discovery of quality markers in the rhizome of *Atractylodes chinensis* using GC-MS fingerprint and network pharmacology. Arab J Chem 16:105114
483. Chen F-L, Liu D-L, Fu J, Yang J, Bai L-P, Zhang W, Jiang Z-H, Zhu G-Y (2022) Atrachinenins A–C, three pairs of caged $C_{27}$ meroterpenoids from the rhizomes of *Atractylodes chinensis*. Chin J Chem 40:460
484. French SA, Sumby CJ, Huang DM, George JH (2022) Total synthesis of atrachinenins A and B. J Am Chem Soc 144:22844
485. Grieco PA, Nunes JJ, Gaul MD (1990) Dramatic rate accelerations of Diels–Alder reactions in 5 *M* lithium perchlorate-diethyl ether: the cantharidin problem reexamined. J Am Chem Soc 112:4595
486. Yan Y-M, Zhang H-X, Liu H, Wang Y, Wu J-B, Li Y-P, Cheng Y-X (2019) (±)-Lucidumone, a COX-2 inhibitory caged fungal meroterpenoid from *Ganoderma lucidum*. Org Lett 21:8523

487. Flower RJ (2003) The development of COX2 inhibitors. Nat Rev Drug Discov 2:179
488. Huang G, Kouklovsky C, de la Torre A (2023) Retro-[4+2]/intramolecular Diels–Alder cascade allows a concise total synthesis of lucidumone. Synlett 34:1195
489. Lyakhova EG, Kolesnikova SA, Kalinovsky AI, Berdyshev DV, Pislyagin EA, Kuzmich AS, Popov RS, Dmitreno PS, Makarieva T, Stonik VA (2017) Lissodendoric acids A and B, manzamine-related alkaloids from the Far Eastern sponge *Lissodendoryx florida*. Org Lett 19: 5320
490. Dixon DJ, Jakubec P (2021) Total synthesis of manzamine alkaloids. In: Kiyota H (ed) Topics in heterocyclic chemistry—marine natural products. Springer, Singapore, p 23
491. Ippoliti FM, Adamson NJ, Wonilowicz LG, Nasrallah DJ, Darzi ER, Donaldson JS, Garg NK (2023) Total synthesis of lissodendoric acid A via stereospecific trapping of strained cyclic allene. Science 379:261
492. Minagawa K, Kouzuki S, Nomura K, Yamaguchi T, Kawamura Y, Matsushima K, Tani H, Ishii K, Tanimoto T, Kamigauchi T (2001) Bisabosquals, novel squalene synthase inhibitors I. Taxonomy, fermentation, isolation and biological activities. J Antibiot 54:890
493. Du X, Liu H, Wu Y, Tang Y (2023) Bio-inspired formal total synthesis of (±)-bisabosqual A. Org Chem Front 10:1042
494. Zhou B, Yue J-M (2022) Natural products are the treasure pool for antimalarial agents. Natl Sci Rev 9:112
495. Kwon OE, Lee HS, Lee SW, Bae K, Kim K, Hayashi M, Rho M-C, Kim Y-K (2006) Dimeric sesquiterpenoids isolated from *Chloranthus japonicus* inhibited the expression of cell adhesion molecules. J Ethnopharmacol 104:270
496. Huang G, Huang Z, Ma X, Feng Z, Yuan F, Qin S, Fu S, Liu B (2023) Asymmetric total synthesis of sarbracholide shizukaol B. Org Chem Front 10:3591
497. Hirasawa Y, Morita H, Kobayashi J (2004) Senepodine F, a new $C_{22}N_2$ alkaloid from *Lycopodium chinense*. Heterocycles 64:515
498. Hirasawa Y, Kobayashi J, Morita H (2009) The *Lycopodium* alkaloids. Heterocycles 77:679
499. Nakashima Y, Inoshita T, Kitajama M, Ishikawa H (2023) Asymmetric total synthesis of senepodine F. Org Lett 25:1151
500. Inoshita T, Goshi K, Moringa Y, Umeda Y, Ishikawa H (2019) Enantioselective construction of octahydroquinolines via trienamine-mediated Diels–Alder reactions. Org Lett 21:2903
501. Achan J, Talisuna AO, Erhart A, Yeka A, Tibenderana JK, Baliraine FN, Rosenthal PJ, D'Alessandro U (2011) Quinine, an old anti-malarial drug in a modern world: role in the treatment of malaria. Malar J 10:144
502. Dubey A, Singh Y (2021) Medicinal properties of cinchona alkaloids—a brief review. Asian J Res Pharm Sci 11:2231
503. Haripriya P, Rai R, Vijayakrishna K (2025) Polymer anchored *Cinchona* alkaloids: synthesis and their applications in organo-catalysis. Top Curr Chem 383:9
504. Lei L, Yang H-F, Ding Q-H, Wei K, Yang Y-R (2023) Total syntheses of *Cinchona* alkaloids via photoredox-catalyzed deoxygenative arylation. Org Lett 25:4586
505. Mikawa T, Takahashi N,, Ohkishi H, Sato Y, Miyadoh S, Sezaki M (1990) Novel antibiotic vermisporin, process for the production thereof and pharmaceutical composition comprising it as an active antibacterial agent. U.S. pat. no. 4933180A
506. Dong H, Hu D, Hong B, Wang J, Lei X (2023) Total synthesis of diverse tetramic acid bearing *cis*-decalin natural products. Angew Chem Int Ed 62:e202301872
507. Bi D, Xia G, Li Y, Liang X, Zhang L, Wang L (2018) Two new cassane diterpene lactams from the fruits of *Caesalpinia mimosoides* Lam. Nat Prod Res 32:875
508. Jing W, Zhang X, Zhou H, Wang Y, Yang M, Long L, Gao H (2019) Naturally occurring cassane diterpenoids (CAs) of *Caesalpinia*: a systematic review of its biosynthesis, chemistry and pharmacology. Fitoterapia 134:226
509. Kinoshita T, Kaneko M, Noguchi H, Kitagawa I (1996) New cassane diterpenes from *Caesalpinia bonduc* (Fabaceae). Heterocycles 43:409

510. Papidocha SM, Bulthaupt HH, Carreira EM (2023) Synthesis of neoceasalpin A, AA, and nominal neocaesalpin K. Angew Chem Int Ed 62:e2022310149
511. Jiang H-Y, Wang W-G, Zhou M, Wu H-Y, Zhan R, Li X-N, Du X, Li Y, Pu J-X, Sun H-D (2014) Diterpenes from *Isodon sculponeatus*. Fitoterapia 93:142
512. Li L-M, Li G-Y, Pu J-X, Xiao W-L, Ding L-S, Sun H-D (2009) *ent*-Kaurane and cembrene diterpenoids from *Isodon sculponeatus* and their cytotoxicity. J Nat Prod 72:1851
513. Cao W, Wang Z, Hao Y, Wang T, Fu S, Liu B (2023) Total synthesis of sculponin U through a photoinduced radical cascade reaction. Angew Chem Int Ed 62:e202305516
514. Xiong J, Tan N-H, Ji C-J, Lu Y, Gong N-B (2008) Vilmoraconitine, a novel skeleton $C_{19}$-diterpenoid alkaloid from *Aconitum vilmorinianum.* Tetrahedron Lett 33:4851
515. Ji J, Chen J, Qin S, Li W, Zhao J, Li G, Song H, Liu X-Y, Qin Y (2023) Total synthesis of vilmoraconitine. J Am Chem Soc 145:3903
516. Ferreira RJ, Spengler G, Orthaber A, dos Santos DJVA, Ferreira M-JU (2021) Pedrolane, a polycyclic diterpene scaffold containing a bicyclo[2.2.1]heptane system, from *Euphorbia pedroi.* Org Lett 23:274
517. Fadel M, Carreira EM (2023) Enantioselective total synthesis of (+)-pedrolide. J Am Chem Soc 145:8332
518. Chen DF, Zhang SX, Wang HK, Zhang SY, Sun QZ, Cosentino LM, Lee KH (1999) Novel anti-HIV lancilactone C and related triterpenes from *Kadsura lancilimba.* J Nat Prod 62:94
519. Kuroiwa H, Suzuki S, Irie K, Tsukano C (2023) Total synthesis and structure revision of (+)-lancilactone C. J Am Chem Soc 145:14587
520. Yu X, Xiao L, Luo T (2024) Enantioselective total synthesis of (−)-vinigrol: the evolution of a transannular Diels–Alder strategy. J Org Chem 89:1709
521. Rashid S, Lone WI, Rashid A, Bhat BA (2024) Inverse electron demand Diels–Alder reaction in total synthesis of bioactive natural products. Tetrahedron Chem 9:100066
522. Du Y-E, Byun WS, Lee SB, Hwang S, Shin YH, Shin B, Jang Y-J, Hong S, Shin J, Lee SK, Oh D (2020) Formicins, *N*-acetylcysteamine-bearing indenone thioesters from a wood ant-associated bacterium. Org Lett 22:5337
523. Khamkar SL, Handore KL, Shinde HM, Reddy DS (2024) Highly stereoselective Diels–Alder-based strategy for the synthesis of 3-*epi*-formicin A and 1-*epi*-formicin B. Org Lett 26:3961
524. Kobayashi J, Tsuda M, Kawasaki N, Matsumoto K, Adachi T (1994) Keramaphidin B, a novel pentacyclic alkaloid from a marine sponge *Amphimedon* sp.: a plausible biogenetic precursor of manzamine alkaloids. Tetrahedron Lett 35:4383
525. Kong F, Andersen RJ (1995) Ingenamine alkaloids isolated from the sponge *Xestospongia ingens*: structures and absolute configurations. Tetrahedron 51:2895
526. Rodríguez J, Peters BM, Kurz L, Schatzman RC, McCarley D, Lou L, Crews P (1993) An alkaloid protein kinase C inhibitor, xestocyclamine A, from the marine sponge *Xestospongia* sp. J Am Chem Soc 115:10436
527. Meng Z, Fürstner A (2020) Total synthesis provides strong evidence: xestocyclamine A is the enantiomer of ingenamine. J Am Chem Soc 142:11703
528. Baldwin JE, Whitehead RC (1992) On the biosynthesis of manzamines. Tetrahedron Lett 33:2059
529. Kurihara Y, Yagi M, Noguchi T, Yasufuku H, Okita A, Yoshimura S, Oishi T, Chida N, Okamura T, Sato T (2024) Total synthesis of keramaphidin B and ingenamine by base-catalyzed Diels–Alder reaction using dynamic regioselective crystallization. J Am Chem Soc 146:11054
530. Wu Z-L, Huang X-J, Xu M-T, Ma X-Y, Li L-R, Shi L, Wang W-J, Jiang R-W, Ye W-C, Wang Y (2018) Flueggeacosines A–C, dimeric securinine-type alkaloid analogues with neuronal differentiation activity from *Flueggea suffruticosa.* Org Lett 20:7703
531. Liu L, Olson TL, Wood JL (2024) Total synthesis of (−)-flueggeacosine C. Org Lett 26:7341–7346

532. Wei W, Wu X, Yang X (2016) Novel phthalide derivatives from the rhizomes of *Ligusticum chuanxiong* and their inhibitory effect against lipopolysaccharide-induced nitric oxide production in RAW 264.7 macrophage cells. RSC Adv 6:61037
533. Zhang Z, Li X, Song Q, Li Y, Tian X, Ali S, Yao Y, Li P, Wang Z, Zheng H (2024) Asymmetric synthesis of (+)-chuanxiongnolide L1 via a stereoselective oxidative dearomatization/Diels–Alder strategy. Org Lett 26:2928
534. Houghton JL, Green KD, Chen W, Garneau-Tsodikova S (2010) The future of aminoglycosides: the end or renaissance? Chembiochem 11:880
535. Okachi R, Takasawa S, Sato T, Sato S, Yamamoto M, Kawamoto I, Nara T (1977) Fortimicins A and B, new aminoglycoside antibiotics. II. Isolation, physico-chemical and chromatographic properties. J Antibiot 30:541
536. Lu Y, You X-Y, Zhang Q, Lu Q-T, Hou J-L, Cai Q (2025) Enantioselective total synthesis of fortimicin B. Angew Chem Int Ed 64:e202424235
537. Wu C, van der Heul HU, Melnik AV, Lübben J, Dorrestein PC, Minnaard AJ, Choi YH, van Wezel GP (2019) Lugdunomycin, an angucycline-derived molecule with unprecedented chemical architecture. Angew Chem Int Ed 58:2809
538. Nuutila A, Xiao X, van der Heul HU, van Wezel GP, Dinis P, Elsayed SS, Metsä-Ketelä M (2024) Divergence of classical and C-ring-cleaved angucyclines: elucidation of early tailoring steps in lugdunomycin and thioangucycline biosynthesis. ACS Chem Biol 19:1131
539. Zhu L, Huang J (2025) Total synthesis of lugdunomycin via sequential photoinduced spiroketalization and isobenzofuran Diels–Alder reactions. Angew Chem Int Ed 64: e202422615
540. Jana D, Khatua A, Kundu S, Noskar S, Nandy M, Bidai A (2025) Enantioselective total synthesis of taiwaniaadducts I, J, and L. J Am Chem Soc Au 5:1376
541. Songthammawat P, Phumjan T, Batsomboon P, Ploypradith P, Ruchirawat S (2025) Total synthesis of morulasin A. Org Lett 27:12325
542. Jeon BS, Wang SA, Ruszczycky MW, Liu HW (2017) Natural [4+2]-cyclases. Chem Rev 117: 5367
543. Zhang B, Wang KB, Wang W, Wang X, Liu F, Zhu J, Shi J, Li LY, Han H, Xu K, Qiao HY, Zhang X, Jiao RH, Kiuk KN, Liang Y, Tan RX, Ge HM (2019) Enzyme-catalysed [6+4] cycloadditions in the biosynthesis of natural products. Nature 568:122
544. Jamieson CS, Ohashi M, Liu F, Tang Y, Houk KN (2019) The expanding world of biosynthetic pericyclases: cooperation of experiment and theory for discovery. Nat Prod Rep 36:698
545. Ohashi M, Jamieson CS, Cai Y, Tan D, Kanayama D, Tang M-C, Anthony SM, Chari JV, Barber JS, Picazo E, Kakule TB, Cao S, Garg NK, Zhou J, Houk KN, Tang Y (2020) An enzymatic Alder-ene reaction. Nature 586:64
546. Watanabe K (2021) Discovery and investigation of natural Diels-Alderases. J Nat Med 75:434
547. Auclair K, Sutherland A, Kennedy J, Witter DJ, Van den Heever JP, Hutchinson CR, Vederas JC (2000) Lovastatin nonaketide synthase catalyzes an intramolecular Diels–Alder reaction of a substrate analogue. J Am Chem Soc 122:11519
548. Wang J, Liang J, Chen L, Zhang W, Kong L, Peng C, Su C, Tang Y, Deng Z, Wang Z (2021) Structural basis for the biosynthesis of lovastatin. Nat Commun 12:867
549. Kasahara K, Miyamoto T, Fujimoto T, Oguri H, Tokiwano T, Oikawa H, Ebizuka Y, Fuji I (2010) Solanapyrone synthase, a possible Diels–Alderase and iterative type I polyketide synthase encoded in a biosynthetic gene cluster from *Alternaria solani*. Chembiochem 11: 1245
550. Kim HJ, Ruszczycky MW, Choi S-H, Liu Y, Liu H (2011) Enzyme-catalysed [4+2] cycloaddition is a key step in the biosynthesis of spinosyn A. Nature 473:109
551. Fage CD, Isiorho EA, Liu Y, Wagner DT, Liu H, Keatinge-Clay AT (2015) The structure of SpnF, a standalone enzyme that catalyzes [4+2] cycloaddition. Nat Chem Biol 11:256
552. Tian Z, Sun P, Yan Y, Wu Z, Zheng Q, Zhou S, Zhang H, Yu F, Jia X, Chen D, Mándi A, Kurtán T, Liu W (2015) An enzymatic [4+2] cyclization cascade creates the pentacyclic core of pyrroindomycins. Nat Chem Biol 11:259

553. Sato M, Kishimoto S, Yokoyama M, Jamieson CS, Narita K, Maeda N, Hara K, Hashimoto H, Tsunematsu Y, Houk KN, Tang Y, Watanabe K (2021) Catalytic mechanism and *endo*-to-*exo* selectivity reversion of an octalin-forming natural Diels–Alderase. Nat Catal 4:223
554. Chang Z, Ansbacher T, Zhang L, Yang Y, Ko TP, Zhang G, Liu W, Huang J-W, Dai L, Guo R-T, Major DT, Chen C-C (2019) Crystal structure of LepI, a multifunctional SAM-dependent enzyme which catalyzes pericyclic reactions in leporin biosynthesis. Org Biomol Chem 17: 2070
555. Sun Q, Hu Y, Gu Y, Huang J, He J, Luo L, Yang Y, Yin S, Dou C, Wang T, Fu X, He L, Qi S, Zhu X, Yang S, Wei X, Cheng W (2019) Deciphering the regulatory and catalytic mechanisms of an unusual SAM-dependent enzyme. Sig Transduct Target Ther 4:17
556. List B, Li Y (2019) Biosynthesis of streptoseomycin involves a [6+4] cycloaddition. Synfacts 15:663
557. Wang KB, Wang W, Zhang B, Wang X, Chen Y, Zhu HJ, Liang Y, Tan RX, Ge HM (2021) A [6+4]-cycloaddition adduct is the biosynthetic intermediate in streptoseomycin biosynthesis. Nat Commun 12:2092
558. Liu SH, Sun JL, Hu YL, Zhang L, Zhang X, Yan ZY, Guo X, Guo ZK, Jiao RH, Zhang B, Tan RX, Ge HM (2022) Biosynthesis of sordarin revealing a Diels–Alderase for the formation of the norbornene skeleton. Angew Chem Int Ed 61:e202205577
559. Ding Q, Guo N, Gao L, McKee M, Wu D, Yang J, Fan J, Weng J-K, Lei X (2024) The evolutionary origin of naturally occurring intermolecular Diels-Alderases from *Morus alba*. Nat Commun 15:2492
560. Gao L, Yang J, Lei X (2022) Enzymatic intermolecular Diels–Alder reactions in synthesis: from nature to design. Tetrahedron Chem 2:100013

**Heinz Falk** has been Professor Emeritus at the Institute of Organic Chemistry of the Johannes Kepler University, Linz, Austria since 2007. He studied chemistry at the University of Vienna starting in 1959 and completed his dissertation in organometallic chemistry in 1966. In 1971, he spent a postdoctoral year at ETH Zürich in the laboratory of Albert Eschenmoser. Upon his return to Vienna in 1972 he attained the Habilitation in Organic Chemistry at the University of Vienna. Starting in 1966 he served as an assistant at the Institute of Organic Chemistry at the University of Vienna before being promoted to Associate Professor of Physical Organic Chemistry in 1975. In 1979, he became Full Professor of Organic Chemistry at the Johannes Kepler University of Linz, where he founded the Institute of Organic Chemistry, serving in this chair position for 28 years. From 1989 through 1991, he was Dean of the Faculty of Engineering and Natural Sciences.

The main research area of Prof. Falk is the structural analysis, synthesis, stereochemistry, and photochemistry of plant and animal photosensitizing and photosensory pigments. Besides fundamental studies of the stereochemistry of planar chiral compounds and the chemistry of bile pigments, the main group of compounds covered in his work are pigments derived from the fundamental phenanthro[1,10,9,8-opqra]perylene-7,14-dione chromophore inclusive of natural pigments like hypericin. Furthermore, he has pursued research on applied problems of industrial relevance, like oxidation, ozonization, non-natural amino acids, and catalysis. During the course of his career, he has supervised nearly one hundred doctoral students and several postdoctoral fellows. He has published two books: (1978) Ausgewählte Übungsbeispiele. Zur Nomenklatur Organischer Verbindungen. Springer-Verlag, Wien New York; ISBN 3211814795, and (1989) The Chemistry of Linear Oligopyrroles and Bile Pigments. Springer-Verlag, Wien New York; ISBN 3211821120. Also, he has published about 300 papers and has several patents.

Prof. Falk has served as Editor-in-Chief of Monatshefte für Chemie/Chemical Monthly for a decade and as Series Editor of Progress in the Chemistry of Organic Natural Products since 1998. He is, in addition to his memberships in the Austrian and German Chemical Societies and the New York Academy of Sciences, an elected member of the Austrian Academy of Sciences and the European Academy of Sciences.

MIX
Papier aus verantwortungsvollen Quellen
Paper from responsible sources
FSC® C105338

If you have any concerns about our products,
you can contact us on
**ProductSafety@springernature.com**

In case Publisher is established outside the EU,
the EU authorized representative is:
**Springer Nature Customer Service Center GmbH**
**Europaplatz 3, 69115 Heidelberg, Germany**

Printed by Libri Plureos GmbH
in Hamburg, Germany